Advances in Intelligent Systems and Computing

Volume 263

Series editor

Janusz Kacprzyk, Polish Academy of Sciences, Warsaw, Poland
e-mail: kacprzyk@ibspan.waw.pl

For further volumes:
http://www.springer.com/series/11156

About this Series

The series "Advances in Intelligent Systems and Computing" contains publications on theory, applications, and design methods of Intelligent Systems and Intelligent Computing. Virtually all disciplines such as engineering, natural sciences, computer and information science, ICT, economics, business, e-commerce, environment, healthcare, life science are covered. The list of topics spans all the areas of modern intelligent systems and computing.

The publications within "Advances in Intelligent Systems and Computing" are primarily textbooks and proceedings of important conferences, symposia and congresses. They cover significant recent developments in the field, both of a foundational and applicable character. An important characteristic feature of the series is the short publication time and world-wide distribution. This permits a rapid and broad dissemination of research results.

Thouraya Bouabana-Tebibel
Stuart H. Rubin
Editors

Integration of Reusable Systems

Springer

Editors
Thouraya Bouabana-Tebibel
Laboratoire de Communication dans les Systèmes Informatiques
Ecole Nationale Supérieure d'Informatique
Algiers
Algeria

Stuart H. Rubin
SPAWAR Systems Center Pacific
San Diego
USA

ISSN 2194-5357 ISSN 2194-5365 (electronic)
ISBN 978-3-319-04716-4 ISBN 978-3-319-04717-1 (eBook)
DOI 10.1007/978-3-319-04717-1
Springer Cham Heidelberg New York Dordrecht London

Library of Congress Control Number: 2014931756

Printed on acid-free paper

Springer is part of Springer Science+Business Media (www.springer.com)

Preface

Software reuse and integration has been described as the process of creating software systems from existing software rather than building software systems from scratch. Whereas reuse solely deals with the artifacts creation, integration focuses on how reusable artifacts interact with the already existing parts of the specified transformation. As a consequence, every integration can be seen as consisting of an analysis of the parts and of their subsequent synthesis into the new whole.

Although significant progress has been made on software reuse and integration, some important issues remain to be fixed. One of these addresses scalability by showing how to make best use of reusable components for very large systems. "Cloud-Based Tasking, Collection, Processing, Exploitation, and Dissemination in a Case-Based Reasoning System" proposes a novel computational intelligence methodology, which can learn to map distributed heterogeneous data to actionable meaning for dissemination. This approach provides a core solution to the tasking, collection, processing, exploitation, and dissemination problem. The expected performance improvements include the capture and reuse of analyst expertise, and, for the user, prioritized intelligence based on the knowledge derived from distributed heterogeneous sensing. "Simulation-Based Validation for Smart Grid Environments: Framework and Experimental Results" describes a simulation-based approach to understanding and examining the behavior of various components of a Smart Grid in the context of verification and validation. To achieve this goal, it adopts the discrete event system specification methodology, which allows the generalization and specialization of entities in the model and supports a customized simulation with specific scenarios.

Another issue is how to do sufficient formal specifications to support reliable construction and functioning of very large and complex systems. High-level representation mechanisms, including rigorous techniques for specification and verification, are needed. "An Institution for Alloy and Its Translation to Second-Order Logic" deals with the Alloy formal method for which it lays out the foundations to fully integrate the formalism in a platform which supports a huge network of logics, logic translators, and provers. This makes possible for Alloy specifications to borrow the power of several, nondedicated proof systems. "A Framework for Verification of SystemC Designs Using SystemC Waiting State Automata" presents the SystemC waiting-state automaton which is a compositional abstract formal model for verifying properties of SystemC. It is first shown how to extract automata for SystemC components. Next, an approach is proposed to infer relations between predicates generated during symbolic execution. The correctness of the abstract analysis is proven by model checking. "Formal MDE-Based Tool Development" proposes a rigorous methodology to create formal tools for GUI-based domain-specific languages. It aims at providing a productive and trustworthy development methodology to safety critical industries. The methodology combines metamodel-based GUI generators with executable backends automatically generated from formal specifications. As for "Formal Modeling and Analysis of Learning-Based Routing in Mobile Wireless Sensor Networks," it presents a formal model for a learning-based routing protocol specific to wireless sensor networks. The model is based on a Bayesian learning method, using a Structural Operational Semantics style. It is analyzed by means of the rewriting logic tool Maude.

Reuse and integration are key concepts in information retrieval and data mining. They structure and configure the stored information in a way to facilitate its extraction and enhance its usefulness. "On the Use of Anaphora Resolution for Workflow Extraction" addresses the problem of workflow extraction from textual process descriptions and presents a framework to support the development of extraction applications. Resolution approaches are presented, and syntactic and semantic evaluation functions are developed. These functions which are based on precision, recall, and F-measure are used to assess the quality of the data-flow. Furthermore, the data mining community has turned a significant fraction of its attention to time series data. Virtually, the availability of plentiful labeled instances is assumed. However, this assumption is often unrealistic. Semi-supervised Learning seems like an ideal paradigm, because it can leverage the knowledge of both labeled and unlabeled instances. "A Minimum Description Length Technique for Semi-Supervised Time Series Classification," first, demonstrates that in many cases a small set of human annotated examples are sufficient to perform accurate classification. Second, it provides a novel parameter-free stopping criterion for semi-supervised learning. The experimental results suggest that the approach can typically construct accurate classifiers even if given only a single annotated instance.

Another key element in the success of reuse is the ability to predict variabilities. "Interpreting Random Forest Classification Models Using a Feature Contribution Method" presents an approach to show how feature contributions measure the influence of variables/features on the prediction outcome and provide explanations as to why a model makes a particular decision. It demonstrates how feature contributions can be applied to understand the dependence between instance characteristics and their predicted classification and to assess the reliability of the prediction.

In reuse, there is also a need for a seamless integration between the models output from domain analysis and the inputs needed to for domain implementations such as components, domain specific languages, and application generators. "Towards a High Level Language for Reuse and Integration" proposes a language which gathers specialization and composition properties. The language is designed in a way to be specific to complex system domains. It supports, on the other hand, a component-based structure that conforms to a user-friendly component assembly. It is conceived in the spirit of SysML concepts and its programs generate Internal Block Diagrams.

Aspect orientation is a promising solution to software reuse. By localizing specific features in code aspects, not only a modular separation of concern is devised, but software development can also be incrementally transitioned to improve productivity and time-to-market. "An Exploratory Case Study on Exploiting Aspect Orientation in Mobile Game Porting" critically examines how aspect orientation is practiced in industrial-strength mobile game applications. The analysis takes into account technical artifacts, organizational structures, and their relationships. Altogether these complementary and synergistic viewpoints allow to formulate a set of hypotheses and to offer some concrete insights into developing information reuse and integration strategies.

Another area of potentially interesting research in reuse and integration is to identify what should be made reusable and which reusable corporate artifacts and processes will give the highest return on investment. "Developing Frameworks from Extended Feature Models" proposes an approach to develop a framework based on features defining its domain. The approach shows developers how to proceed, making them less prone to insert defects and bad smells in the outcome frameworks. It allows that even subjects with no experience in framework development can execute this task correctly and spending less time.

Researchers also argue for better methods to support specification and reasoning on knowledge component depositories. In "About Handling Non-conflicting Additional Information" the focus is on logic-based Artificial Intelligence systems that must accommodate some incoming symbolic knowledge that is not inconsistent with the initial beliefs but that however requires a form of belief change. First, the study investigates situations where the incoming piece of knowledge is both more informative and deductively follows from the preexisting beliefs. Likewise, it considers situations where the new piece of knowledge must replace or amend some previous beliefs, even when no logical inconsistency arises.

Safety and reliability are important issues which may be adequately addressed by reuse and integration. "A Multi-Layer Moving Target Defense Approach for Protecting Resource-Constrained Distributed Devices" proposes a Moving Target Defense approach for protecting resource-constrained mobile devices through fine-grained reconfiguration at different architectural layers. It introduces a coverage-based security metric to identify the configuration that best meets the current requirements. Likewise, in "Protocol Integration for Trust-Based Communication" a secure scheme based on trust is proposed to protect against packet dropping in mobile ad hoc networks. For this purpose, four already existing methods are integrated in a complementary way to the basic routing protocol in order to provide the required security.

Currently, most reuse research focuses on creating and integrating adaptable components at development or at compile time. However, with the emergence of ubiquitous computing, reuse technologies that can support adaptation and reconfiguration of architectures and components at runtime are in demand.

This edited book includes 15 high-quality research papers written by experts in information reuse and integration to cover the most recent advances in the field. These papers are extended versions of the best papers which were presented at IEEE International Conference on Information Reuse and Integration and IEEE International Workshop on Formal Methods Integration, held in San Francisco in August 2013. They have been selected among 111 accepted papers and have been accepted for publication in this book after they have been extended and have undergone a peer review process.

Thouraya Bouabana-Tebibel
Stuart H. Rubin

Contents

Cloud-Based Tasking, Collection, Processing, Exploitation, and Dissemination in a Case-Based Reasoning System

Stuart H. Rubin and Gordon K. Lee

Abstract The current explosion in sensor data has brought us to a tipping point in the intelligence, surveillance, and reconnaissance technologies . This problem can be addressed through the insertion of novel artificial intelligence-based methodologies. The scope of the problem addressed in this chapter is to propose a novel computational intelligence methodology, which can learn to map distributed heterogeneous data to actionable meaning for dissemination. The impact of this approach is that it will provide a core solution to the tasking, collection, processing, exploitation, and dissemination (TCPED) problem. The expected operational performance improvements include the capture and reuse of analyst expertise, an order of magnitude reduction in required bandwidth, and, for the user, prioritized intelligence based on the knowledge derived from distributed heterogeneous sensing. A simple schema example is presented and an instantiation of it shows how to practically create feature search spaces. Given the availability of high-speed parallel processors, such an arrangement allows for the effective conceptualization of non-random causality.

Keywords Boolean features · Case-based reasoning (CBR) · Cloud-based tasking · Data exploitation · Schema instantiation

S. H. Rubin (✉)
SSC-PAC, San Diego, CA 92152-5001, USA
e-mail: stuart.rubin@navy.mil

G. K. Lee (✉)
Department of Electrical and Computer Engineering, San Diego State University,
San Diego, CA, USA
e-mail: glee@mail.sdsu.edu

T. Bouabana-Tebibel and S. H. Rubin (eds.), *Integration of Reusable Systems*,
Advances in Intelligent Systems and Computing 263, DOI: 10.1007/978-3-319-04717-1_1,

1 Introduction

The explosion in sensor data presents a massive challenge in the intelligence, surveillance, and reconnaissance technologies that may only be solved through the insertion of novel artificial intelligence-based methodologies. The nature of the problem is that heterogeneous data (i.e., structured and unstructured) must be reduced in size by an order of magnitude prior to dissemination. The focus of this position chapter is to suggest a novel computational intelligence methodology that can learn to map distributed heterogeneous data to actionable meaning for dissemination. This framework can provide a core solution to the tasking, collection, processing, exploitation, and dissemination (TCPED) problem. Further, it is expected that the approach will demonstrate the viability of the methodologies core concepts and thus justify a scaled-up investment in its development. The expected operational performance improvements include the capture and reuse of analyst expertise, an order of magnitude reduction in required bandwidth, and, for the user, prioritized intelligence based on the knowledge derived from distributed heterogeneous sensing.

To date, there have been many varied approaches to the TCPED problem ([1–7], for example). A common problem with these approaches is that either they employ *NP*-hard technologies that do not scale well (e.g., neural networks), or they attempt to apply logics, which are incomplete to tasks that inherently depend upon the development of quality heuristics—the learning of which is a science in itself.

Consider a schema-definition methodology; could such a strategy be scaled up for cloud-based computing? Will analysts who are non-programmers find it to be user friendly? Can the realized schemas for such applications as weather prediction or web searching find significant causal Boolean features? Boolean features are True/False responses to arbitrarily complex effective questions that collectively (along with at least one situational variable) define a situational context. Can schemas be symmetrically instantiated for greater speed of discovery? Can features be autonomously acquired, which enable correct predictions to be made that could not practically be made in their absence? While uncommon features may well be discovered (e.g., in the weather prediction application, finding changes in the temperature other than crossing the freezing boundary), we are looking to see that some common features are among them (e.g., changes in the barometric pressure).

According to Sam Fusaro [8], we are missing a capability for predictive analysis. The operational effectiveness of the approach presented here will bear proportion to the extent to which our approach can capture causality and thus model the cognitive process.

A fundamental principle of the tasking, collection, processing, exploitation, and dissemination (TCPED) is that intelligence processes must remain operational regardless of the amount of available bandwidth. A cloud-based system is needed to provide decentralized control. It is also advantageous because it is self-organizing and highly survivable. The goal of this chapter is to suggest that schema instantiation and case-based reasoning may be employed to resolve the large data processing issues in cloud computing. The approach presented here contributes to spectrum dominance by evolving a feature-based understanding of the environment. One property of the proposed methodology is that it can be trained to meet most bandwidth constraints by virtue of its learning from skilled analysts. This is important for increasing the scale of the Intelligence, Surveillance, Reconnaissance, and Targeting (ISR&T) picture that will fuse data (and capabilities) from the cloud communities [9]. Furthermore, cloud computing will enable an individual client to utilize sensors regardless of their location.

Cloud computing has fewer problems than a network of heterogeneous machines. A grid computing system is suggested for the cloud computing system's back end. In this way, the cloud system can access the processing power of all available computers on the back end to make our methodologies potentially complex iterative calculations tractable. The cloud also integrates cloud communities that include the intelligence information collection communities.

2 An Illustrative Example

Suppose that we have the following case base, where c_i are cases, w_j are weights, the i_j are situational variables (features), and d is the associated dependency class. Here, an asterisk, "*", represents a situational variable whose value is unknown, or was not recorded. Also, cases are acquired at the logical head, moved to the logical head when fired, and expunged from the logical tail when necessary to release space. Table 1 presents the schema for an arbitrary case base. The cases are shown in logical order, which is used by the uniform or 3-2-1 skew [10, 11]. The use of this skew is optional (i.e., in comparison with uniform weighting) and is useful for domains where the value of the data deteriorates in linear proportion to its time of collection—valuing more recent data more highly. The selection of a particular skew is domain specific. For example, the rate of radioactive decay is known to be proportional to how much radioactive material is left (excluding the presence of certain metals). The nuclear decay equation may be used as a skew for various radioactive materials and is given by $A(t) = A_0 e^{-\lambda t}$. Here, A(t) is the quantity of radioactive material at time t, and A_0 = A(0) is the initial quantity. The term λ is a positive number (i.e., the decay constant) defining the rate of decay for the particular radioactive material. A countably infinite number of other skews may be applicable.

In the following assignment of skew-weights, the skew vector, **S**, favors the logical head of the case base in keeping with Denning's principle of *temporal locality* [12]. Cases, which were most-recently acquired or fired, and thus appear at or nearer to the logical head of a case-base, are proportionately more heavily weighted under the 3-2-1 skew. Of course, this differs from a uniform skew. The closer a case is to the top of its linked list, the greater its weight or importance. A heuristic scheme (i.e., the 3-2-1 skew) for achieving this with a dependency class, d, consisting of $|d|$ cases is to assign the head case a weight of $\frac{2|d|}{|d|(|d|+1)}$. The map just below the head map has a weight of $\frac{2(|d|-1)}{|d|(|d|+1)}$. Finally, the tail map of the segmented case base has a weight of $\frac{2}{|d|(|d|+1)}$. The ith map from the head has a weight of $\frac{2(|d|-i+1)}{|d|(|d|+1)}$, for $i = 1, 2, \ldots, |d|$. For example, using a vector of four weights, the 3-2-1 skew (**S**) is $\mathbf{S} = (0.4, 0.3, 0.2, 0.1)^T$. There are a countably infinite number of possible skews, such that $\sum s_k = 1.0$.

The evaluation of the members of a dependency class is the contiguous weighted sum of its constituent elements (see below). A subsequent example will show how the weights are computed using the uniform and 3-2-1 skews, which again may be selected because they best fit domain characteristics. The weights are uniform if the skew is not known, or if there is no decay in the value of a case once recorded.

Table 1 shows a doubly-linked list. Zero indicates a list end. The list-head of the previous list is m and of the next list is one. The list-head of the free list (i.e., unused array elements) begins with the list-head of the previous list if the rows are fully utilized. Otherwise, the list-head of the free list points to the first row in the list of unutilized rows, in sequence. It simply contains every freed row, in arbitrary order.

Shift values are maintained for each non-Boolean variable (Table 1). These shifts are initialized to one minus the minimum field values, or zero—whichever is greater. If the resultant shift exceeds zero, each non-Boolean variable is initially shifted up by the computed shift value. Whenever a new contextual or situational variable has value less than or equal to the negation of its corresponding shift, then the shift takes the absolute value of that variable plus one. Non-Boolean variables not previously shifted (e.g., the context) will be shifted up by that amount, while all previously shifted ones (e.g., field values) will be shifted up by the values new—old shifts. Whenever a case is expunged, if the expunged non-Boolean variables have values of one, then new field minimums are found (i.e., an O(m) process) and if their values exceed one, the associated shifts and the previously shifted variables are both reduced by the amount that those values exceed one. Thus, all non-Boolean non-asterisk variables will have value of at least one. This prevents divide-by-zero errors in normalization as well as problems in adjusting zero-valued weights.

Next, define a context by c_j for $j = 1, 2, \ldots, n$. The nearness of a pair of cases, c_i and c_j, where the context is taken as c_j, is given by:

$$match(i) = \frac{\sum_{k=1}^{n} w_k \left| c_{i,k} - c_{j,k} \right|}{\left| participating\ situational\ variables \right|}, i \neq j.$$

It follows that since all weights and participating variable differences are normalized, $match(i) \in [0, 1]$. A participating situational variable is one that does not include an "*" in its field. If there are no such fields, then the pair of cases is omitted from the computation. If there are no such pairs of cases, then the match cannot be computed and thus is undefined.

The ranges of non-Boolean variables are normalized using double-precision computations. The non-Boolean vectors must be defined by positive elements. This is necessary to insure that any paired case differential, $|c_{i,k} - c_{j,k}|$, will never exceed unity. There must be at least one non-Boolean vector containing no asterisks. This is necessary to prevent divide-by-zero errors. The sums used for normalization are saved

Table 1 The case base schema

Wts	w_1	w_2	w_3	...	w_n	→
Ind	i_1	i_2	i_3	...	i_n	→
Feature	Non-B	Bool	Non-B	...	Bool	→
Shift	0	-	0	...	-	...
c_1	10	0	5	...	1	→
c_2	15	1	10	...	0	→
c_3	*	*	15	...	0	→
...	...	...	...	...	...	...
c_m	5	1	10	...	1	→
Wts	Dep.	Prv.	Nxt.			
Ind	D	-	-			
Feature	d	-	-			
Shift	-	-	-			
c_1	1	0	2			
c_2	2	1	3			
c_3	2	2	m			
...	...	...	...			
c_m	1	3	0			

Table 2 An arbitrary case base

Wts	w_1	w_2	w_3	w_4		Dep.
Ind	i_1	i_2	i_3	i_4	→	D
Feature	NB	Bool	NB	Bool	→	d
c_1	0.333	0	0.125	1	→	1
c_2	0.5	1	0.25	0	→	2
c_3	*	*	0.375	0	→	2
c_4	0.167	1	0.25	1	→	1

for the subsequent normalization of any context. The sums for situational variables i_1 and i_3 in Table 1 are 30 and 40, respectively (asterisks are skipped). Normalization of these variables is shown in Table 2. Boolean and non-Boolean contextual differences will all be comparable (i.e., resulting in a uniform contribution of their importance, conditioned by their associated weight), since no paired case differential, $|c_{i,k} - c_{j,k}|$, will ever exceed unity.

The dependency class selected to be fired will be the weighted match (i), which has a minimal class value (see below). In the event of a tie, the dependency averaging (i.e., substituting the case dependencies relative position from the logical head for its match (i) value), nearer (at) the logical head of the case base is selected as the winner as a result of temporal locality [12]. The single case dependency, which is nearer (at) the logical head, is selected as the winner in the event of a second tie (e.g., d = 1 in Table 2 because $(1 + 4)/2 = (2 + 3)/2$, but c_1 is the logical head). Using 3-2-1 skew weighting, d = 1 wins again because $(2/3^*1 + 1/3^*4) < (2/3^*2 + 1/3^*3)$.

Relative fused possibilities are produced (e.g., using the uniform or 3-2-1 skew), which evidence that the decision to favor one class dependency over another may be more or less arbitrary. Of course, in some domains it may be more appropriate to present the user with an ordered list of alternative dependency classes, along with their weighted match (i) values, and let the user decide. The decision is necessarily a domain-specific one.

There are domains for which it is useful to know that the case base does not embody the desired matching case(s) and all but perhaps the slightest "guessing" is to be enjoined. This can be achieved by placing a squelch, greater than or equal to zero, on the minimum match (i). Here, if this minimum computed match (i) just exceeds the set squelch, then the system will respond with, "I'm very unsure of the correct action". The correct action dependency (d) will be paired with the context and acquired as a new case, at the earliest opportunity, if the dependency should prove to be incorrect. This dependency may or may not comprise a new action class. The logical tail (LRU'd member) of the case base may be expunged, as necessary, to make room for the new case acquisition. The qualifying phrases are, "very unsure", "somewhat unsure", "somewhat sure", "very sure", depending on the difference, (minimum match (d)—squelch, where d represents a dependency class). Notice that an exact match would have a difference of—squelch. Thus, any difference $< -\text{squelch}/2$ would be associated with the qualifier, "very sure". Any $-\text{squelch}/2 \leq \text{difference} \leq \text{squelch}/2$ would be associated with the qualifier, "somewhat sure". Any $\text{squelch}/2 < \text{difference} \leq \text{squelch}$ would be associated with the qualifier, "somewhat unsure". Finally, any $\text{squelch} < \text{difference}$ would be associated with the qualifier, "very unsure". The most appropriate value for the squelch may be determined experimentally and is domain specific.

Notice that the acquisition of a new case here not only insures that its situational context will be known until, if ever, the case falls off of the logical tail of the case base, but contexts in its immediate field (i.e., having minimal differences with it) will likewise be associatively known. Cases identified as erroneous may be (a) overwritten with the correct dependency, if known, (b) expunged, or (c) bypassed through the acquisition of a correct case at the logical head of the case base so that the erroneous case will eventually fall off of the logical tail of the case base. The choice of methodology here is domain specific in theory. In practice, alternative (a) is the usual favorite, where the domain is deterministic. Alternative (b) is the usual favorite, where correct actions are not forthcoming. Alternative (c) is the usual favorite, where the domain is non-deterministic (assumed herein).

Next, consider the arbitrary case base shown in Table 2. Here, we observe two dependency classes and two case instances mapping to each class. The goal is to find a normalized set of weights, w_j, which will serve in mapping an arbitrary context to the best-matching case situation and thus to the most appropriate action class, if any. We may take the squelch to be 0.1 here on the basis of trial and error for this example. If the minimum computed match (i) >0.1, then the system will reply to the effect that it is very unsure of the correct action. A correct case will be acquired at the logical head, if the correct action is known, should the dependency prove to be incorrect. In this event, the LRU'd case at the logical tail may be expunged to make room.

Ideally, each case situation is to be compared with each other case situation, exactly once, for the purpose of computing the global weight vector, **W**. The number of comparisons here is $m-1+m-2+\ldots+2+1$, or $\frac{m(m-1)}{2}$, which is $O(m^2)$ on a serial machine. This is only tractable for limited values of m. However, an excellent heuristic for reducing the order of magnitude of comparisons on large case bases without significantly reducing the quality of results is to compare the ciel (square root of the number of class members) with that defining number of members in each class including its own, though compared cases must be distinct. This is particularly useful for supporting the evolution of new features because it trades the time required for an exact evaluation of the weights for an (initial) rough estimation of any new features worth—allowing many more candidate features to be explored.

Once the feature set is stable, if ever, the full serial $O(m^2)$ class comparisons may be resumed. If m parallel processors can be utilized, then the runtime complexity here can be reduced to $O(m)$. This is certainly tractable, where we define tractability based on an efficient sort executing on a serial machine (e.g., MergeSort having average and worst case times of $O(N \log N)$ [13]).

The count of comparisons is made in order from the logical head, since these are the most recently acquired/fired cases and thus are intrinsically the most valuable based on temporal locality [12]. For example, in Table 2, cases c_1 and c_4 are members of the first class dependency and cases c_2 and c_3 are members of the second class dependency. Thus, c_1 is compared against $\left\lceil \sqrt{2} \right\rceil = 2$ members of its dependency class as well as two members of the second dependency class. Note that when a class member is being compared against its own class, the initial class member may be counted towards the computed limit in the number of comparisons (though skipped when $i = j$). Furthermore, if each class is dynamically linked based on order from the logical head so that one need not traverse other class members to get to the next member of the same class and each such class maintains a pointer to the next class in order, then, the number of comparisons is $\left\lceil \sqrt{m-1} \right\rceil + \left\lceil \sqrt{m-2} \right\rceil + \ldots + \left\lceil \sqrt{2} \right\rceil + 1$, which is $O(m)$ on a serial machine. (In practice, this is not too difficult to accomplish using two passes through the case base, or $O(m)$ extra storage.) If m parallel processors can be utilized, then the runtime complexity can be reduced to $O(\log m)$. If m^2 parallel processors can be utilized, then this can be reduced to $O(c)$, or constant time. Naturally, the use of m^2 parallel processors would only be practical for limited values of m.

Returning to our example in Table 2, c_1 will be compared against c_2, then against c_3, and next against c_4. Then, c_2 will be compared against c_3, then against c_4. Finally, c_3 will be compared against c_4. Note that c_i is not compared against c_i because this would serve to distort the resultant weights towards uniformity. Uniformity vies against variable (feature) selection. Also, the evolution of the weights is context free in terms of other weights. However, it is actually context sensitive because predicted dependency classes found to be in error are re-acquired as new cases having correct dependencies. The situational part of these new cases defines a new *well* for matching similar cases [14]. Absolute values are not raised to any power on account of the availability of Boolean features, whose differences are immutable under any power. This also provides for faster computation as the power function is relatively slow in comparison with the arithmetic operators.

First, let us compare c_1 against c_2. These two situations belong to distinct classes (D). Here, we want to assign maximal weights, w_j, to the situational variables that are the most quantitatively different. Going across w_1 to w_4 in Table 2 for c_1 and c_2, $|c_1 - c_2|$ is computed as: $|0.333 - 0.5| = 0.167$, $|0 - 1| = 1$, $|0.125 - 0.25| = 0.125$, $|1 - 0| = 1$. We note that the qualitative features are at least as important as any other situational variable. This is evidenced by their more extreme absolute values relative to the non-Boolean situational variables. Here, **W** = (0.167, 1, 0.125, 1).

Next, let us compare c_1 against c_3. These two situations belong to distinct classes (D). Here, we want to assign maximal weights, w_j, to the situational variables that are the most quantitatively different. Going across w_1 to w_4 in Table 2 for c_1 and c_3, $|c_1 - c_3|$ is computed as: *, *, $|0.125 - 0.375| = 0.25$, $|1 - 0| = 1$. Here, **W** = (*, *, 0.25, 1).

Next, let us compare c_1 against c_4. These two situations belong to the same class (D). Here, we want to assign maximal weights, W_j, to the situational variables that are the most quantitatively the same. Going across w_1 to w_4 in Table 2 for c_1 and c_4, $|c_1 - c_4|$ is computed as: $|0.333 - 0.167| = 0.166$, $|0 - 1| = 1$, $|0.125 - 0.25| = 0.125$, $|1 - 1| = 0$. The Boolean situational variable differences need be complimented and the remaining non-asterisk variables subtracted from 1.0 because we need to weight the variables that are most similar most heavily. Here, **W** = (0.834, 0, 0.875, 1).

Next, let us compare c_2 against c_3. These two situations belong to the same class (D). Here, we want to assign maximal weights, w_j, to the situational variables that are the most quantitatively the same. Going across w_1 to w_4 in Table 2 for c_2 and c_3, $|c_2 - c_3|$ is computed as: *, *, $|0.25 - 0.375| = 0.125$, $|0 - 0| = 0$. The Boolean situational variable differences need be complimented and the remaining non-asterisk variables subtracted from 1.0 because we need to weight the variables that are most similar most heavily. Here, **W** = (*, *, 0.875, 1).

Next, let us compare c_2 against c_4. These two situations belong to distinct classes (D). Here, we want to assign maximal weights, w_j, to the situational variables that are the most quantitatively different. Going across w_1 to w_4 in Table 2 for c_2 and c_4, $|c_2 - c_4|$ is computed as: $|0.5 - 0.167| = 0.333$, $|1 - 1| = 0$, $|0.25 - 0.25| = 0$, $|0 - 1| = 1$. Here, **W** = (0.333, 0, 0, 1).

Next, let us compare c_3 against c_4. These two situations belong to distinct classes (D). Here, we want to assign maximal weights, w_j, to the situational variables that are the most quantitatively different. Going across w_1 to w_4 in Table 2 for c_3 and c_4, $|c_3 - c_4|$ is computed as: *, *, $|0.375 - 0.25| = 0.125$, $|0 - 1| = 1$. Here, **W** = (*, *, 0.125, 1).

Now, it is time to compute the normalized weight vectors. First, the raw computed weight vectors, **W**, are given in Table 3. Next, each $c_{i,j}$ is normalized going across for $j = 1$ to n. Table 4 presents the results. The computed weights are then summed

Table 3 The raw computed weights

Wts	w_1	w_2	w_3	w_4	Init. $\sum$
$c_{1,2}$	0.167	1	0.125	1	2.292
$c_{1,3}$	*	*	0.25	1	1.25
$c_{1,4}$	0.834	0	0.875	1	2.709
$c_{2,3}$	*	*	0.875	1	1.875
$c_{2,4}$	0.333	0	0	1	1.333
$c_{3,4}$	*	*	0.125	1	1.125

Table 4 The normalized rows

Wts	w_1	w_2	w_3	w_4	Fin. $\sum$
$c_{1,2}$	0.0729	0.4363	0.0545	0.4363	1.0
$c_{1,3}$	*	*	0.2	0.8	1.0
$c_{1,4}$	0.3079	0	0.323	0.3691	1.0
$c_{2,3}$	*	*	0.4667	0.5333	1.0
$c_{2,4}$	0.2498	0	0	0.7502	1.0
$c_{3,4}$	*	*	0.1111	0.8889	1.0
$\sum$	0.6306	0.4363	1.1553	3.7778	6.0
Avg	0.2102	0.1454	0.1926	0.6296	1.1778
Norm	0.1785	0.1234	0.1635	0.5346	1.0

and divided by the number of non-asterisked, situational variables and normalized to yield the final normalized four weights.

Suppose we had the situational context defined by $\mathbf{c}_1$ in Table 2. Clearly, the correct dependency class is 1. Let us see how this might be predicted.

First, let us compare c_1 against c_1. The raw context, $c_1 = (10, 0, 5, 1)$. Next, we perform a column-normalization by dividing the non-Boolean variables by the previously saved column sums. Thus, $c_1 = (10/30, 0, 5/40, 1) = (0.333, 0, 0.125, 1)$. Going across w_1 to w_4 in Table 2 for c_1, $|c_1 - c_1|$ is computed as: $|0.333 - 0.333| = 0$, $|0 - 0| = 0$, $|0.125 - 0.125| = 0$, $|1 - 1| = 0$. Thus, match $(i = 1) = 0$, which is minimal and thus is correctly matched. (Note that skewed averages are still necessary if non determinism is allowed. Here however, the case situation, c_1, occurs only once in Table 2—insuring that the correct dependency is $d = 1$).

Next, let us compare the situational context = (10, 1, 5, 0) against the four cases in Table 2 to find the best match and thus the predicted value for the dependency. Boolean matches are more significant and thus we would expect the dependency class to be 2 here.

Let us compare this situational context against c_1. Next, we perform a column-normalization by dividing the non-Boolean variables by the previously saved column sums. Thus, $c_1 = (10/30, 1, 5/40, 0) = (0.333, 1, 0.125, 0)$. Going across w_1 to w_4 in Table 2 for c_1, $|c_1 - \text{context}|$ is computed as: $|0.333 - 0.333| = 0$, $|0 - 1| = 1$, $|0.125 - 0.125| = 0$, $|1 - 0| = 1$.

$$\text{Thus, } match(1) = \frac{0.1785(0) + 0.1234(1) + 0.1635(0) + 0.5346(1)}{4} = 0.1645.$$

The denominator is four here because none of the situational variables union the context, in Table 2, has an asterisk and thus are all participating.

Next, let us compare this situational context against c_2. Going across w_1 to w_4 in Table 2 for c_2, $|c_2 - \text{context}|$ is computed as: $|0.5 - 0.333| = 0.167$, $|1 - 1| = 0$, $|0.25 - 0.125| = 0.125$, $|0 - 0| = 0$. Thus,

$$match(2) = \frac{0.1785(0.167) + 0.1234(0) + 0.1635(0.125) + 0.5346(0)}{4} = 0.0126.$$

Next, let us compare this situational context against c_3. Going across w_1 to w_4 in Table 2 for c_3, $|c_3 - \text{context}|$ is computed as: *, *, $|0.375 - 0.125| = 0.25$, $|0 - 0| = 0$.

$$match(3) = \frac{0.1785(0) + 0.1234(0) + 0.1635(0.25) + 0.5346(0)}{2} = 0.0204.$$

The denominator is two here because a total of two situational variables union the context, in Table 2, have asterisks and thus are not participating—being substituted for by zero above. Four situational variables reduced by two non-participating ones, leaves two participating situational variables.

Next, let us compare this situational context against c_4. Going across w_1 to w_4 in Table 2 for c_4, $|c_4 - \text{context}|$ is computed as: $|0.167 - 0.333| = 0.166$, $|1 - 1| = 0$, $|0.25 - 0.125| = 0.125$, $|1 - 0| = 1$. Thus,

$$match(4) = \frac{0.1785(0.166) + 0.1234(0) + 0.1635(0.125) + 0.5346(1)}{4} = 0.1462.$$

Next, we compute the uniform skew value for each class. c_1 and c_4 have d = 1 and c_2 and c_3 have d = 2. Thus, match (class_1) = $(\text{match}(1) + \text{match}(4))/2 = (0.1645 + 0.1462)/2 = 0.1554$. Match ($\text{class}_2$) = $(\text{match}(2) + \text{match}(3))/2 = (0.0126 + 0.0204)/2 = 0.0165$. Clearly, the second class (i.e., d = 2) is the better selection by a factor of almost ten. Here, the minimum match (i) – squelch = 0.0165 – 0.1 = –0.0835 < –squelch/2 and thus we would be "very sure" of it being the correct class. Also, the first class is above the set squelch of 0.1 and thus we would be "very unsure" of it being the correct class.

Had the 3-2-1 skew been used instead of the uniform skew, then the first member of each class would have been weighted more heavily, since it is closer to the logical head. The 3-2-1 skew only considers an element in positional relation to elements in the same class. The 3-2-1 skew for two elements is (2/3, 1/3). Thus, match (class_1) = $(0.6667 * 0.1645 + 0.3333 * 0.1462) = 0.158$. Match ($\text{class}_2$) = $(0.6667 * 0.0126 + 0.3333 * 0.0204) = 0.015$. Here, the use of the 3-2-1 skew has once again selected the second class (i.e., d = 2), as desired. We are slightly more sure of it being the correct class under the 3-2-1 than uniform skew because 0.015 – 0.1 = –0.085 < –0.0835 < –squelch/2.

Table 5 The case base after acquisition

Wts	w_1	w_2	w_3	w_4		Dep.
Ind	i_1	i_2	i_3	i_4	→	D
Feature	NB	Bool	NB	Bool	→	d
c_1	0.333	1	0.125	0	→	3
c_2	0.333	0	0.125	1	→	1
c_3	0.5	1	0.25	0	→	2
c_4	*	*	0.375	0	→	2
c_5	0.167	1	0.25	1	→	1

Suppose however that it was learned that the correct dependency for the context was not d = 2, but rather something else—say d = 3, without loss of generality. Table 5 shows the resulting logical ordering of the updated case base. (Note that Table 5 would depict a nondeterministic case base if the context was an exact match for an existing situation, but the associated action differed.) If this table was limited to the storage of four cases, then case c_5 would be expunged to make room for case c_1. Physical case movement or search is not required. Again, this would involve updating the pointers to a doubly-linked global list. In Table 5, the global head pointer would be updated to point to case c_1 upon acquisition. Case acquisition or firing results in logical movement to create a new list head, where not redundant. The next pointer is updated to point to the previous head, or c_2. The next lower case having the same dependency, or case c_5, is eventually visited through global traversal. Counts of the number of members having the same dependency are maintained for the computation of the uniform or 3-2-1 skew. Case c_5 is last on the global list. Thus, it can be expunged by adding it to the free list and changing the last on the list to point to its immediate predecessor, or case c_4. Thus, the tail is updated to point to c_4.

Suppose that case c_4 were to be fired. It is logically moved to the head of the global list. Thus, the head pointer is set to c_4. Case c_4s next pointer is set to the previous head, or case c_1. Case c_4s predecessor's next pointer (case c_3) is set to point to case c_4s successor (case c_5).

Notice how the chance of generating an erroneous dependency (and its associated possibility) decreases with each case acquisition— given a segmented and relatively stable domain. This is because each case situation induces a proximal matching field for every context. The utility of this matching field is proportionate to the combined degree to which the situational variables were discriminators during training. This completes our example of case acquisition.

Next, consider the situational variables, i_j. Again, using m parallel processors, a runtime complexity of $O(\log m)$ is achievable. This is very fast and such speed can be put to good use in the evolution of new and better features. Table 4 shows the weights, **W** = (0.1785, 0.1234, 0.1635, 0.5346). In order of non-increasing significance they are w_4, w_1, w_3, and w_2. Observe that despite the missing data for the first two weights for one case (i.e., c_3 in Table 2), w_1 is not among the first two weighted-features to be replaced. w_2 is the least-significant non-zero weighted-feature, or the first to be replaced. (Zero-valued weights need time to compute at least one comparative evaluation before being adjudicated.) We have seen that missing data does not affect the value assigned to a weight, as desired. The capability to handle this situation is generally required for feature evolution.

One evolutionary strategy is to run numerous distinct tables, having n situational variables each, in parallel. Each table holds predominantly distinct situational variables (features), though limited duplication is permitted to accelerate feature generation time. After all of the variables have received computed weights, the n greatest weights, and associated distinct variables (features) are concatenated in one table and renormalized. At least one non-Boolean situational variable containing no asterisks is required in the final table. Care must be exercised that no feature is "dangling"—due to reference to a missing situational variable(s), or even other feature(s). The requirement to compute new normalized elements requires that the original data (see Table 1) be the starting point each time—along with the appropriate sums (i.e., in view of a change in the included fields). This strategy is most useful where many parallel/distributed processors are available. More discussion on the use of parallel platforms and its need is provided in [15]. A stable sort, such as the n-way MergeSort [13], is used to sort the weight vectors. A stable sort maintains the relative order of records with equal values. This is important for feature evolution. Here, new situational features are inserted at the right. Thus, in the event of a tie, preexisting situational features will be more highly valued, since they have survived the longer test of time. Again, MergeSort has best and worst case times of $O(\mathrm{N} \log \mathrm{N})$ and MergeSort's best case takes about half as many iterations as its worst case [13].

If no situational variable that is referenced by a feature(s) is allowed to be replaced and no feature that is referenced by another feature(s) is allowed to be replaced, then a second serial-processor strategy may be more efficient. The (non-zero lowest-weight) non-referenced variables and features can be replaced (or augmented) with new variables and/or features. To be replaced, the non-referenced variable or feature must have a weight, which is less than the average weight of all variables and features (except itself) having non-zero weights. Thus, better results can be expected if the number of features referencing other features is minimized (or eliminated). This is necessary to insure that relatively valuable non-referenced variables are not lost. Here, a few of the non-zero lowest-weight features can be found using a single-pass algorithm, which is O(n) on a sequential machine. The computation of Table 6 follows the same process as was illustrated for the computation of Table 3. The sums are unaffected. Tables 6 and 7 contain a replaced feature, i_2. The new feature, though far from perfect, evidences better discrimination than the one it replaced. The theoretical ideal weight vector, **W**, would have minimal cardinality such that all w_j are equal. Ideally, a single situational variable would suffice to determine the dependency class. Table 7 computes the new weight vector, **W** = (0.1163, 0.2995, 0.1411, 0.4431).

Table 8 shows the acquisition of a new case by our arbitrary case base as set forth in Table 1. This case includes a new Boolean feature, i_5. Notice that this feature could not be computed for any but the new case due to the unavailability of data. Nevertheless, the system is designed so that the computed weight, $w_5 = 0$, can co-exist with the

Table 6 The raw computed weights using the new w2

Wts	w_1	w_2	w_3	w_4	Init. $\sum$
$c_{1,2}$	0.167	1	0.125	1	2.292
$c_{1,3}$	*	*	0.25	1	1.25
$c_{1,4}$	0.834	1	0.875	1	3.709
$c_{2,3}$	*	*	0.875	1	1.875
$c_{2,4}$	0.333	1	0	1	2.333
$c_{3,4}$	*	*	0.125	1	1.125

Table 7 The normalized rows using the new w2

Wts	w_1	w_2	w_3	w_4	Fin. $\sum$
$c_{1,2}$	0.0729	0.4363	0.0545	0.4363	1.0
$c_{1,3}$	*	*	0.2	0.8	1.0
$c_{1,4}$	0.2249	0.2696	0.2359	0.2696	1.0
$c_{2,3}$	*	*	0.4667	0.5333	1.0
$c_{2,4}$	0.1428	0.4286	0	0.4286	1.0
$c_{3,4}$	*	*	0.1111	0.8889	1.0
$\sum$	0.4406	1.1345	1.0682	3.3567	6.0
Avg	0.1469	0.3782	0.1780	0.5595	1.2626
Norm	0.1163	0.2995	0.1411	0.4431	1.0

Table 8 An acquisition for the arbitrary case base

Wts	w_1	w_2	w_3	w_4	w_5		Dep.
Ind	i_1	i_2	i_3	i_4	i_5	→	D
Feature	NB	Bool	NB	Bool	Bool	→	d
c_1	20	1	20	0	1	→	2
c_2	10	0	5	1	*	→	1
c_3	15	1	10	0	*	→	2
c_4	*	*	15	0	*	→	2
c_5	5	1	10	1	*	→	1

other weights, **W**, despite the single data point. In fact, having few data points (e.g., the square root of the number of class members), turns out to be an advantage for the evolution of new features. This is because having few data points supports a more uniform coverage of the feature search space. Most importantly, features are generated through the instantiation of schema, which involve situational variables and relational operators. Features may also be taken from case-base dependencies. A simple schema is presented in Fig. 1, discussed in more detail in [16].

```
Define Boolean Weather_Change_Feature (var x, t1, t2; t):
  /* In general, schemas may call other schemas. */
  Randomly Select x ∈ {pressure, humidity, temperature};
  Randomly Select t1, t2  ∈ {t, t-1, t-2, t-3}
    Such That t2 > t1;
  If x[t2] Randomly Select op ∈ {>, <} x[t1]
    Return (1)
Return (0).
```

Fig. 1 A simple weather features schema

3 A Knowledge-Based Solution

The tasking, collection, processing, exploitation, and dissemination problem calls for a knowledge-based approach. The scope of this approach requires the acquisition and generalization of massive amounts of distributed knowledge to succeed. The nature of TCPED problems requires the piecing together of often diverse knowledge segments to arrive at a solution. The search for such solutions can be made tractable through the use of schemata (Figs. 1, 2, 3).

A simple schema is presented in Fig. 1. The search space for this schema is $3 \times 6 \times 2 = 36$ possible instances. The first random select statement allows for three choices (i.e., for pressure, humidity, or temperature). The six combinations for the second random select statement is derived from taking n = 4 things, r = 2 at a time, where the number of combinations, c, is defined by $c = \frac{n!}{r!(n-r)!}$. Finally, we see that there are two random choices for the final set of relational operators, $\{>, <\}$. Figures 2 and 3 show sample features, which are instances of their parent schema, which is found in Fig. 1. Again, they may be automatically discovered and validated through computational search. Figures 2 and 3 present one of 36 possible instances of this schema.

3.1 Case-Based Reasoning

A case base consists of a set of situations and a sequence of actions such that the set is mapped to an appropriate sequence by way of experience, hence the term, experiential knowledge. This knowledge differs from rules in that it generally embeds causality, rather than literally explain it. Thus, cases are far easier to capture, maintain, and select for application.

Cases also differ from rules in that they are mapped to, rather than applied in the form of a logical inference engine. The problem is that it is generally impossible to directly capture causality. Any attempt to do so (e.g., through the use of rules) invariably leads to secondary interactions, which grow to become ever-more difficult to predict with scale. Cases are not associated with this difficulty because they are

```
Define Boolean Pressure_Increase_Feature (pressure, t):
  If pressure[t] > pressure[t-1]
    Return (1)
Return (0).
```

Fig. 2 A “pressure” instance of the schema

```
Define Boolean Humidity_Decrease_Feature (humidity, t):
  If humidity[t-1] < humidity[t-3]
    Return (1)
Return (0).
```

Fig. 3 A “humidity” instance of the schema

limited to the capture of experience, which may differ from the underpinning cause and effect. Case bases are also far less costly to maintain, for this reason, as has been borne out by industrial experience.

Automating TCPED processes requires the scalability and the ease of maintenance found in case-based systems, but needs the causal capture found in rule-based systems. The latter is necessary for the capture of analyst expertise and to minimize the number of case instances and thus the time needed for training the system.

3.2 An Example

Situational knowledge consists of a set of conditional variables and Boolean features, which when satisfied imply a dependency. Dependencies define some system action and are indexed by class membership. For example, a UAV might run client-side software to record its GPS latitude and longitude (two variables), terrain data as reported by its sensors (variables); detect a road (feature), a river (feature), and an overpass (feature) [17]. This data (and potentially much more) is relayed to a cloud, which is running a weighted feature-based Case-Based Reasoning (CBR) system as a service. It maintains and evolves weights on all situational variables and features. If it finds no exact match for this mix of variable data and Boolean features, then in accordance with the novel methodology, it maps the data to the closest matching case.

This case implies a certain action, which is indexed and is a member of a specific (new) class. Moreover, a squelch is set to insure that the system will report when it does not know a proper match for a supplied context, or its level of confidence in the found match is below some preset threshold.

Here, that action class might be, “send text message to coalition cloud partner with UAV situational context + the tag, Boolean feature “Force Status?”. The coalition cloud, running the same algorithm on a distinct case base, has evolved say high weights on the GPS coordinate variables and the Force Status Boolean feature. This triggers its own member of an action class, say, "Friend". This is text messaged back to the cloud that the UAV is communicating with. This action, by definition, advertises a Boolean feature, "Friend", which augments (or substitutes into) the existing set of weighted situational variables and features. Now, the inference engine has the original set of six situational variables and features (five if one is replaced) plus this one. A new best-fit case is matched as before. (Note that the tagged Boolean feature, "Friend" need not be available and if not the firing pattern would be as previously described.) This time the triggered associated action class is say, "do not transmit imagery—all secure". Here, we see that operational bandwidth is conserved. The things to take away from this example is that if the following five conditions are all satisfied, then it will be practical to evolve dynamic solutions to the TCPED problems.

1. The cases can be rapidly acquired by the cloud servers.
2. Erroneous cases can be identified and expunged or updated.
3. The situational weights can be accurately evolved.
4. New features can be evolved, and/or received from fired dependency categories, to replace lower-quality features (i.e., those having the least non-zero weights).
5. The cloud servers can communicate using relatively low (available) bandwidth.

3.3 The Need for an Open Architecture

The payoff from pursuing this approach is that problems can be solved that are not explicitly programmed for. The problem with expert systems is that the knowledge acquisition bottleneck [18] makes it impractical to avoid programming with scale (e.g., NASA's software tool for building expert systems—CLIPS). In the system proposed in this chapter, knowledge truly evolves from cases and the cases derive from analyst expertise. This expertise can be captured in the form of schema and exploited using cloud computing. The languages, used to define the schema, can be bootstrapped using this approach. The risk occurs if an open architecture is not utilized. This is because high-end schema-definition languages are more or less domain specific. Thus, they will improve over time and given an open architecture, the migration to upgrades, in the cloud, will be transparent to analysts on the front end. It is important that knowledge be retained through all upgrades.

3.4 Smart Tagging, Indexing, and Advertising

Several institutions, including the US Navy, are pursuing cloud strategies for ISR/TCPED [19]. Processes are locally written and stored in the cloud. These processes can consist of Boolean features and effective dependencies. For example, suppose a stream of data is sent to the cloud. We need smart and adaptive methodologies to manage mission data so we are not shipping the same product back five different ways and storing it 12 different times in accordance with current practice [20]. Rather, Boolean features are triggered by that stream. The matching set of situational variables and features triggers the best-matching case situation, which triggers the associated dependency if the certainty is above squelch (i.e., above a set minimum numeric quality criterion).

This cloud service does two basic things. First, it tags the data stream with a meaning (e.g., "Force Status?"). Second, it corresponds with a specific Boolean feature, which indicates that it was fired. That is, it advertises that the data was tagged with "Force Status?" indicated. Advertised data may be posted in any calling or called case base (i.e., intra and inter cloud). The risk here occurs where the memory space is too small to allow purchase of the advertised feature when even the least-weighted situational feature is needed and should not be arbitrarily lost. Another payoff of cloud computing is that adequate memory is generally not a problem.

This is potentially far more than "separating the wheat from the chaff" [19]. The approach to the problem in [19] is to process this data in situ so as to limit communication transmissions to mission critical data by autonomously extracting necessary information from the data stream amidst a sea of extraneous material. While this is helpful, we take the process to the next logical step. That is, we iteratively tag data streams and use the collective sequence of tags to direct further tagging. This is semantic scaffolding (also known as bootstrapping).

One rarely knows what data is needed so the iterative indexing of features effectively enables what is known in computer science as, "data-directed translation". Here, the various tags provide a context for subsequent tags. In human terms, this is building a cognitive picture. This feature, "Force Status?" always resets upon the next iteration of the inference engine. Tagging and advertising data are accomplished through the use of index tables, which are maintained in RAM for rapidity of access using limited communication paths. The resulting system rapidly evolves a correct complex behavior. This is attributed to the case interactions through smart tagging and indexing capabilities to advertise relevant data. The complexity of this process will be quadratic if each of m cases is compared against each other on a serial architecture, or linear in the cloud, where at least m distributed/parallel processors are available. However, if each case is compared against the square root of the number of cases in each class, then the complexity is reduced to linear on a serial architecture, or logarithmic in the cloud, where at least m distributed/parallel processors are available. This order of magnitude reduction in the number of ideal dependency category comparisons speeds up the discovery of new features by speeding up their evaluation and ranking. There will be insignificant impact on the quality of the resultant features because the cases, in each class taken for comparison, will be the most-recently acquired or fired (i.e., logically ordered). These cases will have the greatest utility.

4 On Boolean Features

Networks of case bases allow one bases' fired dependency category to serve as another's situational feature. Other features follow from the instantiation of domain-specific schema. For example, a temperature variable might need to distinguish the three natural states of water—solid, liquid, and gas. Here are a few tuples that serve to illustrate a simple schema and its instantiations, Schema: (Temperature°F, Freezing?, Boiling?), (32°, 1, 0), (72°, 0, 0), and (212°, 0, 1). The use of Boolean features is very common and effective as a descriptor.

As few as 20 Boolean features can provide universal situational identifications. For example, a parlor game exists called, "20 questions". Players take turns asking questions, which can be answered with a simple "Yes" or "No". The information, as measured by Shannon's entropy statistic, needed to identify an arbitrary object, from among 220 objects, is at most 20 bits [21]. The best questions will divide the field of remaining choices by two each time. Indeed, this is how binary search algorithms operate. Transforming this result, if binary decision points can differentiate an arbitrary object, then Boolean features can similarly characterize it. The better the questions, or equivalently the better the Boolean features the fewer that will be needed on average.

While these Boolean features are simplistic, they can be far more complex. For example, consider boosted/bagged neural networks, trained to recognize a particular face over streaming video. Note that if a network of architecturally distinct subsystems is trained, it is called, boosting. However, if identical subsystems are given distinct training, it is called, bagging. Schemas define alternative weight vectors, which program the neural networks recognition capabilities. Such weight vectors are obtained on the basis of external training. Furthermore, Zero Instruction Set Computer (ZISC) hardware can provide more than enough speed for real- time full-motion HD video Boolean feature recognition (e.g., less than 10 microseconds) [22]. Additional discussion on the use of Boolean features is given in [23].

5 The Role of Analysts

A new case is formed from the triggered features (i.e., both schemata instances and advertised dependencies) and the analyst-supplied action dependency. This machine training makes more efficient use of existing analysts through the capture and reuse of their expertise by the system as mediated through cloud-based architectures. The ISR tipping point continues to increase the demand for additional intelligence analysts to perform TCPED mission functions in traditional ways [20]. The expertise for the human piece of the intelligence analysis and exploitation that will be needed must be effectively captured for reuse, or we will not have the manpower to reduce the volumes of data that must be moved around.

Notice the advantage provided by iterative processing using an inference engine. That is, each iteration of the inference engine can transform the data stream by transforming it with the results of its processing (i.e., tagging). Such tagging can result in the iterative expansion and contraction of the grammatical object. This defines the universal, or Type 0, grammar. It then follows that no semantics is theoretically too sophisticated for it to effectively capture and process [24]. Intelligent systems that will not be found to be lacking in capabilities are needed in many applications.

Tags can also be formed by reaching back in the data stream and using previous processing results to tag the current stream. For example, once the data was tagged with “Force Status?” the set of features that is triggered on the next iteration of the inference engine recognizes that a request for information was sent to a federated cloud. What would a trained analyst do at this point? (S)he might notice from the reach-back that no hostile forces are in the vicinity. Knowing this and that no problems are indicated is sufficient to enable the analyst to tag the data stream with “Friend”. This message is subsequently routed back to the cloud that made the request. This is defined as an executable tag, meaning that it is transmitted.

Notice that this methodology provides for the iterative fusion and reduction of multiple data streams using federated clouds running software as a service. Processing is done using the clouds platform as a service, which can execute in real time. One key to the success of the full- scale methodology is providing the analysts with an easy to use front end that (1) can capture their observations and decisions in software for reuse and (2) can execute as a service in the cloud environment. The creation of dependencies and most Boolean features does not require excessively complex software. Such software is said to be lightweight or thin-client because it can be composed exclusively on the client’s platform.

5.1 Case Specification and Schema Definition

The qualitative statement of intelligence needs derives transparently from the inference engine in finding the best match for the specified contextual situation. The user’s intent and system objectives are embedded in a replay of analyst actions. They are not literally defined per se, but rather are an emergent property of system complexity.

Once schemas and cases are specified by the analysts, the evolution of better features is transparent. This is possible because the cases are partitioned into classes on the basis of their dependencies. Cases having the same dependency are assigned greater weights on their variables (features) that have near similar normalized values. Conversely, cases having distinct dependencies are assigned higher weights on their variables (features) that have the most different normalized values. This supports the evolution of the ideal feature set, occurs in the cloud, and converges on the capture of causality. Causality takes the form of a weighted set of appropriate Boolean features. Predicted dependency categories, found to be in error, are reacquired as new (nondeterministic) cases having correct dependencies. The situational part of these new cases defines a contextual well for matching similar cases [14].

The unique case-based weighted inference engine iteratively tags and removes tags from the data to evolve a simulation of our cognitive processes. Traditional CBR systems were designed to do this, but our novel CBR system autonomously evolves better features and more optimal weights. It also embodies data management to rank the utility of each acquired and fired case through logical movement. This also enables the square-root comparison method to work using relatively little data. Moreover, it can work with multiple alternative actions (i.e., non-determinism).

Knowledge acquisition is far easier than it would be using an expert system. Analysts could not work with the latter due to the inherent complexity of capturing causality. Moreover, unlike conventional programming, analysts need not be exact when specifying schemas. That is, the crispness associated with conventional computer programming is mollified to form a fuzzy space of alternatives. Our complex system may pass messages back and forth internally and between clouds before allowing a recommendation to emerge in satisfaction of the commander's intent and mission objectives. One property of the output of such a system is that it can be trained to meet most bandwidth constraints by virtue of its exercise by the analyst(s).

5.2 *A Revolutionary Answer to an Evolutionary Need*

A recent TCPED study estimated a future need for over 300 additional intelligence analysts [20]. Their knowledge and skills need to be captured for replay in training the cloud-based CBR systems. Otherwise, adding more analysts is an evolutionary answer to a revolutionary need. Analysts may opt to use sophisticated, but easy to work with agent based modeling toolkits to parse the data and insert the proper tags (e.g., ABLE, SOAR 6 in JAVA, SimPlusPlus in C++, but no programming required, et al.). They will also write top-down and/or bottom-up schema (see below), which are used to constrain the feature space. A treatise on the use of experts and analysts is given in [15].

6 On Unsupervised Feature Learning

While schema definition effectively provides for semi-supervised learning because the machine is dependent on the human and vice versa, the question arises if the unsupervised learning of features is practical. First, in order to be tractable in the large, knowledge is required to guide search. Thus, our question at once is transformed into, "What is the most efficient representation(s) for knowledge?" and, "What is the paradigm(s) for knowledge acquisition?"

The answer to the first question is to employ more constrained schema, more parallel computation, and a better way to associate the problem definition with an associated schema. Top-down and bottom-up methods have been given for schema generalization. These methods depend upon having a symbolic representation.

This representation needs to be composable to create a search space of alternatives. The selection among alternatives is knowledge based (e.g., using rules or preferably cases).

The answer to the second question is to use the same case-based weighted feature acquisition system. In other words, the dependencies at this level may be feature schema, which are more general than the independent features leading to their selection. These dependencies capture greater knowledge than embodied by the independent features leading to their selection.

Next, we consider the basis for feature-based schema selection. Features are defined by the absence of noise [14, 17–30]. In other words, features are randomizable. Such randomization can be lossless (i.e., invertible), or with greater or lesser degrees of loss (e.g., in the "real" world). For example, an upper right angle and a lower right angle can be learned by a neural network and associated with the concept of right angle. However, the learned information is not compressed and in this sense it is not randomized. Hence, we say that a feature has not been extracted. All manner of methods have been used to extract the concept (e.g., positive and negative examples, grammatical inference, etc.). However, they all fall short in what we refer to as conceptualization.

Patterns may be stored and recursively adapted within limits to match meta-patterns. New patterns may be learned. In essence, such mapping procedures can facilitate recognition. But, randomization is not limited to structural randomization. It may be functional as well. For example, a large flat rock and a chair are functionally randomizable. Again, we see the need for symbolic representations of knowledge. While symmetric knowledge can be discovered through the use of an inference engine (e.g., a sandy beach is like a bed), random knowledge cannot. It must be limited to the product of exhaustive search. Such search must involve conceptual tokens (i.e., words and phrases) because only they are randomizations of the intended concept (e.g., neural weights are not randomizations). The compression of fundamental memories in a neural network requires context (e.g., is it that they are right angles or that they are sharp?). As such, context must be separable. This can only be accomplished through the use of symbolic representations. This suggests the use of neural networks as a frontend for symbolic representation and the symbolic composition of schemata using system dependencies (i.e., knowledge-based composition of schemata).

Sometimes a single schema is sufficient for an instance of it to capture a general concept. At other times, a more or less linear independent association of schema instances is necessary to capture a general concept. This is because randomization is an iterative process. The end point of randomization is recursively enumerable, but not recursive. Thus, while fewer schemata instances are necessary over time, one never truly knows if a relative minimum has been attained. That is why conceptualization is time dependent [23].

7 Five Research Challenges

There are five research challenges, amongst others, which need to be addressed [16].

1. Develop a schema definition methodology suitable for small-scale testing and realizable software to facilitate the implementation of self-modifying code.
2. Investigate and report on how to best constrain the implied search.
3. Investigate tagging using advertised features (i.e., self-referential feedback in machine learning)—this will also prove the utility of intra and inter cloud communication if successful.
4. Evaluate the quality of the evolved features as determined by the evolution of their associated weights towards a set having minimal statistical variance (i.e., a surrogate criterion for high utility features). Also, if the number of descriptive features is allowed to vary then a successful feature evolution will usually, but not always reduce the number of maintained features. (While having a minimal number of maximally-descriptive features is desired, that minimum is domain specific and cannot be known in the general case.)
5. Evaluate the subjective quality of the feature evolution process—including the number and relevance of the features evolved.

With research challenges, one must also ask some questions which may lead to potential reinforcements of proposed approaches or lead to other approaches not initially considered.

1. Can the schema-definition methodology utilize the massively distributed environment provided by cloud computing? Can libraries of schema templates be defined and used as is common for object-oriented programming languages? Will schemas be easy for analysts to use to constrain the search space? Can random instances of analyst-specified weather prediction schemas and/or symmetric instances of analyst-specified weather prediction features yield significant causal Boolean features?
2. That random instantiation of schemas works was empirically demonstrated in 1998 in an unpublished computer program. Will the random instantiation of schemas, or the symmetric extension of features, if either, prove to be a more productive methodology for analyst use? This topic is suitable for a patent disclosure(s) and/or research paper(s). It is likely to be best realized through economies of scale. Can Bernstein's concept of multiple analogies, where having multiple derivational paths increases the likelihood of having a valid result, be empirically confirmed.
3. Can correct (i.e., meaningful) weather predictions be made using two or more cycles of the inference engine such that at least one advertised Boolean feature tag, which was not present in the first cycle, and is subsequently posted, leads to the firing of a correct action dependency? This demonstrable event, which implies successful schema definition and instantiation, provides clear indication that the methodology can scale, be cloud-based, and address TCPED intelligence processes. That is the impact. The needed bandwidth will be relatively low

because text, not say HD video, will be streamed, for the most part, over the network.

4. Ideally, a minimum number of features will all have about the same weight and suffice to determine the proper dependencies. Statistical variance, in the weights, will be computed along a path from the initial weight vector and associated feature set to an evolved weight vector and associated feature set. A plotted approximation of a decreasing exponential curve will evidence that the quality of the Boolean features is successfully improving. Furthermore, if the number of maintained features is allowed to vary based on the magnitude of the associated evolved weights, then a decreasing number of maintained features, in general, will likewise support the improving quality of the evolving features.
5. The features evolved for the application domain (e.g., weather prediction) will be evaluated by their number (less is better), their relevance as defined by their associated weights (greater is better), and most importantly by their use in the actual domain. While uncommon meteorological features may well be discovered here (e.g., changes in the relative humidity), we are looking to see that some common meteorological features are among them (e.g., changes in the barometric pressure). Freely available weather history reports can be obtained from the Web, from <http://weathersource.com/past-weather/weather-history-reports/free>. This data can be used to support feature evolution. Successful evaluation of the five categorical questions collectively provides clear evidence that the proposed methodology works as described. Each loosely coupled CBR system in the cloud will benefit from each other's quality improvements. The features and their weights are continuously evolved on the back end (e.g., the cloud's grid computing system) to dynamically optimize performance.

The approach suggested here advances beyond others research by providing the following ten fundamental capabilities.

1. It evolves causal behavior—it does not require the user to find rules to specify it.
2. It utilizes schema-definition languages to define and evolve candidate features.
3. Candidate features are also automatically derived from every fired dependency. This is also supported by low bandwidth intra and inter-cloud networking.
4. It compares cases against the same as well as distinct classes of case dependencies to evolve near optimal features and associated weight sets for the supplied TCPED problems.
5. The weight sets may be rapidly evolved, using a novel "square-root class comparison method" to more quickly converge upon the best features.
6. Cases are acquired, fired, and expunged to free space using dynamic memory management and logical (i.e., not physical) movement for speed.
7. System training is transparently provided by teams of distributed analysts, using intelligent agents so that their expertise is readily captured for replay.
8. Every action is associated with an evaluation of its certainty. Actions having low certainties are subject to being squelched (i.e., filtered).

9. It learns from analyst training to iteratively find meaning in multiple data streams. Dependency actions can disseminate this meaning, in the cloud, using an order of magnitude less bandwidth. This is made possible through the use of smart tagging and advertising.
10. All of the novel advancements above utilize the massive parallelism available through platform as a service using federated cloud architectures.

A successful approach for conceptualizing causality, and thus modeling the cognitive process, will define a solution for the entire intelligence process, or TCPED. There is a small risk of failure here. However, the methodology overviewed herein shows that TCPED intelligence processes can be automated to a much greater extent than is presently possible. That is the payoff.

8 Conclusion

Determining causal relationships from observations and experiments is fundamental to human reasoning, decision making, and the advancement of science. In the final analysis, this methodology converges on effectively capturing causality—to the extent that it is not randomly based. It shows that the evolution of features and the machine learning of cases provide a unified framework for the capture of intelligence. Furthermore, the definition of schemas, for the production of novel features, is defined by randomness and symmetry [11, 25, 27].

The methodology is robust and can handle limited noise and/or missing data. In particular, evolution can occur on top of an existing case base—even if the data needed to evaluate the new variables (features) cannot be had for the preexisting cases. A squelch is set to insure that the system will report when it does not know a proper match for a supplied context. Cases are logically acquired at the head of the base and moved there whenever fired. The least-recently used (LRU'd) cases are expunged from the tail when necessary to free space

Weights are evolved by processing each case against each other in a segmented case base and partitioning the cases on the basis of having the same or distinct class dependencies. Greater weights are used if the situational variables being compared are different and the associated action dependencies belong to distinct classes. Similarly, greater weights are used if the situational variables being compared are similar and the associated action dependencies belong to the same class. Situational variables are normalized and unified with Boolean features. The methodology can be O(log m) in runtime complexity if m parallel processors are utilized and the square root of the number of class members is used for comparison purposes. This allows for the rapid evolution of new features—decreasing the entropy of the supplied data. Contexts are mapped to the class whose components have minimal weighted error (i.e., evaluate closest to zero). These components are typically uniformly weighted, or weighted using the 3-2-1 skew. The latter reflects the age-weighted properties of data, or a domain-specific alternative skew may be designed. Methods are supplied for acquiring new variables (features) and replacing those found to be less relevant.

This methodology shows that the evolution of features and the machine learning of cases provide a unified framework for the capture of intelligence. Furthermore, the definition of schemas, for the production of novel features, is defined by randomness and symmetry. Finally, networks of segmented case bases allow one bases fired dependency class to serve as another's situational feature. Such networks define generalized and-or graphs (GAGs). These graphs have the representational power of Type 0 grammars. Successful situational features are more likely to be taken from segmented case base dependencies, where they were previously discovered. Here, new cases and segmented bases can be formed through the association of triggered dependency classes and the most-recent distinct actions.

Acknowledgments The authors thank SSC-PAC for financial support. This research document was produced, in part, by a U.S. government employee as part of his official duties.

References

1. Kumar, S., Dharm, R.: A contemporary approach to hybrid expert systems: case-based reasoning. In: International Conference on Computer and Communication Technology, ICCCT-2010, pp. 736–740 (2010)
2. Maurer, M., Brandic, I., Sakellariou, R.: Simulating autonomic SLA enactment in clouds using case-based reasoning. In: Lecture Notes in Computer Science. LNCS, vol. 6481, pp. 25–36. Springer, Heidelberg (2010)
3. Ben Mustapha, N., Zghal, H.B., Aufaure, M.A., Ben Ghezala, H.: Semantic search using modular ontology learning and case-based reasoning. In: Proceedings of the International Conference on Extending Database Technology/International Conference on Database Theory. ACM (2010)
4. Smiti, A., Elouedi, Z.: Using clustering for maintaining case-based reasoning systems. In: Proceedings of the 5th IEEE International Conference on Modeling, Simulation and Applied Optimization, ICMSAO (2013)
5. Pawlish, M., Varde, A.S., Robila, S.A.: Cloud computing for environment-friendly data centers. In: Proceedings of the 3rd International Workshop on Cloud Data Management, ACM (2012)
6. Bahga, A., Madisetti, V.K.: Analyzing massive machine maintenance data in a computing cloud. IEEE Trans. Parallel Distrib. Syst. **23**(10), 1831–1843 (2012)
7. Horn, G.: A vision for a stochastic reasoner for autonomic cloud deployment. In: Proceedings of the 2nd Nordic Symposium on Cloud Computing and Internet Technologies, ACM (2013)
8. Rosenberg, B.: Harnessing the full power of sensor fusion. In: Defense Systems. http://defensesystems.com/Articles/2009/09/02/C4ISR1-Sensor-Fusion.aspx (2009)
9. Costlow, T.: Military pushes for smaller and capable sensor inputs for UAVs. Defense Syst. **5**(11), 26–27 (2011)
10. Rubin, S.H.: System and method for geodesic data mining. US Patent No. 7,840,506 B1, 23 Nov 2010
11. Rubin, S.H.: Is the kolmogorov complexity of computational intelligence bounded above? In: Proceedings of the IEEE International Conference on Information Reuse and Integration (IRI), Las Vegas, pp. 455–461 (2011)
12. Deitel, H.M.: An Introduction to Operating Systems. Prentice Hall, Inc., Upper Saddle River (1984)

13. Merge sort, Wikipedia, the free encyclopedia, http://en.wikipedia.org/wiki/Merge_sort
14. Rubin, S.H., Murthy, S.N.J., Smith, M.H., Trajkovic, L.: KASER: knowledge amplification by structured expert randomization. IEEE Trans. Syst. Man Cybern. B Cybern. **34**(6), 2317–2329 (2004)
15. Rubin, S., Lee, G.: Requirements for an intelligent system for experts. In: Proceedings of the ISCA International Conference on Computers and Their Applications, Honolulu (2010)
16. Rubin, S., Lee, G.: Cloud-based tasking, collection, processing, exploitation, and dissemination. In: Proceedings of the IEEE International Conference on Information Reuse and Integration, San Francisco (2013)
17. Kim, Z.: Real-time road detection by learning from one example. In: Proceedings of the IEEE Workshop on Application of Computer Vision, pp. 455–460 (2005)
18. Feigenbaum, E.A., McCorduck, P.: The Fifth Generation: Artificial Intelligence and Japan's Computer Challenge to the World. Addison-Wesley Pub. Co., Reading (1983)
19. Davis, S.A.: Information dominance, agile acquisition, and intelligence integration, Q&A with Terry Simpso. PEO C4I's Principal Deputy for Intelligence. United States Navy, Space and Naval Warfare Systems Command, Office of Public Affairs and Corporate, Communications (Feb 25, 2011) www.public.navy.mil/spawar/Press/Documents/Publications/2.23.11_TerrySimpson.pdf
20. CHIPS Magazine, Interview with J. Terry Simpson, PEO C4I Principal Deputy for Intelligence. www.doncio.navy.mil/CHIPS/ArticleDetails.aspx?ID=2289 (April-June 2011)
21. Twenty Questions, Wikipedia. http://en.wikipedia.org/wiki/Twenty_Questions#cite_note-1
22. Deck, W.C.: Target tracking with the zero instruction set computer, VDM, Saarbrücken (2010)
23. Rubin, S., Lee, G.: Predictor-corrector equations for feature extraction. In: Proceedings of the ISCA International Conference on Computers and Their Applications, Honolulu (2013)
24. Honavar, V., Slutzki, G. (eds.): Grammatical Inference. Lecture Notes in Artificial In-telligence, vol. 1433. Springer, Berlin (1998)
25. Chaitin, G.J.: Randomness and mathematical proof. Sci. Am. **232**(5), 47–52 (1975)
26. Rubin, S.H.: System and method for knowledge amplification employing structured expert randomization (KASER), Patent No. US 7,047,226. 16 May 2006
27. Rubin, S.H.: On randomization and discovery. Inf. Sci. **177**(1), 170–191 (2007)
28. Rubin, S.H.: Chapter 17—Knowledge amplification by structured expert randomization—KASERs in SoS design. CRC book. System of Systems—Principles and Applications, pp. 421–450. Taylor & Francis Group, Boca Raton (2009)
29. Rubin, S.H.: Chapter 13—On creativity and intelligence in computational systems. In: Advances in Reasoning-Based Image Processing, Analysis, and Intelligent Paradigms. pp. 383–421. Springer, ISRL 29 (2011)
30. Rubin, S.H.: Multilevel constraint-based randomization adapting case-based learning to fuse sensor data for autonomous predictive analysis. NC **101614**, 06 Feb 2012

Simulation-Based Validation for Smart Grid Environments: Framework and Experimental Results

Wonkyu Han, Mike Mabey, Gail-Joon Ahn and Tae Sung Kim

Abstract Large and complex systems, such as the Smart Grid, are often best understood through the use of modeling and simulation. In particular, the task of assessing a complex system's risks and testing its tolerance and recovery under various attacks has received considerable attention. However, such tedious tasks still demand a systematic approach to model and evaluate each component in complex systems. In other words, supporting a formal validation and verification without needing to implement the entire system or accessing the existing physical infrastructure is critical since many elements of the Smart Grid are still in the process of becoming standardized for widespread use. In this chapter, we describe our simulation-based approach to understanding and examining the behavior of various components of the Smart Grid in the context of verification and validation. To achieve this goal, we adopt the discrete event system specification (DEVS) modeling methodology, which allows the generalization and specialization of entities in the model and supports a customized simulation with specific variables. In addition, we articulate metrics

---

A preliminary version of this chapter appeared under the title "Simulation-Based Validation for Smart Grid Environments," in Proceedings of the 14th IEEE International Conference on Information Reuse and Integration, San Francisco, USA, August 14–16, 2013. All correspondences should be addressed to Dr. Gail-Joon Ahn at gahn@asu.edu.

W. Han (✉) · M. Mabey · G.-J. Ahn (✉)
Laboratory of Security Engineering for Future Computing (SEFCOM), Arizona State University, Phoenix, AZ, USA
e-mail: whan7@asu.edu

M. Mabey
e-mail: mmabey@asu.edu

G.-J. Ahn
e-mail: gahn@asu.edu

T. S. Kim
Chungbuk National University, Cheongju-si, South Korea
e-mail: kimts@chungbuk.ac.kr

T. Bouabana-Tebibel and S. H. Rubin (eds.), *Integration of Reusable Systems*, Advances in Intelligent Systems and Computing 263, DOI: 10.1007/978-3-319-04717-1_2,

for supporting our simulation-based verification and validation and demonstrate the feasibility and effectiveness of our approach with a real-world use case.

Keywords Smart grid · Discrete event system specification · Risk assessment · Simulation · Validation

1 Introduction

The Smart Grid is a pervasive new concept intended to provide sophisticated features to the electrical grid, including energy resource sharing, distribution, and load balancing [1–4]. A wide variety of research has been conducted to determine what technological aspects and risks should be considered in the creation of the Smart Grid, such as smart metering technology [5], information system development [6], future standards, and so on (Table 1).

As for the future standards, the National Institute of Standards and Technology (NIST) released the final version of their Smart Grid "Framework 2.0" roadmap in February of 2012 [7]. In this version, they provide a conceptual model to describe the overall Smart Grid system, and propose eight research areas which should be standardized with high priority. The most significant difference between this release and their previous one (i.e., Release 1.0) is the emphasis on improving interoperability among various distributed systems and reducing the number of threats in the Smart Grid. In addition, there exist several functional and non-functional requirements associated with the Smart Grid. For instance, the Energy Power Research Institute (EPRI) published the Integrated Energy and Communication Systems Architecture (IECSA), which describes many functional requirements and scenarios and is helpful for understanding specific domains of the Smart Grid [8]. Also, the Organization for the Advancement of Structured Information Standards (OASIS) and Zigbee published the Energy Market Information Exchange (EMIX) and Smart Energy profile (SEP) 2.0, respectively, which are additional specifications with the goal to develop common object models which can be applied in a Smart Grid system [9, 10]. Also, EPRI releases various use cases (or scenarios) that still need to be verified and validated by scientists and engineers.[1]

Even though the common interest of these research groups clearly expresses the growing risks to the Smart Grid, there exists no systematic method to leverage use cases and articulate critical flaws in a dynamic and large-scale system in the Smart Grid. Since it is more difficult to discover vulnerabilities and threats in a large system, a simulation-based verification and validation process is indispensable. Also, a simulation-based approach helps perform verification and validation without requiring considerable time and resources, needing to implement the entire system, or accessing the existing physical infrastructure, which could hamper its operations

[1] As of January 2013, 213 use cases are available at http://smartgrid.epri.com/Repository/Repository.aspx.

Table 1 Domains in the smart grid conceptual model [7]

Domain	Actors in the domain
Customers	The end users of electricity. May also generate, store, and manage the use of energy. Traditionally, three customer types are discussed, each with its own sub-domain: home, commercial/building, and industrial
Markets	The operators and participants in electricity markets
Service provider	The organizations providing services to electrical customers and utility companies
Operations	The managers of the movement of electricity
Bulk generation	The generators of electricity in bulk quantities. May also store energy for later distribution
Transmission	The carriers of bulk electricity over long distances. May also store and generate electricity
Distribution	The distributors of electricity to and from customers. May also store and generate electricity

or cause failures on running systems. Also, such an effective approach is critical since many elements of the Smart Grid are still in the process of becoming standardized for widespread use. In this chapter, we propose a novel framework to harness the power of simulation in the verification and validation processes for Smart Grid environments. Our framework leverages use case repositories to change its form to identifiable simulation entities and performs automatic validation tasks with corresponding assessment library. We also adopt one of the well-known formal modeling methodology, discrete event system specification (DEVS), to achieve scalability and ease of use. The DEVS modeling methodology allows modelers to articulate the states of each entity so that our framework can easily identify and trace all activities during the simulation.

The rest of this chapter is organized as follows. We give an overview of the related work in Sect. 2 including NIST's conceptual model and validation work. Section 3 describes our framework, called the Simulation-Based Validation Framework, along with DEVS-based model validation. In Sect. 4, we discuss details of our design and implementation of a specific use case to verify that our framework is capable of performing validations against system models as expected, along with details of our evaluation results. Section 5 concludes the chapter and addresses several future directions.

2 Related Work

This section presents the NIST conceptual model and existing validation approaches for the Smart Grid.

2.1 NIST Conceptual Model

The NIST conceptual model divides the Smart Grid into seven domains,[2] each of which contains various actors and applications. Actors can be physical devices, software programs, or organizations which own those devices. Applications are designated tasks performed by actors. Domains consist of actors who have the same objectives and maintain similar characteristics when they are communicating within the same domain. In the Customer domain, all customers are not just consuming electricity, but managing their energy usage and generating Distributed Energy Resources (DER). The Market domain consists of all operators and participants including commercial service providers, energy brokers, and end users. Actors in the Operation domain deliver electricity from generators to end users. The Service Provider domain shares information to cooperate with other domains such as the Market, Operation, and Customer domains. Organizations in the Service Provider domain provide energy installation, facility maintenance, billing services, and account management. Companies in the Bulk Generation domain generate electricity for customers and transmit/distribute energy via the Transmission and Distribution domains, respectively. During these various domain activities, each domain exchanges information with each other to operate their tasks.

2.2 Validation Approaches in the Smart Grid

Various concerns in the Smart Grid have received attention for several years. One of the leading research groups, called the Cyber Security Working Group (CSWG), made a three year plan (beginning April 2011) to develop a standardized framework which consists of examining use cases, evaluating threats, and suggesting countermeasures [11]. In their plan, a use case is first selected for the threat evaluation. Risk assessment is then performed to identify what vulnerabilities the use case is associated with and how they would impact the overall Smart Grid system. From the risk assessment, high-level requirements and mitigation solutions can be specified. After the risk assessment, either a new architecture is developed to prevent the identified risks or existing standards are assessed for possible flaws. These procedures describe how early-stage validation is critical to the next generation of Smart Grid standards. Although the CSWG framework is well-organized, it lacks details on how use cases should be examined and evaluated in their framework.

Another relevant evaluation model is the Electricity Subsector Cybersecurity Capability Maturity Model (ES-C2M2) [12]. This model includes 10 domains and 4 maturity indicator levels that are used to measure how secure each system is. However, this approach uses its own conceptual model which makes it hard to perform the evaluation tasks in a standardized manner. Also, measuring the indicator

[2] Ericsson et al. [6] suggested four domains: Generation, Transmission, Distribution and Markets, respectively, which is mostly covered in NIST model.

level may be subjective. Other researchers have taken different approaches such as agent-based [13, 14], model-based [15, 16], and attack-scenario-based [17] evaluation. Even though these approaches demonstrated interesting evaluation results, their work omitted real use cases. To assess the assurance of each component in the Smart Grid, it is necessary to have a comprehensive but generic framework for considering real use cases systematically.

3 Simulation-Based Validation Framework

This section describes our framework, called the Simulation-Based Validation Framework, which leverages the benefits of simulation with the validation process.

3.1 Overview

As mentioned in Sect. 2, threats in the Smart Grid continue to gain attention; however, there still lacks a systematic, comprehensive, and repeatable framework with which to validate a wide variety of use cases. To accommodate these goals, our framework consists of three core components: (i) *Entity Generator* initiates a simulation by generating a number of entities described in an existing use case; (ii) *Simulation Execution Block* establishes relations between the entities and executes the assessment based on specified requirements and model definitions; and (iii) *Viewer* displays messages that are exchanged between entities during the state transition. Our systematic process allows for the validation to be repeated. Figure 1 shows how three components cooperate with each other.

3.2 Entity Generator

The most important role of the *Entity Generator* component is to create entities that are identifiable by the *Simulation Execution Block*. To achieve this goal, the *Use Case Representation* module modifies the original use case into a composition of entities, actors, and activities. Each entity is then defined using either a certain expression or formal language and entered into the *Model Definition*, which allows the *Simulation Execution Block* to understand what the entity is. For example, most use cases utilize UML diagrams to illustrate functional/non-functional features of actors and activities in the entity. In particular, exchanging messages between actors plays a major role to describe relations of actors and their activities.

We noticed that exchanging messages between actors can be a key criterion to make state-based diagrams, which are message-based state derivations that define the number of states with regard to the number of messages. A use case has a

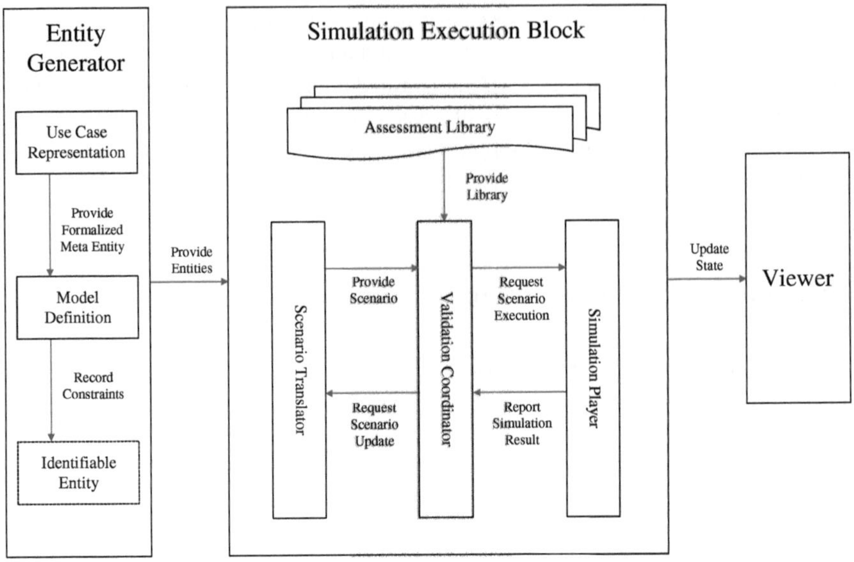

Fig. 1 Simulation-based validation framework

number of incoming messages, which are denoted by $M^I = \{M_1^I, M_2^I, \ldots, M_U^I\}$, and outgoing messages, which are denoted by $M^O = \{M_1^O, M_2^O, \ldots, M_V^O\}$ where $\mid M^I \mid= U$ and $\mid M^O \mid= V$. The entire state set of this entity is defined as $S^E = \{(S_1^E, S_2^E, \ldots, S_i^E, \ldots, S_k^E) \mid S_i^E = (x, y)$, where $x \in M^I$, $y \in M^O$ and $k = U * V\}$. The total number of states in the entire set is $\mid S^E \mid= U * V$, which has too many states because some states may not be used if, according to the use case definition, certain pairs of incoming and outgoing messages cannot be coupled. To accommodate this, we minimize $|S^E|$ by serializing messages which yields the reduced set, denoted by $S^R = M^I \cup M^O$. Note that the controlling logic of the entity is created along with the states during message-based state derivation. With message-based state derivation, most flow charts and sequence diagrams can be translated to the state diagram easily and can be recognized by the *Simulation Execution Block*.

While translating use cases into identifiable entities, conditions and constraints, called *Meta Entities*, are added to the entity in the *Model Definition*. Another item of note is that the entity made by the *Entity Generator* is not connected to any other entities in this phase. Identifying entities from the use case and making a formalized *Meta Entity* are labor intensive, and it may require load balancing and optimization modules to save any substantial amount of time while completing this process.

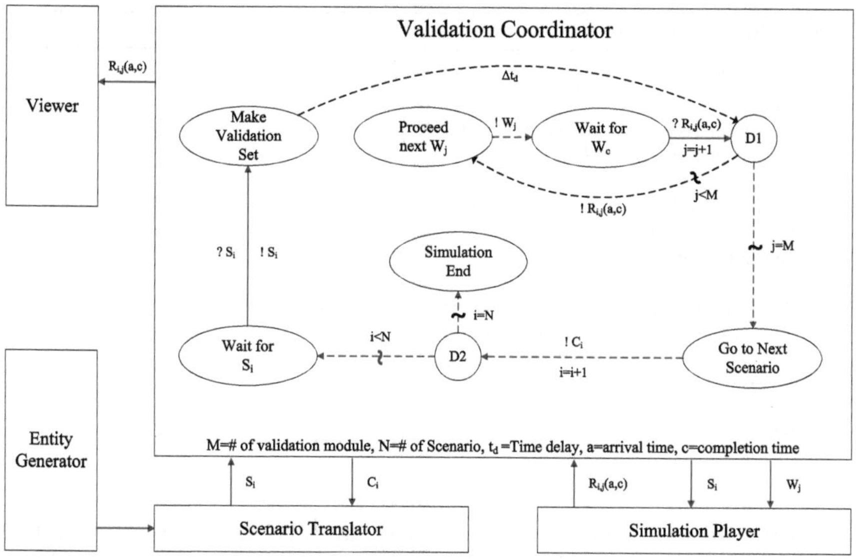

Fig. 2 Validation coordinator using DEVS formalism

3.3 Simulation Execution Block

Once entities are generated, establishing relations between entities should be carried out. The *Scenario Translator* creates these relations by deciding what messages are exchanged between all unique pairs of entities. A scenario is a distinct combination of entities, denoted by $S = \{S_1, S_2, \ldots, S_N\}$. When *Scenario Translator* receives *Request Scenario Update*, *Scenario Translator* provides the next scenario (S_{i+1}). The *Validation Coordinator* is a core element of our framework. It searches all possible validation methods in the *Assessment Library* which maintains requirements and testing modules. The *Validation Coordinator* makes selections from the *Assessment Library* and with those selections creates a validation set, denoted as $W = \{W_1, W_2, \ldots, W_M\}$, and sends a *Request Scenario Execution* message to the *Simulation Player*. Once it receives a *Report Simulation Result* message from the *Simulation Player*, the same action is continued until W_M is finished, which concludes one round of scenario validation. The *Simulation Execution Block* repeats the process again until the last scenario S_N is completed. This automated validation procedure can be easily expanded by adding another library, allowing for the evaluation of numerous use cases by simply changing scenarios.

Since all the components in our framework cooperate interactively each other, providing an adequate description of these interactions using a single algorithm would be prohibitively difficult. Instead, we adopt the DEVS modeling methodology to describe such dynamic interactions [18]. Figure 2 shows the internal and external structures of the *Validation Coordinator* (VC). When the execution begins, the VC

Algorithm 1: Simulation Execution Block

Input: A set of entities, E.
Output: A set of results, R.
/* Simulation Translator: *generate set of scenarios* */
$S \longleftarrow$ ScenarioTranslator(E);
foreach $s_i \in S$ **do**
  /* Validation Coordinator: *generate set of validation* */
  $W \longleftarrow$ AssessmentLibrary(s);
  foreach $w_j \in W$ **do**
    $a \longleftarrow$ arrival time;
    /* Simulation Player: *run* i^{th} *scenario and* j^{th} *validation* */
    $r \longleftarrow$ SimulationPlayer(s_i, w_j);
    $c \longleftarrow$ completion time;
    $R.Append(r, a, c)$;
return R;

stays in the *Wait for* S_i state until a scenario is sent from the *Scenario Translator*. Once the scenario is received, the VC sends S_i to the *Simulation Player* and then transitions to the *Make Validation Set* state. The *Simulation Player* then generates atomic models and establishes relations between atomic models. In the *Make Validation Set* state, the VC obtains a validation set $W = \{W_1, W_2, \ldots, W_M\}$, where $|W| = M$, from the *Assessment Library* with the time delay Δt.

With an initial j value equal to zero, the VC moves to the *Proceed next* W_j state. Before moving to the next state, the VC sends W_j (jth validation at the ith scenario) to the *Simulation Player*. Then, the VC waits until it receives $R_{i,j}(a, c)$ (result of the jth validation at the ith scenario, arrival at time a and completion at time c) at the *Wait for* W_c state. When received, the VC updates $j \longleftarrow j + 1$ and compares the values of j and M. If $j < M$, the VC repeats the process; otherwise ($j = M$) the VC moves to the *Go to Next Scenario* state. After comparing the values of i and N, if $i < N$ then the VC goes through another round of simulation; otherwise ($i = N$) the VC moves to the *Simulation End* state. During simulation, the *Viewer* updates simulation results periodically. To sum up, the overall *Simulation Execution Block* algorithm is shown in Algorithm 1, including all the sub modules of the *Simulation Execution Block*.

Note that although we adopt the DEVS modeling methodology to describe the VC in this work, any other methodology can be leveraged in our framework.

3.4 Viewer

The *Viewer* enables the user to monitor what events occur and what results the simulation generates. Its functionality is not only to display the results of a simulation, but also to educate the user what risks are involved and how they can be resolved. Through the *Viewer*, the effectiveness and reliability of countermeasures can be evaluated.

4 Case Study: Implementation Details and Evaluation Results

To demonstrate the feasibility and reliability of our framework, this section starts with a use case from a real-time pricing scenario and articulates critical components in this use case. Next, we describe how requirements specified in this use case can be realized in our framework. Also, we elaborate upon the results from our evaluation.

4.1 Requirements for Real-Time Pricing

In the *Real-Time Pricing* (RTP) scenario detailed in [8], each of the domain stakeholders correspond with each other to circulate pricing information and exchange their constraints, such as power outage, ancillary services, etc. The motivation for RTP arises from the disparity between the amount of electricity generated by power plants and the amount of energy demanded by customers. Ideally, power companies would be able to accurately predict exactly how much demand there would be at any given time, but the reality is that sporadic usage spikes and ebbs create energy surpluses and shortages all the time, resulting in either wasted energy production or shortages in the amount of electric power delivered to customers.

Intuitively, there is a direct relationship between the demand for electricity and its price, increasing during periods of peak usage,[3] and decreasing when demand is low, such as during the night. However, Service Providers typically use what is called a *fixed price list* or *fixed tariff*, which does not reflect a fine-grained view of market circumstances. Hence, the RTP approach, which updates prices hourly, provides greatly improved price data and is able to vitalize the energy market. In other words, it would significantly contribute to the fulfillment of the business continuity objective that is one of the important requirements for critical infrastructure, including the Smart Grid. Hourly price calculation models have been proposed by many researchers [2, 19, 20]. For our case study, we selected Allcott's model [21] since this approach formulates accurate price changes according to customers' demand. The following is a slightly modified RTP calculation equation.[4] Based on Allcott's model, we additionally introduce a distributed energy resource (DER) factor, $\sum_i d_{it}$, in Eq. 1 and an ancillary service cost, P_a, in Eq. 2.

$$Q_t^s(P_t) = \sum_j k_{jt} + \sum_i d_{it} \tag{1}$$

---

[3] Peak usage times may vary for each Energy Service Provider, but are generally weekday afternoons from 2 pm to 6 pm in Arizona. The relevant reference is available at http://www. azenergy.gov/SavingTips/TimeOfUse.aspx.

[4] $\alpha = \frac{RTP_{users}}{All_{users}}$, P_t = real-time price, $\bar{P}$ = fixed tariff price, P_c = capacity market cost, P_a = ancillary service cost, η = elasticity of demand variable, ϵ_t =error fixing variable.

$$Q_t^d(P_t, \bar{P}, P_c) = \{\alpha(P_t + P_c + P_a)^\eta + (1-\alpha)(\bar{P} + P_c + P_a)^\eta\} \cdot \epsilon_t \tag{2}$$

$$Q_t^s(P_t) = (1+m)Q_t^d(P_t, \bar{P}, P_c) \tag{3}$$

Equation 1 is the total generation function that sums power plants' generation and DER generation where t is a specific time period, j is a power plant instance, and i is a customer. Equation 2 defines the expected customer's demand. Allcott used three kinds of prices which are P_t, $\bar{P}$ and P_c and we added one more price factor P_a. Equation 3 shows the equilibrium equation which determines the real-time price. For this calculation, we adopt a reserve margin index m which can be obtained at [22].

In order to identify the target requirements for our case study, we first provide a summary of the decision process: the Bulk Generation company announces the initial raw prices at which it will sell energy in the energy market. After adding transmission and distribution costs, each company finalizes their base price. At the same time, each energy service provider gathers customers' estimated energy demand and sends an aggregated demand amount to the energy market, where the real-time price is calculated.

Based on this decision process, we notice that protecting customers' privacy and maintaining price data integrity are essential in RTP. However, since the latter requirement is closely coupled with RTP model, our simulation mainly focuses on how an RTP scenario can be realized in our framework and how our simulation can detect key components involved in the RTP decision process.

4.2 Design and Implementation

To design an RTP use case in our framework, we adopt the conceptual model from NIST. We use only four of the domains by making the assumption that there is zero cost incurred by the Transmission and Distribution domains. In the Bulk Generation domain, there are five types of power plants according to their energy source: coal, natural gas, nuclear, hydro electric, and renewable. Customers' residency styles in the Customer domain (represented in our model by the Customer Building Automation System) can be one of four types: detached, semi-detached, apartment, and terraced [23]. Data types of our case study are shown in Table 2.

Once electricity is generated, the next step performs load-balancing and pricing for the electricity. The energy scheduler balances total supply and expected demand by mediating between the Bulk Generation entity and the Service Provider entity (equivalent to the Energy Service Provider). The real-time pricing decision is made by the Base RTP Calculator in the Market domain, but prices may fluctuate since customers' energy usage may be affected by the set price. Once the real-time price is calculated, pricing information is delivered to the customers. Figure 3 illustrates these components and its relationships along with the real-time price decision process.

Table 2 Input data types

CBAS domain		Bulk generation domain	
Variable name	Data type	Variable name	Data type
CustomerID	int	BulkID	int
CustomerType	String	BulkType	String
Period	int	BulkCapacity	double
CorrespondingESP	int	BulkDestination	double
EnergyDemand	double[]	Period	int
DERCapacity	double	LoadElectricity	double
DERLoad	double	RawPrice	double
AncillaryService	String	Constraints	String
Constraints	String		

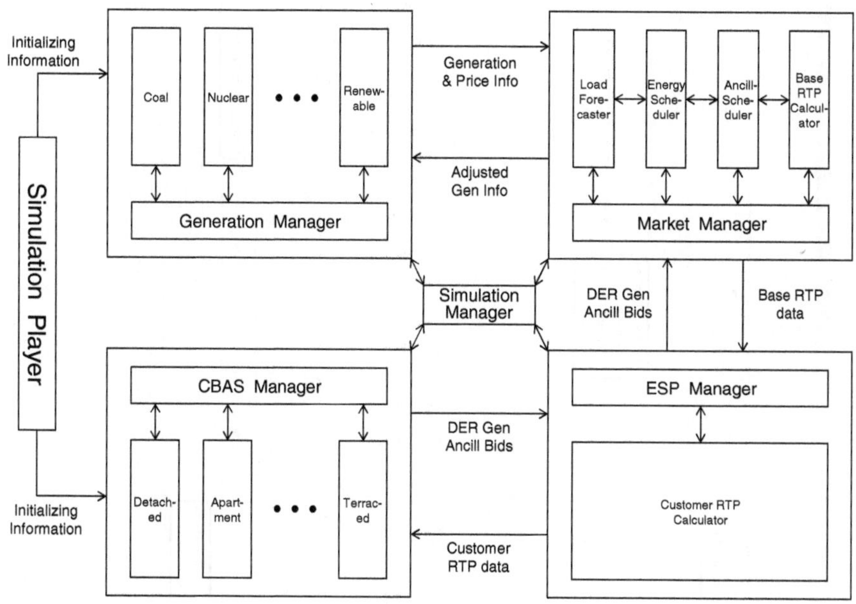

Fig. 3 RTP scenario generation

Based on our framework, to realize the RTP use case with the DEVS modeling methodology, we utilize a DEVS supporting simulator called MS4.[5] By adopting the DEVS supporting simulator, we realize the procedures illustrated in Fig. 2. One of the advantages of using MS4 is that it provides a simulation viewer, eliminating the need to construct our own specialized viewer.

As shown in Fig. 4, the overall appearance is quite similar to the RTP design. To support our framework, four simulation entities representing the four domains

[5] MS4 software is available at http://www.ms4systems.com/pages/ms4me.php.

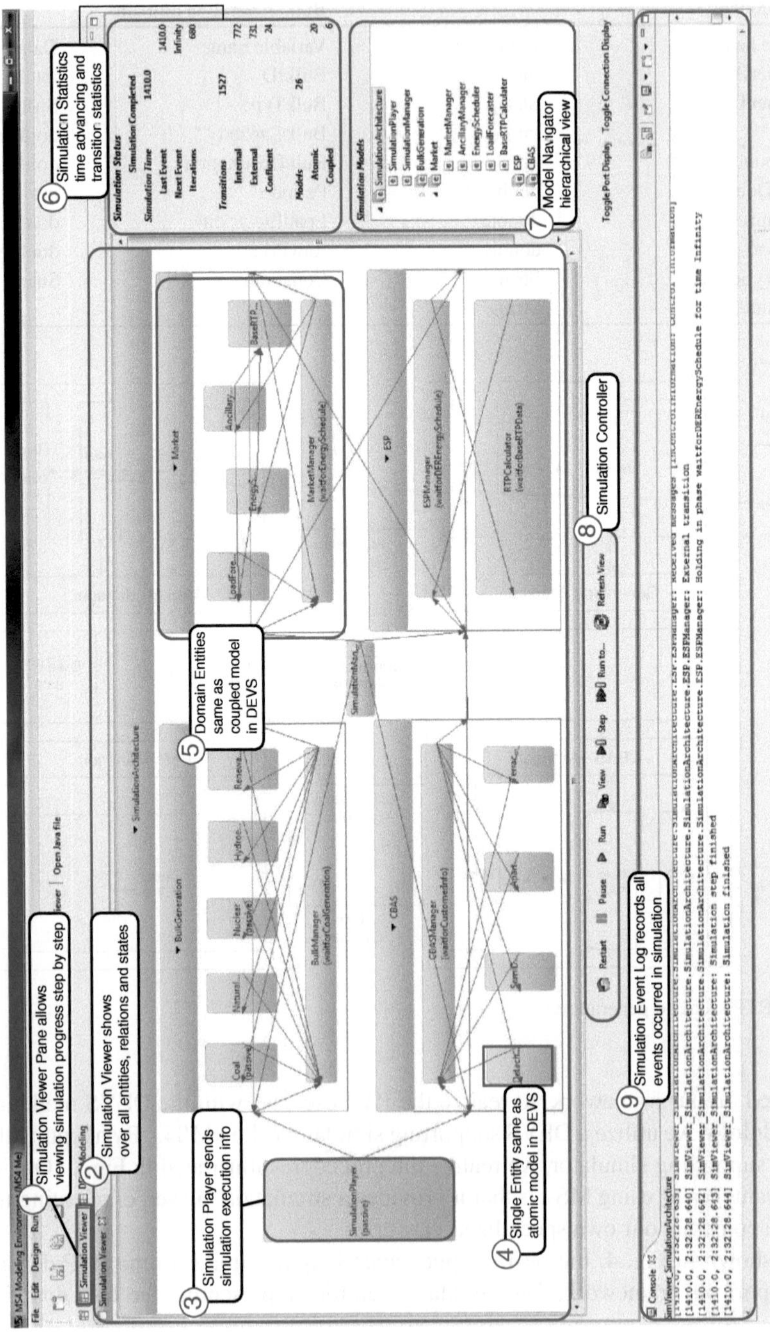

Fig. 4 Simulation entities in MS4 simulator

were implemented. State transitions in each entity and message exchanges among entities were analyzed for each step (see simulation controller). After the simulator completely executed the use case, we produced simulation statistics. Moreover, result graphs were generated for further analyzing the simulation results.[6]

4.3 Simulation Results

To perform a realistic simulation, we considered two Energy Service Providers in the state of Arizona (SRP and APS)[7] and used production information and retail energy prices for their power plants. In addition, we took the customers' daily energy usage behavior available from [23]. By applying real-world data, simulating an RTP use case is more reliable and meaningful.

Table 3 shows the number of exchanged messages when the number of power plants is 9 and the number of customers is 200 (100 for each ESP). It shows that 56 % of all intra-domain messages (1,046/1,881) are exchanged within the Market Operation (MO) domain, which means MO is the key infrastructure to protect for supporting a reliable RTP decision process. Furthermore, 64 % of all inter-domain messages (800/1,243) are generated between the ESP and the Customer Building Automation System (CBAS). Hence, the network between the ESP and the CBAS needs to be carefully supervised to prevent potential data leakage.

The next set of results depicts the simulation under different scenarios. The RTP ratio, denoted by α, represents the percentage of customers that have elected to use the RTP model for their service. Three test scenarios with different values of α were considered with diverse residency types as follows.

$$S = \{S_{\alpha=0.1}, S_{\alpha=0.3}, S_{\alpha=0.5}\}$$

$$W = \{W_{Detached}, W_{SemiDetached}, W_{Apartment}, W_{Terraced}\}$$

As shown in Table 4, the standard deviation of each residency type is considerably reduced when the value of α increases. This result shows how radical price changes can be produced when the value of α is small, which can cause severe distrust in the RTP system. In Figs. 5 and 6, detached residency type shows how α value impacts overall RTP system's safety. When $\alpha = 0.1$, RTP fluctuation is huge compared to $\alpha = 0.3$ or $\alpha = 0.5$. This means that RTP may face unexpected, huge fluctuations when RTP is in its early stages of customer adoption. Moreover, SRP's maximum real-time price is 5.63 times its fixed tariff price, as shown in Fig. 7, and APS's maximum real-time price is 5.15 times its fixed tariff price, as shown in Fig. 6.

[6] The simulation viewer also provides state updates, message exchange animations, as well as a mechanism for advancing time.

[7] The information of each energy service provider is available at https://www.srpnet.com and http://www.aps.com/en/residential/Pages/home.aspx, respectively.

Table 3 Number of exchanged messages

Intra-domain		Inter-domain	
Bulk generation (BG)	18	Simulation player ↔ Domains	6
Market operation (MO)	1,046	BG ↔ MO	19
Energy service provider (ESP)	409	MO ↔ ESP	418
Customer building automation system (CBAS)	408	ESP ↔ CBAS	800
Total	1,881	Total	1,243

Table 4 Standard deviation comparison

Scenario	Detached		SemiDetached		Apartment		Terraced	
	SRP	APS	SRP	APS	SRP	APS	SRP	APS
$\alpha = 0.1$	0.049313	0.172181	0.077588	0.107786	0.103452	0.057526	0.053144	0.1214
$\alpha = 0.3$	0.02106	0.027133	0.019465	0.02707	0.023043	0.028997	0.016388	0.02779
$\alpha = 0.5$	0.011554	0.015463	0.011423	0.016457	0.008256	0.020822	0.012499	0.016702
Total	0.031302	0.105675	0.046573	0.064329	0.061451	0.03866	0.032567	0.072146

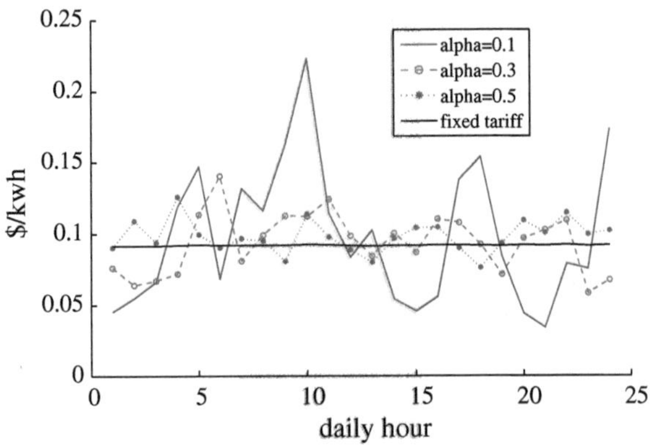

Fig. 5 SRP RTP fluctuation (detached type only)

These radical price changes may cause further distrust in RTP, hence countermeasures to mitigate these changes, such as modifying the elasticity constant or revising high prices, should be considered.[8]

We conducted another experiment to test the elasticity of price. As presented in Eq. 2, η is a crucial factor which determines the fluctuation of real-time prices. Typically η is between -0.04 and -0.15 [20, 24–26]. In order to test which η value is valid for our case study, we changed η value at intervals of 0.04 and measured the deviation of prices by taking $|\text{PRICE}_{RTP} - \text{PRICE}_{Fixed}|$. As shown in Fig. 8, the result at $\eta = -0.12$ deviates much less than $\eta = -0.04$ and $\eta = -0.08$. Due to the nature of Eq. 2 (expected customer's demand), we can reduce RTP fluctuation by

[8] In order to reduce redundancy, we mainly address compulsive cases from our evaluation results in this chapter.

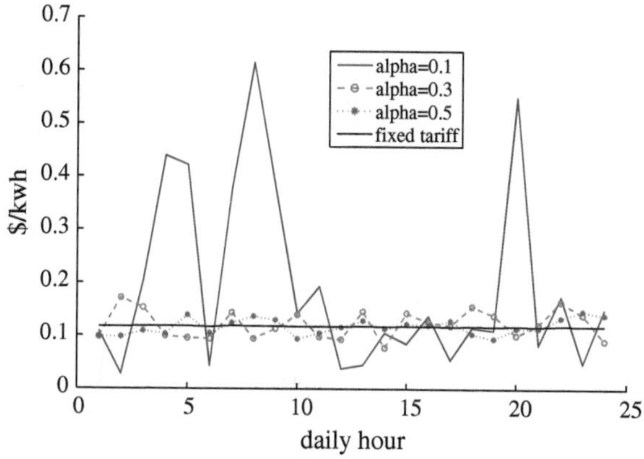

Fig. 6 APS RTP fluctuation (detached type only)

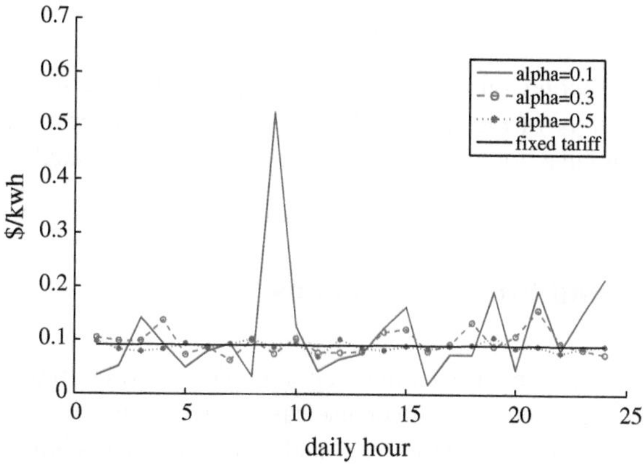

Fig. 7 SRP RTP fluctuation (apartment type only)

setting a small value for η when making a decision on a real-time price. However, finding an appropriate fluctuation level is our goal for RTP simulation, instead of mainly reducing RTP fluctuation. Hence, we measured a break-even η value that satisfies the same gross sales amount of electricity between two cases: under fixed-tariff scenario and under RTP scenario. This helps us understand what price is acceptable for both fixed tariff users and RTP users. We assumed that customers' demand has not been changed by the real-time price. We found that when $\eta = -0.083$, gross sales have the same value for all scenarios. This result leads us to determine that an appropriate η value is also important for RTP, and improper values of η may cause distrust of the system because of the large fluctuations in the simulation results.

Through our case study of RTP, we evaluated the feasibility of our framework. First, our framework provides an easy transformation from entity design to simulation execution, which enables us to understand RTP use case without requiring

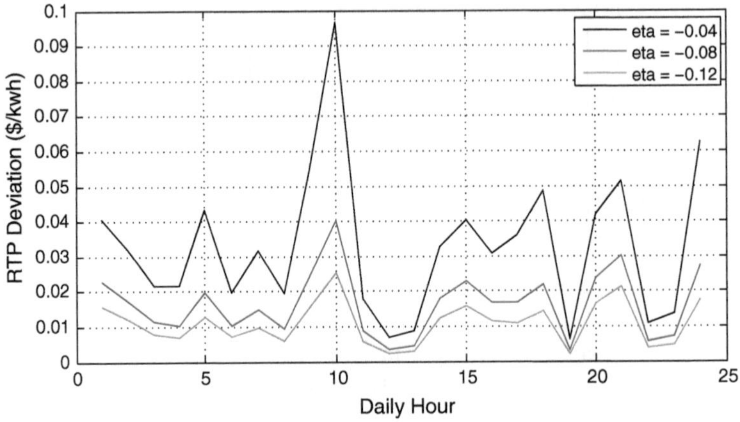

Fig. 8 Changes in elasticity of price (ESP = SRP, Type = Detached, $\alpha = 0.3$)

significant effort. Second, our simulation-based approach using the DEVS modeling methodology gives us various practical results that can answer the critical requirements of RTP that have not been previously identified or validated.

5 Conclusion and Future Directions

As the Smart Grid system has become more complex, its validation and verification process depends heavily upon realistic use cases, such as new requirements, energy resource relocation, and so on. Moreover, such a critical process is a tedious and difficult task without the support of an appropriate systematic approach. To resolve these problems, we have proposed the Simulation-Based Validation Framework using the DEVS modeling methodology. Our framework consists of three core components: Entity Generator, Simulation Execution Block and Viewer. We demonstrated how the Entity Generator can create individual entities in an identifiable format, and the Simulation Execution Block can generate a number of scenarios (with entities made by the Entity Generator) and execute the simulation while the Viewer provides updates on simulation's progress. Also, by performing various simulation experiments on a real-time pricing use case, we showed how critical issues in use cases can be simulated and discovered based on the proposed framework.

In our future work, we will articulate various requirements with our framework and further enhance our approach to support use case generation and validation intuitively, particularly focusing on security requirements. In particular, the following areas can be further studied:

Business intelligence Simulation-based approaches utilize various kinds of use cases to search for and find any possible risks that can impact running systems. As James [27] claimed that "acknowledging the business impact of cyber,...leveraging timely business intelligence,...[and] broaden[ing] awareness" are crucial to establishing resilient cyber systems, such characteristics of simulation-based approach can be also extended to the area of business intelligence. For instance, our framework can provide useful data for producing various scenarios that need to be investigated. Moreover, predictive analysis can be performed by our framework. Watson et al. [28] introduced the importance of a business intelligence system that fosters "the use of information and analytics." As shown in Sect. 4.3, forecasting variables of interest and testing scenarios with different conditions in our framework would be tremendously helpful to make business decisions in an effective manner.

Risk management As our RTP use case illustrates the impact of economical issues and the price of electricity on Smart Grid systems, our framework can be further extended to assess potential risks in large-scale distributed systems. Varaiya et al. [29] pointed out that the price of electricity is the biggest risk with respect to the economic challenges of Smart Grid systems. Chao [30] even claimed that fixed uniform price policies remain a considerable barrier and may prevent the success of the Smart Grid. Although RTP is an essential factor to realize the Smart Grid, our study showed that fluctuations of RTP should be restricted due to potential risks and it is necessary to mitigate such fluctuations to a manageable and acceptable risk level for preventing customers from evading the use of RTP systems. Consequently, our approach would help discover potential risks and evaluate diverse mitigation methods to minimize potential risks. In addition, our approach can be further extended to support other domains by articulating uses cases for those target domains.

Acknowledgments This work was partially supported by grants from the National Science Foundation and the Department of Energy.

References

1. Garcia, R.C., Contreras, J., van Akkeren, M., Garcia, J.B.C.: A garch forecasting model to predict day-ahead electricity prices. IEEE Trans. Power Syst. **20**(2), 867–874 (2005)
2. Mohsenian-Rad, A.-H., Leon-Garcia, A.: Optimal residential load control with price prediction in real-time electricity pricing environments. IEEE Trans. Smart Grid **1**(2), 120–133 (2010)
3. Arora, M., Das, S.K., Biswas, R.: A de-centralized scheduling and load balancing algorithm for heterogeneous grid environments. In: Proceedings of the International Conference on Parallel Processing Workshops (ICPPW'02), pp. 499–505 (2002)
4. Molderink, A., Bakker, V., Bosman, M.G.C., Hurink, J.L., Smit, G.J.M.: Domestic energy management methodology for optimizing efficiency in smart grids. In: Proceedings of the IEEE Bucharest PowerTech 2009, 1–7 July 2009
5. Metke, A.R., Ekl, R.L.: Security technology for smart grid networks. IEEE Trans. Smart Grid **1**(1), 99–107 (2010)
6. Ericsson, G.N.: Cyber security and power system communication—essential parts of a smart grid infrastructure. IEEE Trans. Power Delivery **25**(3), 1501–1507 (2010)
7. Nist framework and roadmap for smart grid interoperability standards. http://www.nist.gov/public_affairs/releases/upload/smartgrid_interoperability_final.pdf (2012). Accessed Feb 2012
8. Energy power research institute, real-time pricing—top level. http://smartgrid.epri.com/Repository/Repository.aspx (2012). Accessed Feb 2012

9. Cox, W., Holmberg, D., Sturek, D.: Oasis collaborative energy standards, facilities, and zigbee smart energy. In: Grid-Interop Forum 2011 (2011)
10. Zigbee smart energy 2.0 draft 0.9 public application profile. http://www.zigbee.org/Standards/ZigBeeSmartEnergy/ZigBeeSmartEnergy20PublicApplicationProfile.aspx (2012). Accessed July 2012
11. Cybersecurity working group final three-year plan. http://collaborate.nist.gov/twikisggrid/bin/view/SmartGrid/CSWGRoadmap (2011). Accessed Apr 2011
12. Electricity subsector cybersecurity capability maturity model (es-c2m2). http://energy.gov/oe/services/cybersecurity/electricity-subsector-cybersecurity-capability-maturity-model (2012). Accessed May 2012
13. Lin, J., Sedigh, S., Miller, A.: Modeling cyber-physical systems with semantic agents. In: Proceedings of the IEEE 34th Annual Computer Software and Applications Conference Workshops (COMPSACW) 2010, 13–18 July 2010
14. Pipattanasomporn, M., Feroze, H., Rahman, S.: Multi-agent systems in a distributed smart grid: design and implementation. In: IEEE/PES Power Systems Conference and Exposition (PSCE'09), pp. 1–8 Mar 2009
15. Stevens, F., Courtney, T. Singh, S., Agbaria, A., Meyer, J.R., Sanders, W.H., Pal, P.: Model-based validation of an intrusion-tolerant information system. In: Proceedings of the 23rd IEEE International Symposium on Reliable Distributed Systems (SRDS'04), pp. 184–194 Oct 2004
16. Nicol, D.M., Sanders, W.H., Trivedi, K.S.: Model-based evaluation: from dependability to security. IEEE Trans. Dependable Secure Comput. **1**(1), 48–65 (2004)
17. Jonsson, E., Olovsson, T.: A quantitative model of the security intrusion process based on attacker behavior. IEEE Trans. Softw. Eng. **23**(4), 235–245 (1997)
18. Zeigler, B.P., Praehofer, H., Kim, T.G.: Theory of Modeling and Simulation: Integrating Discrete Event and Continuous Complex Dynamic Systems. Academic Press, San Diego (Feb 2000)
19. Allcott, H.: Real time pricing and electricity markets. http://www-prd-0.gsb.stanford.edu/facseminars/events/applied_microecon/documents/ame_03_09_allcott.pdf (2009). Accessed Jan 2009
20. Taylor, Thomas N., Schwarz, Peter M., Cochell, James E.: 24/7 hourly response to electricity real-time pricing with up to eight summers of experience. J. Regul. Econ. **27**, 235–262 (2005)
21. Allcott, H.: Real-time pricing and electricity market design. https://files.nyu.edu/ha32/public/research/Allcott-Real-TimePricingandElectricityMarketDesign.pdf (2013). Accessed Mar 2013
22. Levelized cost of new generation resources in the annual energy outlook 2013. http://www.eia.gov/forecasts/aeo/electricity_generation.cfm (2013). Accessed Jan 2013
23. Ghaemi, S., Brauner, G.: User behavior and patterns of electricity use for energy saving. Internationale Energiewirtschaftstagung an der TU Wien, IEWT (2009)
24. Patrick, R.H., Wolak, F.A.: Estimating the customer-level demand for electricity under real-time market prices. Technical report, National Bureau of Economic Research, Washington (Apr 2001)
25. Herriges, J.A., Baladi, S.M., Caves, D.W., Neenan, B.F.: The response of industrial customers to electric rates based upon dynamic marginal costs. Rev. Econ. Stat. **75**(3), 446–454 (1993)
26. Boisvert, R.N., Cappers, P., Goldman, C., Neenan, B., Hopper, N.: Customer response to rtp in competitive markets: a study of niagara mohawk's standard offer tariff. Energy J. **28**(1), 53–74 (2007)
27. White, J.: 12 steps toward cyber resilience. InfoSecurity Professional INSIGHTS 2(2). https://www.isc2.org/infosecurity-professional-insights-archives.aspx?terms=12-Steps-toward-Cyber-Resilience (2013)
28. Watson, H.J., Wixom, B.H.: The current state of business intelligence. Computer **40**(9), 96–99 (2007)
29. Varaiya, P.P., Wu, F.F., Bialek, J.W.: Smart operation of smart grid: risk-limiting dispatch. Proc. IEEE **99**(1), 40–57 (2011)
30. Chao, Hung-po: Price-responsive demand management for a smart grid world. Electr. J. **23**(1), 7–20 (2010)

An Institution for Alloy and Its Translation to Second-Order Logic

Renato Neves, Alexandre Madeira, Manuel Martins and Luís Barbosa

Abstract Lightweight formal methods, of which ALLOY is a prime example, combine the rigour of mathematics without compromising simplicity of use and suitable tool support. In some cases, however, the verification of safety or mission critical software entails the need for more sophisticated technologies, typically based on theorem provers. This explains a number of attempts to connect ALLOY to specific theorem provers documented in the literature. This chapter, however, takes a different perspective: instead of focusing on one more combination of ALLOY with still another prover, it lays out the foundations to fully integrate this system in the HETS platform which supports a huge network of logics, logic translators and provers. This makes possible for ALLOY specifications to "borrow" the power of several, non dedicated proof systems. The chapter extends the authors' previous work on this subject by developing in full detail the semantical foundations for this integration, including a formalisation of ALLOY as an institution, and introducing a new, more general translation of the latter to second-order logic.

Keywords Model finding · Theorem proving · Second–order logic

---

R. Neves (✉) · L. Barbosa (✉)
INESC TEC (HASLab), University of Minho, Braga, Portugal
e-mail: nevrenato@gmail.com

L. Barbosa
e-mail: lsb@di.uminho.pt

A. Madeira
Department of Mathematics, University of Aveiro, Aveiro, Portugal
e-mail: madeira@ua.pt

M. Martins
Center for Research and Development in Mathematics and Applications—Department of Mathematics, University of Aveiro, Aveiro, Portugal
e-mail: martins@ua.pt

T. Bouabana-Tebibel and S. H. Rubin (eds.), *Integration of Reusable Systems*, Advances in Intelligent Systems and Computing 263, DOI: 10.1007/978-3-319-04717-1_3,

1 Introduction

In [17] the authors discussed the integration of ALLOY [9] in HETS platform of logics, logic translators and provers, by sketching its formalisation as an institution [6, 7] and defining its encoding into CASL [14], an extension of multi-sorted first order logic with partiality and free types. The motivation was clear: to offer a systematic way to connect ALLOY to a huge network of logics and logical systems in order to complement the model finder strategies of the former with suitable theorem provers already linked into the latter.

Actually, ALLOY, based on a single sorted relational logic whose models can be automatically tested with respect to bounded domains, is a most successful tool for the working software engineer. Its simple but powerful language combined with an analyser which can promptly give counter-examples depicted graphically, makes it increasingly popular both in academia and industry: Successful stories report on the discovery of faults in software designs, supposedly faultless, by taking advantage of ALLOY. The tool, however, may also bring a false sense of security, as absence of counter-examples does not imply model's correctness. Therefore, in the project of critical systems our research claims the use of ALLOY should be framed into broader toolchains involving more general, even if often less friendly, theorem provers. In such a toolchain properties can be first tested within the ALLOY analyser; if no counter-examples are found, a theorem prover would be asked to generate a proof, at least in what concerns critical design fragments.

There is a number of results on connecting ALLOY to specific theorem provers. The perspective taken in [17], however, and complemented in this chapter goes a step further by "plugging" ALLOY into the HETS network, providing a number of effective and sound connections to several logical systems and tools at once. Currently, HETS integrates several world-class reasoners, namely ISABELLE [18], LEO- II [4], SPASS [21], VAMPIRE [19], DARWIN [2], among many others. Plugging ALLOY to HETS, makes thus possible the translation of its models to a number of languages in the network, and naturally, borrowing for free the corresponding proof support. Experiments can then be carried out in different tools, typically tuned to specific application areas.

Actually, HETS [15] may be regarded as a "motherboard" for logics where different "expansion cards" can be plugged. The latter are individual logics (with associated analysers and proof tools) as well as logic translations to "transport" properties and proofs between them. To make them *compatible*, logics are formalised as *institutions* [6] and logic translations as *comorphisms*. This is the price to be paid: the integration of ALLOY in this network entails the need for formalising ALLOY as an institution and providing a translation to a relevant logic in the HETS network.

The present chapter addresses this challenge by developing in full detail the semantical foundations for this integration, including a formalisation of ALLOY as an institution, and introducing a new, more general translation of it to second-order

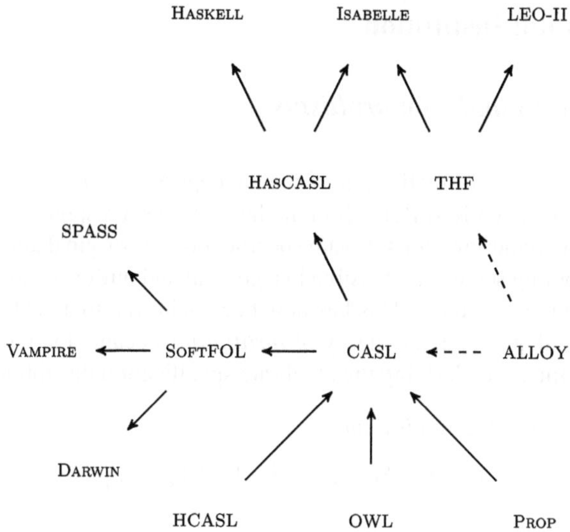

Fig. 1 HETS sub-network extended with ALLOY

logic (SOL) [12]. This new translation allows a more natural embedding of ALLOY, when compared to the previous, essentially first-order translation to CASL introduced in [17]. Moreover, this translation establishes a connection between ALLOY and higher-order provers, such as LEO- II and ISABELLE, using the logic THF [3] which integrates second-order features. The original translation to CASL, presented in [17], is re-framed in this context, because, in practice, it opens doors to a broad number of theorem provers, namely for first-order systems. Both translations are depicted in Fig. 1 as dashed arrows.

Related work. The idea of connecting ALLOY to a theorem prover is not new—see, for example, references [1, 10, 20]. The usual approach is to translate ALLOY models into the input language of a given theorem prover and (re-)formulate the proof targets accordingly. For instance, [20], one of the most recent proposals in this trend, translates models into a first-order dialect supported by the KEY theorem prover.

Paper structure. The formalisation of ALLOY as an institution and the definition of suitable comorphisms is presented in Sects. 3, 4 and 5. Before that, in Sect. 2, a brief overview of the theory of institutions is provided as a background for the chapter. Section 6 reports on a (fragment of a) case study in the medical domain on the combined use of ALLOY and HETS, to illustrate the potential and limits of the approach proposed here. Finally, Sect. 7 concludes.

2 Background: Institutions

2.1 Institutions and Comorphisms

An *institution* [7] is a formalisation of the concept of a logical system, introduced by Joseph Goguen and Rod Burstall in the late 70's, as a response to the increasing number of logics emerging for software specification. Its original aim was to develop as much computing science as possible in a general and uniform way, independently of particular logical systems. This has now been achieved to an extent even greater than originally thought, as the theory of *institutions* became the most fundamental mathematical theory underlying the algebraic specification discipline.

Definition 1. *An* institution *is a tuple*

$$(Sign^{\mathcal{I}}, Sen^{\mathcal{I}}, Mod^{\mathcal{I}}, \{\models^{\mathcal{I}}_{\Sigma}\}_{\Sigma \in |Sign^{\mathcal{I}}|})$$

where

- $Sign^{\mathcal{I}}$ *is a category of signatures and signature morphisms,*
- $Sen^{\mathcal{I}} : Sign^{\mathcal{I}} \to \mathbb{S}et$ *is a functor relating signatures to the corresponding sentences, where* $\mathbb{S}et$ *is the category of Sets,*
- $Mod^{\mathcal{I}} : (Sign^{\mathcal{I}})^{op} \to \mathbb{C}at$ *is a functor giving for each signature* Σ*, the category of its models, where* $\mathbb{C}at$ *is the category of categories,*
- $\models^{\mathcal{I}}_{\Sigma} \subseteq |Mod^{\mathcal{I}}(\Sigma)| \times Sen^{\mathcal{I}}(\Sigma)$ *is the satisfaction relation between models and sentences such that, for each morphism* $\varphi : \Sigma \to \Sigma'$ *in* $Sign^{\mathcal{I}}$*, and for any* $M' \in |Mod^{\mathcal{I}}(\Sigma')|$ *and* $\rho \in Sen^{\mathcal{I}}(\Sigma)$*,*

$$M' \models^{\mathcal{I}}_{\Sigma'} Sen^{\mathcal{I}}(\varphi)(\rho) \text{ iff } Mod^{\mathcal{I}}(\varphi)(M') \models^{\mathcal{I}}_{\Sigma} \rho$$

The reduct of M' through a signature morphism φ is defined by $Mod^{\mathcal{I}}(\varphi)(M')$, and denoted by $M' \upharpoonright \varphi$. Dually, M' is called a model φ–expansion of $M' \upharpoonright \varphi$.

Example 1. To illustrate the concept of an institution, consider, in this example, the construction of an institution for first-order logic (FOL).

Signatures. $Sign^{\mathrm{FOL}}$ is a category whose objects are triples (S, F, P), where S is the set of sort symbols, F a family of function symbols indexed by their arity, $F = \{F_{w \to s} | w \in S^*, s \in S\}$ and P a family of relational symbols also indexed by their arity, $P = \{P_w | w \in S^*\}$. A signature morphism in this category is also a triple $(\varphi_{st}, \varphi_{op}, \varphi_{rl})$ such that for $\varphi : (S, F, P) \to (S', F', P')$, if $\sigma \in F_{w \to s}$, then $\varphi_{op}(\sigma) \in F'_{\varphi_{st}(w) \to \varphi_{st}(s)}$, and if $\pi \in P_w$ then $\varphi_{rl}(\pi) \in P'_{\varphi_{st}(w)}$.

Sentences. For each signature object $(S, F, P) \in |Sign^{\mathrm{FOL}}|$, $Sen^{\mathrm{FOL}}(S, F, P)$ is the smallest set of first order sentences:

$t \approx t'$,	for $t, t' \in terms$
$\pi(t_1, \ldots, t_n)$,	for $t_1, \ldots, t_n \in terms$ and $\pi \in P_w$
$\neg\rho$,	for $\rho \in Sen^{\mathrm{FOL}}(S, F, P)$
$\rho \Rightarrow \rho'$,	$\rho, \rho' \in Sen^{\mathrm{FOL}}(S, F, P)$
$\forall x : s \,.\, \rho$,	$s \in S, \rho \in Sen^{\mathrm{FOL}}(S, F \uplus \{x\}_{\to s}, P)$

where a term of sorts is a syntactic structure $\sigma(t_1, \ldots, t_n)$, such that $\sigma \in F_{s_1,\ldots,s_n \to s}$ and $t_1, \ldots, t_n$ are terms of sort $s_1, \ldots, s_n$, respectively. A signature morphism φ defines a term translation function $terms(\varphi)$, given by

$$terms(\varphi)(\sigma(t_1, \ldots, t_n)) = \varphi_{op}(\sigma)(terms(\varphi)(t_1), \ldots, terms(\varphi)(t_n)).$$

Given a signature morphism φ in $Sign^{\mathrm{FOL}}$, sentences are mapped in the following way:

$$\begin{aligned}
Sen^{\mathrm{FOL}}(\varphi)(t \approx t') &= terms(\varphi)(t) \approx terms(\varphi)(t') \\
Sen^{\mathrm{FOL}}(\varphi)(\pi(t_1, \ldots, t_n)) &= \varphi_{rl}(\pi)(terms(\varphi)(t_1), \ldots, terms(\varphi)(t_n)) \\
Sen^{\mathrm{FOL}}(\varphi)(\neg\rho) &= \neg Sen^{\mathrm{FOL}}(\varphi)(\rho) \\
Sen^{\mathrm{FOL}}(\varphi)(\rho \Rightarrow \rho') &= Sen^{\mathrm{FOL}}(\varphi)(\rho) \Rightarrow Sen^{\mathrm{FOL}}(\varphi)(\rho') \\
Sen^{\mathrm{FOL}}(\varphi)(\forall x : s \,.\, \rho) &= \forall x : \varphi_{st}(s) \,.\, Sen^{\mathrm{FOL}}(\varphi')(\rho), \\
&\quad \text{where } \varphi' \text{ canonically extends } \varphi \text{ with } \varphi'_{op}(x) = x
\end{aligned}$$

Models. For each signature $(S, F, P) \in |Sign^{\mathrm{FOL}}|$, $Mod^{\mathrm{FOL}}(S, F, P)$ is a category whose objects are models with the following components : a carrier set $|M_s|$, for each $s \in S$; a function $M_\sigma : |M_w| \to |M_s|$, for each $\sigma \in F_{w \to s}$; a relation $M_\pi \subseteq |M_w|$, for each $\pi \in P_w$.
For any signature morphism $\varphi : (S, F, P) \to (S', F', P')$, and any (S', F', P')–model M', $Mod^{\mathrm{FOL}}(\varphi)(M')$, or $M' \restriction \varphi$, is defined as:

- for any $s \in S$, $|(M' \restriction \varphi)_s| = |M'_{\varphi_{st}(s)}|$
- for any $\sigma \in F_{w \to s}$, $(M' \restriction \varphi)_\sigma = M'_{\varphi_{op}(\sigma)}$
- for any $\pi \in P_w$, $(M' \restriction \varphi)_\pi = M'_{\varphi_{rl}(\pi)}$

Satisfaction. For any Σ–model $M \in |Mod^{\mathrm{FOL}}(\Sigma)|$, with $\Sigma \in |Sign^{\mathrm{FOL}}|$, the satisfaction relation is inductively defined in the following way:

$$\begin{aligned}
&M \models^{\mathrm{FOL}}_{\Sigma} t \approx t' && \text{iff } M_t = M_{t'} \\
&M \models^{\mathrm{FOL}}_{\Sigma} \pi(t_1, \ldots, t_n) && \text{iff } (t_1, \ldots, t_n) \in M_\pi \\
&M \models^{\mathrm{FOL}}_{\Sigma} \neg\rho && \text{iff } M \not\models^{\mathrm{FOL}}_{\Sigma} \rho \\
&M \models^{\mathrm{FOL}}_{\Sigma} \rho \Rightarrow \rho' && \text{iff } M \models^{\mathrm{FOL}}_{\Sigma} \rho' \text{ whenever } M \models^{\mathrm{FOL}}_{\Sigma} \rho \\
&M \models^{\mathrm{FOL}}_{\Sigma} \forall x : s \,.\, \rho && \text{iff for any model } x\text{–expansion } M' \text{ of } M,\ M' \models^{\mathrm{FOL}}_{\Sigma'} \rho
\end{aligned}$$

A comorphism is a mapping that "embeds" a structurally "simpler" institution into a more "complex" one.

Definition 2. *Formally, given two institutions* $\mathcal{I}, \mathcal{I}'$, *a comorphism from* $\mathcal{I}$ *to* $\mathcal{I}'$ *is a triple* (Φ, α, β) *consisting of*

- *a functor* Φ: $Sign^{\mathcal{I}} \to Sign^{\mathcal{I}'}$,
- *a natural transformation* α: $Sen^{\mathcal{I}} \Rightarrow Sen^{\mathcal{I}'}.\Phi$,
- *a natural transformation* β: $Mod^{\mathcal{I}'}.\Phi^{op} \Rightarrow Mod^{\mathcal{I}}$,

such that, for any $\Sigma \in |Sign^{\mathcal{I}}|$, $M' \in |Mod^{\mathcal{I}'}(\Phi(\Sigma))|$ *and* $\rho \in Sen^{\mathcal{I}}(\Sigma)$,

$$\beta_{\Sigma}(M') \models^{\mathcal{I}}_{\Sigma} \rho \text{ iff } M' \models^{\mathcal{I}'}_{\Phi(\Sigma)} \alpha_{\Sigma}(\rho)$$

Definition 3. *Let* (Φ, α, β) *be a comorphism. We say that* (Φ, α, β) *is conservative whenever, for each* Σ*–model* M *in* $\mathcal{I}$, *there exists a* $\Phi(\Sigma)$*–model* M' *in* $\mathcal{I}'$ *such that* $M = \beta_{\Sigma}(M')$.

The notion of conservative comorphism is typically used to "borrow" proof support from an institution in a sound way. In this chapter we resort to such notion to complement ALLOY's proof environment by "borrowing" the proof support from other logical systems within HETS.

Definition 4. *Given an* institution $\mathcal{I}$, *one defines the institution of presentations over* $\mathcal{I}$ *by extending signatures* $\Sigma \in |Sign^{\mathcal{I}}|$ *to pairs* (Σ, Γ), *where* $\Gamma \subseteq Sen^{\mathcal{I}}(\Sigma)$, *signature morphisms to presentation morphisms and restricting models* $M \in |Mod^{\mathcal{I}}(\Sigma)|$ *to the ones in which* Γ *is satisfied, i.e., such that* $M \models^{\mathcal{I}}_{\Sigma} \Gamma$.

The latter definition (from [6]) is very useful to deal with comorphisms where the source institution is too "complex" to be transformed into the target one in a straightforward way. Actually, as discussed later in the chapter, due to a sort of "hidden" rules in semantics of ALLOY, each comorphism defined in Sects. 4 and 5 does not go to the institution of the target logic, but to the corresponding institution of presentations.

The notion of an amalgamation square is often essential for proving the satisfaction conditions of institutions and comorphisms. This justifies recalling it here.

Definition 5. *A commuting square of functors,*

$$\begin{array}{ccc} A' & \xrightarrow{G_2} & A_2 \\ {\scriptstyle G_1}\downarrow & & \downarrow{\scriptstyle F_2} \\ A_1 & \xrightarrow[F_1]{} & A \end{array}$$

is a weak amalgamation square *if and only if, for each* $M_1 \in |A_1|$, $M_2 \in |A_2|$, *such that* $F_1(M_1) = F_2(M_2)$, *there is an object* $M' \in |A'|$ *such that* $G_1(M') = M_1$ *and* $G_2(M') = M_2$. *When* M' *is unique the amalgamation square is called strong.*

The model amalgamation of an institution consists of the model amalgamation of the diagrams in $Mod^{\mathcal{I}}$ (external square) induced by the pushout of signatures in $Sign^{\mathcal{I}}$ (internal square),

$$\begin{array}{ccc}
Mod^{\mathcal{I}}(\Sigma') & \xrightarrow{Mod^{\mathcal{I}}(\theta_2)} & Mod^{\mathcal{I}}(\Sigma_2) \\
{\scriptstyle Mod^{\mathcal{I}}(\theta_1)}\downarrow & \begin{array}{ccc} \Sigma' & \xleftarrow{\theta_2} & \Sigma_2 \\ {\scriptstyle \theta_1}\uparrow & & \uparrow{\scriptstyle \varphi'} \\ \Sigma_1 & \xleftarrow[\varphi]{} & \Sigma \end{array} & \downarrow{\scriptstyle Mod^{\mathcal{I}}(\varphi')} \\
Mod^{\mathcal{I}}(\Sigma_1) & \xrightarrow[Mod^{\mathcal{I}}(\varphi)]{} & Mod^{\mathcal{I}}(\Sigma)
\end{array}$$

i.e., for each $M_1 \in |Mod^{\mathcal{I}}(\Sigma_1)|$, $M_2 \in |Mod^{\mathcal{I}}(\Sigma_2)|$, such that $Mod^{\mathcal{I}}(\theta_1)(M_1) = Mod^{\mathcal{I}}(\theta_2)(M_2)$, there is an object $M' \in |Mod^{\mathcal{I}}(\Sigma')|$ called the amalgamation of M_1 and M_2, such that $Mod^{\mathcal{I}}(\varphi)(M') = M_1$ and $Mod^{\mathcal{I}}(\varphi')(M') = M_2$. The square is strong if M' is unique.

Consider comorphism (Φ, α, β). The model amalgamation of β–transformations and functor reducts of $Mod^{\mathcal{I}}$ consists of the weak model amalgamation of the following commutative square,

$$\begin{array}{ccc}
Mod'^{\mathcal{I}}(\Phi(\Sigma')) & \xrightarrow{\beta_{\Sigma'}} & Mod^{\mathcal{I}}(\Sigma') \\
{\scriptstyle Mod^{\mathcal{I}}(\theta_1)}\downarrow & & \downarrow{\scriptstyle Mod^{\mathcal{I}}(\varphi')} \\
Mod^{\mathcal{I}'}(\Phi(\Sigma)) & \xrightarrow[\beta_{\Sigma}]{} & Mod^{\mathcal{I}}(\Sigma)
\end{array}$$

i.e., for each $M_\Phi \in |Mod^{\mathcal{I}'}(\Phi(\Sigma))|$, $M' \in |Mod^{\mathcal{I}}(\Sigma')|$, such that $\beta_\Sigma(M_\Phi) = Mod^{\mathcal{I}}(\varphi)(M')$, there is a model $M'_\Phi \in |Mod^{\mathcal{I}}(\Phi(\Sigma'))|$, such that $Mod^{\mathcal{I}'}(\varphi')(M'_\Phi) = M_\Phi$ and $\beta_{\Sigma'}(M'_\Phi) = M'$. When M' is unique, the amalgamation square is called strong.

3 Alloy as an Institution

ALLOY [9] is based on a single sorted relational language extended with a transitive closure operator. Roughly speaking, an ALLOY specification is divided into declarations, of both relations and signatures, and sentences. Signatures will be called *kinds* from now on to distinguish them from signatures in an institution. Actually, kinds are nothing more than unary relations whose purpose is to restrict other relations. This is in line with the *motto* of ALLOY which regards *everything as a relation*. Additionally, kinds may be given parents by an annotation with the keyword `extends`, establishing the obvious inclusion relation. When two kinds are in different subtrees (i.e. one is not a descendant of the other) they are supposed to be mutually disjoint. Finally, kinds may be of type *Abstract*, i.e., included in the union of its descendants, *Some*, i.e., required to have at least one element, or *One*, i.e., exactly with one element. The ALLOY analyser checks an assertion against a specification by seeking for counter-examples within bounded domains.

In this section we define an institution for ALLOY, $\mathcal{A} = (Sign^{\mathcal{A}}, Sen^{\mathcal{A}}, Mod^{\mathcal{A}}, \models^{\mathcal{A}})$. We proceed as follows:

Signatures. Objects in $Sign^{\mathcal{A}}$ are tuples, (S, m, R, X), composed by:

- A family of sets containing kinds and indexed by a type, $S = \{S_t\}_{t \in \{All, Abs, Som, One\}}$. S_{All} is the set of all kinds, S_{Abs} the set of the abstract ones, S_{Som} the non-empty ones, and, finally, S_{One} collects the kinds containing exactly one element. Notice that, for all S_t, $S_t \subseteq S_{All}$.
- $m : S_{All} \to S_{All}$ is a function that gives the parent of each kind, *i.e.*, $m(s) = s'$ means that s' is the parent of s. Top level kinds are considered the parents of themselves, and therefore, m takes the form of a forest structure.
- A family of relational symbols $R = \{R_w | w \in (S_{All})^+\}$.
- A set of unary relational symbols X, representing the variable symbols declared on quantified expressions. Despite being the same than the elements in S_{One}, once encoded they must be treated differently.

Morphisms $\varphi : (S, m, R, X) \to (S', m', R', X')$ in this category are triples $\varphi = (\varphi_{kd}, \varphi_{rl}, \varphi_{vr})$ such that:

- $\varphi_{kd} : S \to S'$ is a function that, for any $S_t \in S$, if $\pi \in S_t$ then $\varphi_{kd}(\pi) \in S'_t$, and the following diagram commutes:

$$\begin{array}{ccc} S & \xrightarrow{\varphi_s} & S' \\ {\scriptstyle m}\downarrow & & \downarrow{\scriptstyle m'} \\ S & \xrightarrow[\varphi_s]{} & S' \end{array}$$

- φ_{rl} is a family of functions such that, $\varphi_{rl} = \{\varphi_{kd} : R_w \to R'_{\varphi_{kd}(w)}\}_{w \in (S_{All})^+}$;
- $\varphi_{vr} : X \to X'$ is a function.

Sentences. Consider Exp a functor of the same type of $Sen^{\mathcal{A}}$. Given a signature $\Sigma = (S_\Sigma, m_\Sigma, R_\Sigma, X_\Sigma) \in |Sign^{\mathcal{A}}|$, the set of expressions $Exp(\Sigma)$ is the smallest one containing

$$\begin{array}{ll} \pi, & \pi \in (S_\Sigma)_{All} \cup (R_\Sigma)_w \cup X_\Sigma \\ \hat{}e, & e \in Exp(\Sigma) \text{ and } |e| = 2 \\ \sim e, & e \in Exp(\Sigma) \\ e \text{ -> } e', & e, e' \in Exp(\Sigma) \\ e \odot e', & e, e' \in Exp(\Sigma), |e| = |e'|, \text{ and } \odot \in \{+, -, \&\} \\ e \,.\, e', & e, e' \in Exp(\Sigma), \text{ and } |e| + |e'| > 2 \end{array}$$

where the length $|e|$ of an expression e is computed as follows:

$$\begin{array}{ll}
|\pi| & = |w|,\ \text{for } \pi \in (R_\Sigma)_w \\
|\pi| & = 1,\ \text{for } \pi \in (S_\Sigma)_{All} \cup X_\Sigma \\
|\hat{}e| & = |e| \\
|\sim e| & = |e| \\
|e \odot e'| & = |e|, \text{for } \odot \in \{+, -, \&\} \\
|e\,.\,e'| & = (|e| + |e'|) - 2 \\
|e \mathbin{\texttt{->}} e'| & = |e| + |e'|
\end{array}$$

Finally, the set of sentences, $Sen^{\mathcal{A}}(\Sigma)$, is the smallest one containing:

$$\begin{array}{ll}
e\ \texttt{in}\ e' & e, e' \in Exp(\Sigma), \text{for } |e| = |e'| \\
\texttt{not}\ \rho & \rho \in Sen^{\mathcal{A}}(\Sigma) \\
\rho\ \texttt{implies}\ \rho' & \rho, \rho' \in Sen^{\mathcal{A}}(\Sigma) \\
(\texttt{all}\ x : e)\ \rho & e \in Exp(\Sigma),\ |e| = 1, \text{and } \rho \in Sen^{\mathcal{A}}(\Sigma'), \text{where} \\
& \Sigma' = (S_\Sigma, m_\Sigma, R_\Sigma, X_\Sigma \uplus \{x\})
\end{array}$$

Note that other standard boolean connectives may be built from the above. For instance, the conjunction, denoted in ALLOY with symbol and, is usually defined with the implication and negation constructors.

Given a signature morphism $\varphi : \Sigma \to \Sigma'$ in $Sign^{\mathcal{A}}$, expressions and sentences are mapped as follows:

$$\begin{array}{ll}
Exp(\varphi)(\pi) & = \varphi_{kd}(\pi), \text{for } \pi \in (S_\Sigma)_{All} \\
Exp(\varphi)(\pi) & = \varphi_{rl}(\pi), \text{for } \pi \in (R_\Sigma)_{All} \\
Exp(\varphi)(\pi) & = \varphi_{vr}(\pi), \text{for } \pi \in X_\Sigma \\
Exp(\varphi)(\hat{}e) & = \hat{}Exp(e) \\
Exp(\varphi)(e \mathbin{\texttt{->}} e') & = Exp(\varphi)(e) \mathbin{\texttt{->}} Exp(\varphi)(e') \\
Exp(\varphi)(e \odot e') & = Exp(\varphi)(e) \odot Exp(\varphi)(e') \\
Exp(\varphi)(e\,.\,e') & = Exp(\varphi)(e)\,.\,Exp(\varphi)(e')
\end{array}$$

$$\begin{array}{ll}
Sen^{\mathcal{A}}(\varphi)(e\ \texttt{in}\ e') & = Sen^{\mathcal{A}}(\varphi)(e)\ \texttt{in}\ Sen^{\mathcal{A}}(\varphi)(e') \\
Sen^{\mathcal{A}}(\varphi)(\texttt{not}\ \rho) & = \texttt{not}\ Sen^{\mathcal{A}}(\varphi)(\rho) \\
Sen^{\mathcal{A}}(\varphi)(\rho\ \texttt{implies}\ \rho') & = Sen^{\mathcal{A}}(\varphi)(\rho)\ \texttt{implies}\ Sen^{\mathcal{A}}(\varphi)(\rho') \\
Sen^{\mathcal{A}}(\varphi)((\texttt{all}\ x : e)\ \rho) & = (\texttt{all}\ x : Exp(\varphi)(e))\ Sen^{\mathcal{A}}(\varphi')(\rho), \text{where } \varphi'
\end{array}$$

canonically expands φ with $\varphi'_{vr}(x) = x$.

Models. For each signature $(S, m, R, X) \in |Sign^{\mathcal{A}}|$, a model $M \in |Mod^{\mathcal{A}}((S, m, R, X))|$ has,

1. a carrier set $|M|$;
2. an unary relation $M_\pi \subseteq |M|$, for each $\pi \in S_{All}$;
3. a relation $M_\pi \subseteq M_w$, for each $\pi \in R_w$;
4. a singleton relation, $M_\pi \subseteq |M|$, for each $\pi \in X$,

satisfying the following properties for any $\pi, \pi' \in S_{All}$,

1. $M_\pi \subseteq M_{m(\pi)}$
2. if $\pi \in S_{Som}$, then $M_\pi \not\subseteq \emptyset$
3. if $\pi \in S_{One}$, then $\#M_\pi = 1$
4. if $\pi \in S_{Abs}$, then $M_\pi \subseteq \bigcup_{\tau \in m^\circ(\pi)} M_\tau$
5. if π, π' are not related by the transitive closure of m, then $M_\pi \cap M_{\pi'} = \emptyset$

Evaluation of expressions in such models, is done in the following way:

$$\begin{aligned}
M_{\sim e} &= (M_e)^\circ \\
M_{e\,+\,e'} &= M_e + M_{e'} \\
M_{e\,-\,e'} &= M_e - M_{e'} \\
M_{e\,\&\,e'} &= M_e \cap M_{e'} \\
M_{e\,.\,e'} &= M_e \,.\, M_{e'} \\
M_{e\,\text{->}\,e'} &= M_e \times M_{e'} \\
M_{\wedge e} &= (M_e)^+
\end{aligned}$$

A signature morphism, $\varphi : \Sigma \rightarrow \Sigma'$, is mapped to $Mod^{\mathcal{A}}(\varphi) : Mod^{\mathcal{A}}(\Sigma') \rightarrow Mod^{\mathcal{A}}(\Sigma)$, giving for each $M' \in |Mod^{\mathcal{A}}(\Sigma')|$, its φ-reduct, $M' \upharpoonright \varphi \in |Mod^{\mathcal{A}}(\Sigma)|$. The latter is defined as follows:

$$\begin{aligned}
|(M' \upharpoonright \varphi)| &= |M'| \\
(M' \upharpoonright \varphi)_\pi &= M'_{\varphi_{kd}(\pi)}, \text{ for any } \pi \in (S_\Sigma)_{All} \\
(M' \upharpoonright \varphi)_\pi &= M'_{\varphi_{rl}(\pi)}, \text{ for any } \pi \in (R_\Sigma)_w \\
(M' \upharpoonright \varphi)_\pi &= M'_{\varphi_{vr}(\pi)}, \text{ for any } \pi \in X_\Sigma
\end{aligned}$$

Satisfaction. Given a Σ-model M, for $\Sigma \in |Sign^{\mathcal{A}}|$, the satisfaction relation is defined for each Σ-sentence as follows:

$$\begin{aligned}
M &\models^{\mathcal{A}}_{\Sigma} e \texttt{ in } e' && \text{iff } M_e \subseteq M_{e'} \\
M &\models^{\mathcal{A}}_{\Sigma} \texttt{not } \rho && \text{iff } M \not\models^{\mathcal{A}}_{\Sigma} \rho \\
M &\models^{\mathcal{A}}_{\Sigma} \rho \texttt{ implies } \rho' && \text{iff } M \models^{\mathcal{A}}_{\Sigma} \rho' \text{ whenever } M \models^{\mathcal{A}}_{\Sigma} \rho \\
M &\models^{\mathcal{A}}_{\Sigma} (\texttt{all } x : e)\rho && \text{iff } M' \models^{\mathcal{A}}_{\Sigma'} (x \texttt{ in } e) \texttt{ implies } \rho,
\end{aligned}$$

for all model x–expansions M' of M, with Σ' canonically extending Σ with the variable x.

The next step is to prove that $\mathcal{A} = (Sign^{\mathcal{A}}, Sen^{\mathcal{A}}, Mod^{\mathcal{A}}, \models^{\mathcal{A}})$ forms an institution. The proof will proceed through a number of auxiliary results

Lemma 1. *On the conditions above the commuting diagram below is a strong amalgamation square.*

$$\begin{array}{ccc}
(S, m, R, X) & \xrightarrow{\varphi} & (S', m', R', X') \\
{\scriptstyle x}\downarrow & & \downarrow{\scriptstyle x^{\varphi}} \\
(S, m, R, X + \{x\}) & \xrightarrow[\varphi']{} & (S', m', R', X' + \{x\})
\end{array}$$

Proof. Proof in Appendix A.1. □

Lemma 2. *For any signature morphism $\varphi : \Sigma \to \Sigma'$ in $Sign^{\mathcal{A}}$, any Σ–expression e, and any Σ'–model M',*

$$(M' \restriction \varphi)_e = M'_{Exp(\varphi)(e)}$$

Proof. Proof in Appendix A.2. □

Finally, we can prove the satisfaction condition for $\mathcal{A}$ and conclude its characterisation as an institution.

Theorem 1. *The satisfaction condition holds for $\mathcal{A}$.*

Proof. Let $\varphi : \Sigma \to \Sigma'$ be a $Sign^{\mathcal{A}}$ morphism, ρ a Σ–sentence, and M' a Σ'–model:

Consider first the case $\rho := e \;\texttt{in}\; e'$:

$$M' \restriction \varphi \models^{\mathcal{A}}_{\Sigma} e \;\texttt{in}\; e'$$

$\Leftrightarrow$ $\{\models$ defn. $\}$

$$(M' \restriction \varphi)_e \subseteq (M' \restriction \varphi)_{e'}$$

$\Leftrightarrow$ $\{$Lemma 2 $\}$

$$M'_{Exp(\varphi)(e)} \subseteq M'_{Exp(\varphi)(e')}$$

$\Leftrightarrow$ $\{\models$ defn. $\}$

$$M' \models^{\mathcal{A}}_{\Sigma'} Exp(\varphi)(e) \;\texttt{in}\; Exp(\varphi)(e')$$

$\Leftrightarrow$ $\{ Sen^{\mathcal{A}}$ defn. $\}$

$$M' \models^{\mathcal{A}}_{\Sigma'} Sen^{\mathcal{A}}(\varphi)(e \;\texttt{in}\; e')$$

Proofs for the negation and implication are analogous.
When $\rho := (\texttt{all}\ x : e)\ \rho$:

$$M' \upharpoonright \varphi \models^{\mathcal{A}}_{\Sigma} (\texttt{all}\ x : e)\ \rho$$

$\Leftrightarrow$ $\{ \models \text{defn.} \}$

$$\text{For all model } x\text{–expansions } (M' \upharpoonright \varphi)' \text{ of } M' \upharpoonright \varphi, \\ (M' \upharpoonright \varphi)' \models^{\mathcal{A}}_{\Sigma^x} (x\ \texttt{in}\ e)\ \texttt{implies}\ \rho$$

$\Leftrightarrow$ $\{ \text{Lemma 1; I.H.} \}$

$$\text{For all model } x\text{–expansions } M'' \text{ of } M', \\ M'' \models^{\mathcal{A}}_{\Sigma'^x} Sen^{\mathcal{A}}(\varphi')((x\ \texttt{in}\ e)\ \texttt{implies}\ \rho)$$

$\Leftrightarrow$ $\{ \models \text{defn.} \}$

$$M' \models^{\mathcal{A}}_{\Sigma'} (\texttt{all}\ x : Sen^{\mathcal{A}}(\varphi)(e))\ Sen^{\mathcal{A}}(\varphi')(\rho)$$

$\Leftrightarrow$ $\{ Sen^{\mathcal{A}} \text{ defn.} \}$

$$M' \models^{\mathcal{A}}_{\Sigma'} Sen^{\mathcal{A}}(\varphi)((\texttt{all}\ x : e)\ \rho)$$

□

We have proved that $\mathcal{A} = (Sign^{\mathcal{A}}, Sen^{\mathcal{A}}, Mod^{\mathcal{A}}, \models^{\mathcal{A}})$ is an institution.

4 From Alloy to SOL

Second-order logic (SOL), extends first–order logic with quantification over functions and predicates, a feature required to canonically encode the notion of transitive closure primitive in ALLOY. The corresponding institution, SOL $= (Sign^{\text{SOL}}, Sen^{\text{SOL}}, Mod^{\text{SOL}}, \models^{\text{SOL}})$, extends the one in example 1, by allowing quantification over functions and predicates in Sen^{SOL}.

This section describes a conservative *comorphism* from ALLOY to the institution of presentations over SOL, where the notion of presentations is used as a key tactic for dealing appropriately with ALLOY's implicit rules over kinds.

We proceed as follows: Let us define (Φ, α, β) : ALLOY→ SOL^{pres}.

Signature functor. For any signature $(S, m, R, X) \in |Sign^{\mathcal{A}}|$, Φ gives a tuple $((S', F, P), \Gamma)$ where,

- $S = \{U\}$
- $F_w = \begin{cases} \{\pi | \pi \in X\} & \text{if } w = U \\ \emptyset & \text{otherwise} \end{cases}$

- $P_w = \begin{cases} \{\pi \mid \pi \in S_{All}\} \cup \{\pi \mid \pi \in R_s, |s| = 1\} & \text{if } w = U \\ \{\pi \mid \pi \in R_s, |s| > 1\} & \text{if } w = (U, \ldots, U), |w| > 1 \\ \emptyset & \text{otherwise} \end{cases}$

and Γ is the least set containing the following sets of axioms:

1. $\{(\forall u : U)\, \pi(u) \Rightarrow \pi'(u) \mid \pi \in S_{All}, \pi' = m(\pi)\}$
2. $\{(\exists u : U)\, \pi(u) \mid \pi \in S_{One} \cup S_{Som}\}$
3. $\{(\forall u, u' : U)\, (\pi(u) \wedge \pi(u')) \Rightarrow u = u' \mid \pi \in S_{One}\}$
4. $\{(\forall u : U)\, \pi(u) \Rightarrow (\bigvee_{\pi' \in m^{\circ}(\pi)} \pi'(u)) \mid \pi \in S_{Abs}\}$
5. $\{\neg(\exists u : U)\, \pi(u) \wedge \pi'(u) \mid \pi, \pi' \in S_{All}$, π,π' not related by the transitive closure of $m\}$
6. $\{(\forall u_1, \ldots, u_n : U)\, \pi(u_1, \ldots, u_n) \Rightarrow \bigwedge_{i=1}^{n} s_i(u_i) \mid \pi \in R_{s_1,\ldots,s_n}\}$

Sentence transformation. Given any signature $\Sigma = (S_\Sigma, m_\Sigma, R_\Sigma, X_\Sigma) \in |\,Sign^{\mathcal{A}}|$, $\alpha_\Sigma : Sen^{\mathcal{A}}(\Sigma) \to Sen^{\mathrm{SOL}^{pres}}(\Phi(\Sigma))$ is defined as:

$$\begin{aligned}
\alpha_\Sigma(\texttt{not}\ \rho) &= \neg\alpha_\Sigma(\rho) \\
\alpha_\Sigma(\rho\ \texttt{implies}\ \rho') &= \alpha_\Sigma(\rho) \Rightarrow \alpha_\Sigma(\rho') \\
\alpha_\Sigma((\texttt{all}\ x : e)\ \rho) &= (\forall x : U)\ \alpha_{\Sigma^x}((x\ \texttt{in}\ e)\ \texttt{implies}\ \rho), \text{ where} \\
&\quad\ \Sigma^x = (S_\Sigma, m_\Sigma, R_\Sigma, X_\Sigma + \{x\}) \\
\alpha_\Sigma(e\ \texttt{in}\ e') &= (\forall V : U_1 \ldots U_n)\ \eta_V(e) \Rightarrow \eta_V(e'), \text{ such that} \\
&\quad\ V = (v_1, \ldots, v_n), \text{ and } |V| = |e|,
\end{aligned}$$

with η_V[1] being defined as follows:

$$\begin{aligned}
\eta_V(\pi) &= \pi(V), \pi \in (S_\Sigma)_{All} \cup (R_\Sigma)_w \\
\eta_v(\pi) &= \pi = v, \pi \in X_\Sigma \\
\eta_{(v1,v2)}(\hat{}e) &= (\exists R : U_1, U_2) R(v_1, v_2) \wedge \alpha_{\Sigma^R}(e\ \texttt{in}\ R\ \texttt{and}\ e.R\ \texttt{in}\ R) \wedge \\
&\quad\ (\forall S : U_1, U_2)\alpha_{\Sigma^{R,S}}((e\ \texttt{in}\ S\ \texttt{and}\ e.S\ \texttt{in}\ S)\ \texttt{implies}\ R\ \texttt{in}\ S), \\
&\quad\ \text{where } \Sigma^R = (S_\Sigma, m_\Sigma, R_\Sigma + R, X_\Sigma) \text{ and} \\
&\quad\ \Sigma^{R,S} = (S_\Sigma, m_\Sigma, R_\Sigma + R + S, X_\Sigma) \\
\eta_V(\sim e) &= \eta_{V'}(e), \text{ such that } V' = (v_n, \ldots, v_1) \\
\eta_V(e + e') &= \eta_V(e) \vee \eta_V(e') \\
\eta_V(e\ -\ e') &= \eta_V(e) \wedge \neg\eta_V(e') \\
\eta_V(e\ \&\ e') &= \eta_V(e) \wedge \eta_V(e') \\
\eta_V(e \texttt{->} e') &= \eta_{V'}(e) \wedge \eta_{V''}(e'), \text{ where } V' \text{ is the prefix of } V \text{ such that} \\
&\quad\ |V'| = |e| \text{ and } V'' \text{ the suffix of } V \text{ such that } |V''| = |e'| \\
\eta_V(e\ .\ e') &= (\exists y : U)\eta_{(V',y)}(e) \wedge \eta_{(y,V'')}(e'), \text{ where } V' \text{ is the prefix of } V \\
&\quad\ \text{such that } |V'| + 1 = |e|, V'' \text{ the suffix of } V \text{ such that} \\
&\quad\ |V''| + 1 = |e'|
\end{aligned}$$

---

[1] To represent a tuple of n elements, $v_1, \ldots, v_n$, we use notations $(v_1, \ldots, v_n)$ and $v_1, \ldots, v_n$ interchangeably, the latter being usually chosen if potential ambiguity is ruled out by the context.

Model transformation. Consider a signature $\Sigma = (S_\Sigma, m_\Sigma, R_\Sigma, X_\Sigma) \in |Sign^{\mathcal{A}}|$, and a $\Phi(\Sigma)$–model M.

Then $\beta_\Sigma : |Mod^{\mathrm{SOL}^{pres}}(\Phi(\Sigma))| \to |Mod^{\mathcal{A}}(\Sigma)|$ is defined as:

$$\begin{aligned}
|\beta_\Sigma(M)| &= |M_U|, \text{ where } |M_U| \text{ is the carrier of } U \text{ in } M,\\
\beta_\Sigma(M)_\pi &= M_\pi, \text{ for any } \pi \in (S_\Sigma)_{All} \cup (R_\Sigma)_w,\\
\beta_\Sigma(M)_\pi &= \{M_\pi\}, \text{ for any } \pi \in X_\Sigma
\end{aligned}$$

Lemma 3. *On the conditions above, the commuting diagram below is a strong amalgamation square,*

$$\begin{array}{ccc}
Mod^{\mathcal{A}}(S, m, R, X) & \xleftarrow{\;\beta_{(S,m,R,X)}\;} & Mod^{\mathrm{SOL}^{pres}}(S', F', P') \\
x^\beta \big\downarrow & & \big\downarrow x \\
Mod^{\mathcal{A}}(S, m, R, X + \{x\}) & \xleftarrow[\;\beta_{(S,m,R,X+\{x\})}\;]{} & Mod^{\mathrm{SOL}^{pres}}(S', F' + \{x\}, P')
\end{array}$$

Proof. Proof in Appendix A.3 □

Note that Lemma 3 says that, for any (S', F', P')–model M the x^β–expansion of its transformation by $\beta_{(S,m,R,X)}$, is equal to the transformation by $\beta_{(S,m,R,X+\{x\})}$ of its x–expansion, whenever the value taken by both x's in the corresponding expansions is the same.

Lemma 4. *Let $\Sigma = (S_\Sigma, m_\Sigma, R_\Sigma, X_\Sigma)$ be a signature in $Sign^{\mathcal{A}}$, M' a $\Phi(\Sigma)$-model M', and e a Σ-expression. Then, for any tuple $(v_1, \ldots, v_n) \in X_\Sigma$ with $n = |e|$, we have*

$$M' \models^{\mathrm{SOL}^{pres}}_{\Phi(\Sigma)} \eta_{(v_1,\ldots,v_n)}(e) \text{ iff } \beta_\Sigma(M') \models^{\mathcal{A}}_\Sigma (v_1 \texttt{->} \ldots \texttt{->} v_n)\ \texttt{in}\ e$$

Proof. Proof in Appendix A.4 □

Theorem 2. *On the conditions above the satisfaction condition holds in (Φ, α, β). I.e., given a signature $\Sigma \in |Sign^{\mathcal{A}}|$, a $\Phi(\Sigma)$-model M', and a Σ-sentence ρ*

$$M' \models^{\mathrm{SOL}^{pres}}_{\Phi(\Sigma)} \alpha_\Sigma(\rho) \text{ iff } \beta_\Sigma(M') \models^{\mathcal{A}}_\Sigma \rho$$

Proof. Proof in Appendix A.5 □

Theorem 3. *The triple (Φ, α, β) defined above is a conservative comorphism (in $\mathcal{SOL}^{pres}$).*

Proof. Let M be an (S, m, R, X)-model. Now, lets consider the $\Phi(S, m, R, X)$-model M' defined by:

1. $|M'_U| = |M|$
2. $M'_\pi = M_\pi$, for any $\pi \in S_{All} \cup R_w$
3. $M'_\pi = a$ where $\{a\} = M_\pi$, for any $\pi \in X$

Clearly $\beta_{(S,m,R,X)}(M') = M$ and M' satisfies Γ. Hence β is surjective and, therefore, (Φ, α, β) is conservative. □

5 From Alloy to Casl

CASL, the *Common Algebraic Specification Language* [14], was developed within the CoFI initiative with the purpose of creating a suitable language for specifying requirements and to design conventional software packages. CASL specifications extend multi-sorted first order logic with partial functions, subsorting and free types, i.e., types whose elements are restricted to be generated by the corresponding constructors and whose distinct constructor terms must denote different elements; we use free types and the notion of presentations to encode ALLOY's transitive closure in CASL. Signatures in CASL are as in SOL, but extended with a family of partial functions symbols PF indexed by their arity, and a partial order $\leq$ over the symbols in S. As usual, $PF \cap F = \emptyset$.

Currently, CASL is regarded as the *de facto* standard language for algebraic specification. It is integrated into HETS along with many of its expansions, acting, as suggested in Fig. 1, as a glue language inside the HETS network of logics.

A comorphism $(\Phi', \alpha', \beta') : \text{ALLOY} \rightarrow \text{CASL}^{pres}$ may be defined in a very similar way to the comorphism defined in the Sect. 4. Let us, then, analyse the things that change in each component.

Signature functor. For any signature $(S, m, R, X) \in |Sign^{A}|$, Φ' gives a tuple $((S', F, PF, P), \Gamma)$ where

$$\begin{aligned} S' &= \{U, Nat\} \\ PF &= \emptyset \end{aligned}$$

$$F_w = \begin{cases} \{\pi | \pi \in X\} & \text{if } w = U \\ \{0\} & \text{if } w = Nat \\ \{suc\} & \text{if } w = Nat, Nat \\ \emptyset & \text{for the other cases} \end{cases}$$

$$P_w = \begin{cases} \{\pi | \pi \in S_{All}\} \cup \{r | r \in R_s, |s| = 1\} & \text{if } w = U \\ \{r | r \in R_s, |s| > 1\} & \text{if } w = (U, \ldots, U), |w| > 1 \\ \{\pi_r | \pi \in R_s, |s| = 2\} & \text{if } w = Nat, U, U \\ \emptyset & \text{for the other cases} \end{cases}$$

and Γ contains two additional rules:

1. { free type $Nat ::= (0 \mid suc(Nat))$ }
2. $\{(\forall u, v : U)\ \pi_r(0, u, v) \Leftrightarrow r(u, v) \wedge (\forall n : Nat)\ \pi_r(suc(n), u, v) \Leftrightarrow (\exists x : U)$ $\pi_r(0, u, x) \wedge \pi_r(n, x, v) | \pi_r \in R_s, |s| = 2\}$.

Sentence transformation. Given any signature $\Sigma \in |Sign^{\mathcal{A}}|$, where

$\Sigma = (S_\Sigma, m_\Sigma, R_\Sigma, X_\Sigma)$, $\alpha'_\Sigma : Sen^{\mathcal{A}}(\Sigma) \to Sen^{\mathrm{CASL}^{pres}}(\Phi(\Sigma))$ is defined in the same way as α (introduced in Sect. 4), with the following replacing the case of transitive closure over expressions:

$$\eta'_V(\hat{}r) = (\exists n : Nat)\ \pi_r(n, V)$$

Note that only the transitive closure of atomic relations is considered. This is done, however, without loss of generality: for an arbitrary expression we just declare an extra binary relation and state that the latter is equal to the former.

Model transformation. Nothing changes in the model transformation component, i.e., for each $\Sigma \in |Sign^{\mathcal{A}}|$, $\beta_\Sigma = \beta'_\Sigma$.

Theorem 4. *On the conditions above the satisfaction condition holds in* (Φ', α', β').

Proof. We can prove the satisfaction condition by using the following treatment in the case of transitive closure:
When $e := \hat{}r$, $r \in R_w$, with $|w| = 2$:

$$M' \models^{\mathrm{CASL}^{pres}}_{\Phi(\Sigma)} \eta_V(\hat{}r)$$

$\Leftrightarrow$ $\{\alpha$ definition$\}$

$$M' \models^{\mathrm{CASL}^{pres}}_{\Phi(\Sigma)} (\exists n : Nat)\ \pi_r(n, V)$$

$\Leftrightarrow$ $\{\models$ defn. and π_r in Γ defn.$\}$

$$M'_V \in M'_{(r^+)}$$

$\Leftrightarrow$ $\{$elements of V are constants, β definition$\}$

$$\beta_\Sigma(M')_V \subseteq \beta_\Sigma(M')_{(r^+)}$$

$\Leftrightarrow$ $\{$Expression evaluation and $\models$ definition$\}$

$$\beta_\Sigma(M') \models^{\mathcal{A}}_\Sigma V \texttt{ in } \hat{}r$$

Then for all other cases the proof is analogous to the one performed in the last section. □

Theorem 5. *The above comorphism is conservative.*

Proof. We have just to define a model in the same way as in Theorem 3. In addition, the following must also be included:

(a) $M'_{Nat} = \mathbb{N}$;
(b) For any π in $R_{s,s}$, M' has a relation, π_r, defining the transitive closure of r.

Clearly, M' satisfies the additional rules 1, 2 in Γ. □

We have proved that (Φ', α', β') is a comorphism, and furthermore that is conservative. This means that (Φ', α', β') is a sound method for validating ALLOY's specifications through the proof environment of CASL.

6 ALLOY and HETS at Work

6.1 An Introduction to DCR Graphs

DCR graphs, short for *Distributed Condition Response Graphs*, were introduced in [8] to specify workflow models in an implicit way through a number of conditions. A functional style and precise semantics make DCR graphs excellent candidates for modelling critical workflows.

Formally, a DCR graph consists of a set E of events and two relations `condition`, `response` $\subseteq E \times E$ which restrict control flow, regarded as a sequence of event executions. In detail,

- $(e, e') \in$ `condition` iff e' can only be executed after e;
- $(e, e') \in$ `response` iff whenever e is executed the control flow may only come to terminal configuration after the execution of e'.

A mark, or execution state, in a DCR G, is a tuple $(Ex, Res) \in \mathbb{P}(E) \times \mathbb{P}(E)$, where Ex is the set of the events that already occurred and Res the set of events scheduled for execution. A valid execution step in G is a triple (M, M', e) where $M, M' \in \mathbb{P}(E) \times \mathbb{P}(E)$ and $e \in E$ such that, for $M = (Ex, Res)$, $M' = (Ex', Res')$,

1. $\{e' | \texttt{condition}(e', e)\} \subseteq Ex$
2. $Ex' = Ex \cup \{e\}$
3. $Res' = (Res \backslash \{e\}) \cup \{e' | response(e, e')\}$.

Mukkamala [16] suggests a translation of DCR graphs to PROMELA so that the specification of workflows can be checked with the SPIN model checker. The encoding, however, is not easy. For example, the language has only arrays as a basic data structure, thus events and relations have to be encoded as arrays, relations becoming two-dimensional bit arrays. Moreover, SPIN based verification is limited by possible state explosion.

An encoding into ALLOY, on the other hand, seems an attractive alternative. Not only it comes out rather straightforwardly, due to the original relational definition of DCR graphs, but also the ALLOY analyser is eager to avoid potential state space explosion by restricting itself to bounded domains. This restricts, of course, the scope of what can be verified in a specification. However, as illustrated below, ALLOY plugged into the HETS family offers a really interesting alternative to the verification of DCR based workflows.

6.2 DCR Graphs in ALLOY

DCR graphs are encoded in ALLOY as follows,

```
abstract sig Event {
   condition : set Event,
   response : set Event
}

sig Mark {
   executed : set Event,
   toBeExecuted : set Event,
   action : set Mark −> set Event
}

fact {
   all m,m' : Mark, e : Event |
      (m −> m' −> e) in action <=>
         (condition.e in m.executed and
         m'.executed = m.executed + e and
         m'.toBeExecuted = (m.toBeExecuted - e) + e.response )
}
```

This includes the declaration of two kinds (`sig`), one of events and another to define markings. Relations are declared in an object oriented style as fields of kinds (objects). For example, what the declaration of `action` entails is, as expected, a subset of the product Mark × Mark × Event. Finally note how the invariant for valid execution steps is directly captured in the `fact` above. Other DCR properties can be directly checked in ALLOY. For example,

```
all m,m' : Mark, e : Event |
   (m −> m' −> e) in action and e in m'.toBeExecuted
      implies e in e.response
```

formalises the claim that 'after executing an event *e*, if in the next mark *e* is still to be executed, then `response` contains a reflexive pair at *e*".

Of course, this property cannot be proved in ALLOY for an arbitrary domain. To do it another tool inside the network has to be called, provided that ALLOY is already plugged there. Applying the comorphism to CASL defined in the Sect. 6 we get the following encoding of the property:

```
forall m : U . Mark(m) =>
forall m' : U . Mark(m') =>
forall e : U . Event(e) =>
(forall v1,v2,v3 : U . v1 = m /\ v2 = m' /\ v3 = e => action(v1,v2,v3)) /\
(forall v : U . v = e => exists y : U . y = m' /\ toBeExecuted(y,v)) =>
(forall v : U . v = e => exists y : U . y = e /\ response(y,v))
```

which, after a few reduction steps simplifies to

```
forall m,m',e : U .
   Mark(m) /\ Mark(m') /\ Event(e) =>
      (action(m,m',e) /\ toBeExecuted(m',e) => response(e,e))
```

which is can then be verified by the SPASS theorem prover.

6.3 A Medical Workflow

Consider now the following example of a DCR graph representing a medical workflow as introduced in [16]. It concerns the administration of a medicine to a patient. The workflow diagram obtained from the ALLOY analyser is depicted in Fig. 2.

As mentioned in the introduction, ALLOY may give a false sense of security as the scope set for a simulation session may not be wide enough to produce a counter example. To illustrate this situation consider the following property in which we assume transRun = ^(action.Event). In English it reads: "starting with an empty mark (∅, ∅), if by continuously executing events a mark is reached where SecEffect was executed and no further events are to be executed, then this mark has no executed events". In ALLOY,

```
all m,m' : Mark |
   (no m.(executed+toBeExecuted) and
   m' in m.transRun and
   SecEffect in m'.executed and
   no m'.toBeExecuted)
      implies no m'.executed
```

An analysis of the workflow diagram shows the property is false. Actually, if the left side of the implication is true, it may happen that the right hand side is false:

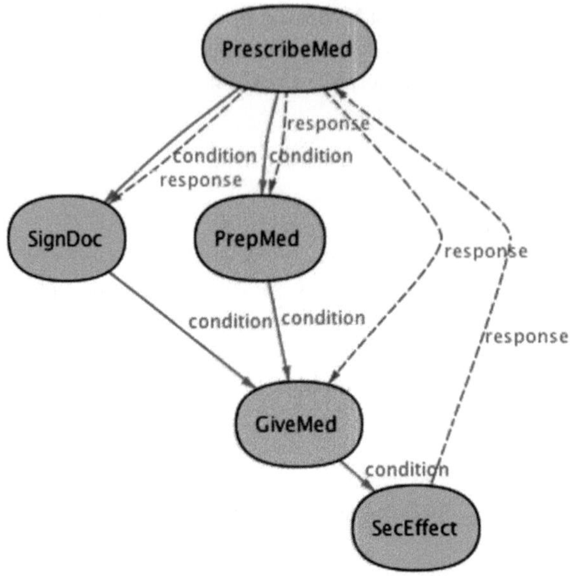

Fig. 2 A medical workflow diagram

the former says there are executed events while the latter contradicts it. The ALLOY analyser, however, is unable to find a counter-example within a scope below 15 (recall the default scope is 3). The problem of this, is that with a scope smaller than 15 (10 marks +5 events) the ALLOY analyser can never reach a mark where the left side of the implication is true, and therefore no counter examples are found.

On the other hand, after encoding into CASL and calling another prover in the HETS network, such as VAMPIRE, the result pops out in a few seconds. A HETS session for this example is reproduced in Fig. 3. In general the ALLOY analyser has difficulties when dealing with similar properties and diagrams with just two more events. In some cases the search, if successful, may exceed 30 min.

We have checked several other properties[2] using both ALLOY, with scope 15, and automatic theorem provers available in HETS, namely SPASS and EPROVER, through the second encoding proposed in this chapter. The experimental results seem to confirm the advantages of the hybrid approach proposed here, with automatic theorem provers taking the job whenever ALLOY is unable to proceed or requires an excessive processing time. In some cases, namely when dealing with encodings of ALLOY models that make heavy use of transitive closure, another member of the HETS network—an interactive theorem prover—has to be called.

[2] Full models at `github.com/nevrenato/IRI_FMI_Annex`.

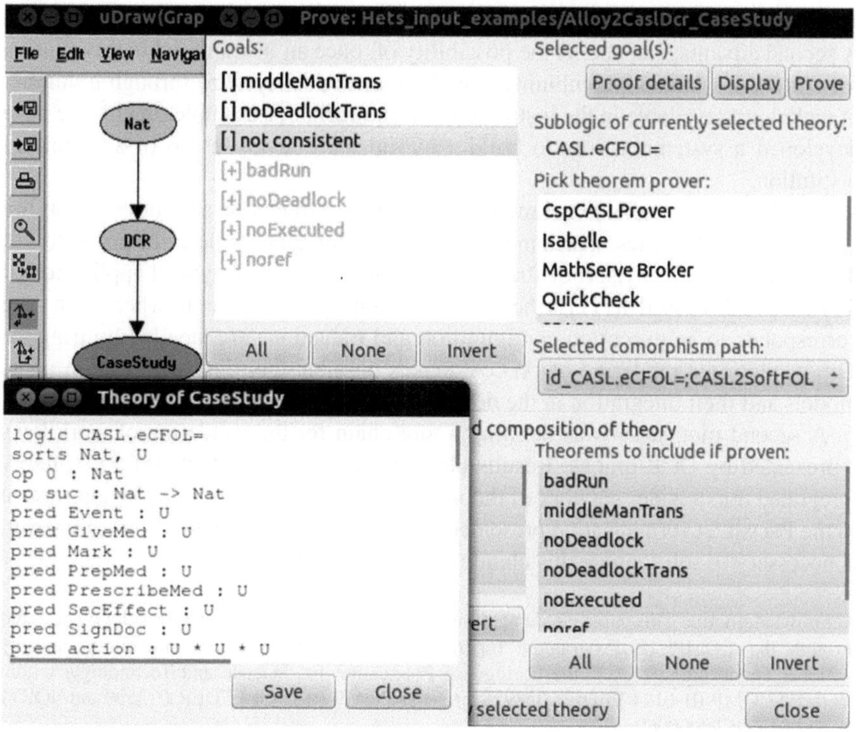

Fig. 3 A HETS session

7 Discussion and Conclusions

The chapter laid the first steps toward establishing a rigorous methodology for modelling and validating software designs by connecting ALLOY to a network of logics and logical tools, rather than, to a single one.

Going generic has, as one could expect, a price to be paid. In our case, this was the development of a proper formalisation of the ALLOY underlying logical system as an institution, together with conservative comorphisms into institutions of presentations over SOL and CASL as entry points in the HETS network. The work reported here extends [17] in working out all the proof details and, mainly, providing a new, sound translation to SOL.

Adopting an institutional framework brings to scene a notational burden the working software engineer may find hard to bear. It should be noted, however, this is done once and for all: our results, once proved, provide a simple method to translate ALLOY models not only into both SOL and CASL specifications. On the other hand, following this approach has a number of advantages. First of all this is a sound way to integrate systems based on a formal relationship between their underlying logical systems. This contrasts with *ad hoc* combinations, often attractive at first sight but not always consistent, which abound in less careful approaches to Software Engineering. A second advantage concerns the possibility of, once an institutional representation for ALLOY is obtained, combining it with other logical systems through a number of techniques available in the institutional framework. For example, in [13] we have developed a systematic way to build a hybrid logic layer on top of an arbitrary institution.

Hybrid logic [5] adds to the modal description of transition structures the ability to refer to specific states, which makes it a suitable language to describe properties of individual states in any sort of structured transition system. A typical application of this method discussed in [11] is the design of reconfigurable systems, where each state corresponds to an execution configuration and transitions are labelled by triggers. The institutional rendering of ALLOY makes possible that the hybridisation of its models and their integration in the development cycle of reconfigurable software.

A second motivation was defining a tool chain for the validation of workflows represented by DCR graphs. Results obtained so far suggest that ALLOY, suitably integrated into a wider network of theorem provers, provides an intuitive alternative to the PROMELA formalisation presented in [16]. More experimental work, however, is necessary to substantiate this claim on general grounds.

Acknowledgments This work is funded by ERDF—European Regional Development Fund through the COMPETE Programme (operational programme for competitiveness) and by National Funds through FCT, the Portuguese Foundation for Science and Technology, within projects FCOMP-01-0124-FEDER-028923, project FCOMP-01-0124-FEDER-022690 and NORTE-01-0124-FEDER-000060.

Appendix A: Proofs

A.1 Lemma 1

The following commuting diagram of ALLOY signature morphisms is a strong amalgamation square.

$$\begin{array}{ccc} (S, m, R, X) & \xrightarrow{\varphi} & (S', m', R', X') \\ \Big\downarrow x & & \Big\downarrow x^{\varphi} \\ (S, m, R, X + \{x\}) & \xrightarrow[\varphi']{} & (S', m', R', X' + \{x\}) \end{array}$$

Proof. Let M_1 be an $(S, m, R, X + \{x\})$–model, and M_2 an (S', m', R', X')–model, such that $Mod^{\mathcal{A}}(x)(M_1) = Mod^{\mathcal{A}}(\varphi)(M_2)$. Thus, for any $\pi \in S_t$, $M_{1\pi} = M_{2\varphi_{kd}(\pi)}$, for any $\pi \in R_w$, $M_{1\pi} = M_{2\varphi_{rl}(\pi)}$, for any $\pi \in X$, $M_{1\pi} = M_{2\varphi_{vr}(\pi)}$, and $|M_1| = |M_2|$.

Then, let us define an $(S', m', R,', X' + \{x\})$-model M' by stating that: For all $\pi \in S'_t$, $M_{2\pi} = M'_\pi$; for all $\pi \in R'_w$, $M'_\pi = M_{2\pi}$; for all $\pi \in X'$, $M_{2\pi} = M'_\pi$; $|M_2| = |M'|$, and $M_{1x} = M'_x$. Clearly, $M_1 = Mod^{\mathcal{A}}(\varphi')(M')$ and $M_2 = Mod^{\mathcal{A}}(x^\varphi)(M')$. Also it is not difficult to show that M' is unique. Therefore the diagram above is a strong amalgamation square. □

A.2 Lemma 2

For any signature morphism $\varphi : \Sigma \to \Sigma'$ in $Sign^{\mathcal{A}}$, any Σ–expression e, and any Σ'–model M',

$$(M' \upharpoonright \varphi)_e = M'_{Exp(\varphi)(e)}$$

Proof. Consider first the case $e := \pi$, for $\pi \in (R_\Sigma)_w$:

$$(M' \upharpoonright \varphi)_\pi$$

$=$ {Reduct defn.}

$$M'_{\varphi_{rl}(\pi)}$$

$=$ {*Exp* defn.}

$$M'_{Exp(\varphi)(\pi)}$$

Proofs for when $\pi \in (S_\Sigma)_{All}$ or $\pi \in X_\Sigma$ are analogous.
When $e := e + e'$:

$$(M' \upharpoonright \varphi)_{e+e'}$$

$=$ {Expression evaluation}

$$(M' \upharpoonright \varphi)_e + (M' \upharpoonright \varphi)_{e'}$$

$=$ {Induction hypothesis}

$$M'_{Exp(\varphi)(e)} + M'_{Exp(\varphi)(e')}$$

$=$ {Expression evaluation}

$$M'_{Exp(\varphi)(e)+Exp(\varphi)(e')}$$

$=$ {*Exp* defn.}

$$M'_{Exp(\varphi)(e+e')}$$

Proofs for the remaining operators are analogous. □

A.3 Lemma 3

The following commuting square of model morphisms and model transformations is a strong amalgamation square,

$$\begin{array}{ccc}
Mod^{\mathcal{A}}(S,m,R,X) & \xleftarrow{\beta_{(S,m,R,X)}} & Mod^{\mathrm{SOL}^{pres}}(S',F',P') \\
{\scriptstyle x^{\beta}}\downarrow & & \downarrow{\scriptstyle x} \\
Mod^{\mathcal{A}}(S,m,R,X+\{x\}) & \xleftarrow[\beta_{(S,m,R,X+\{x\})}]{} & Mod^{\mathrm{SOL}^{pres}}(S',F'+\{x\},P')
\end{array}$$

Proof. Let M_1 be an $(S, m, R, X + \{x\})$–model and M_2 a (S', F', P')–model such that $Mod^{\mathcal{A}}(x^{\beta})(M_1) = \beta_{(S,m,R,X)}(M_2)$. *I.e.*, for any $\pi \in S_{All} \cup R_w \cup X$, $M_{1\pi} = M_{2\pi}$ and $|M_1| = |M_2|$.

Then let us define an $(S', F'+\{x\}, P')$–model M' such that for any $s \in S'$, $|M'_s| = |M_{2s}|$; for any $\sigma \in F'_w$, $M'_\sigma = M_{2\sigma}$; for any $\pi \in P'_w$, $M'_\pi = M_{2\pi}$; $M'_x = \{M_{1x}\}$. Clearly, we have $M_1 = \beta_{(S,m,R,X+\{x\})}(M')$ and $M_2 = Mod^{\mathrm{SOL}^{pres}}(x)(M')$. Moreover, it is not difficult to show that M' is unique. Therefore the diagram above is a strong amalgamation square. □

A.4 Lemma 4

Let $\Sigma = (S_\Sigma, m_\Sigma, R_\Sigma, X_\Sigma)$ be a signature in $Sign^{\mathcal{A}}$, M' a $\Phi(\Sigma)$-model M', and e a Σ-expression. Then, for any tuple $(v_1, \ldots, v_n) \in X_\Sigma$ with $n = |e|$, we have

$$M' \models^{\mathrm{SOL}^{pres}}_{\Phi(\Sigma)} \eta_{(v_1,\ldots,v_n)}(e) \text{ iff } \beta_\Sigma(M') \models^{\mathcal{A}}_{\Sigma} (v_1 \texttt{->} \ldots \texttt{->} v_n) \ \texttt{in}\, e$$

Proof. When $e := \pi$, $\pi \in (R_\Sigma)_w \cup (S_\Sigma)_{All}$:

$$M' \models^{\mathrm{SOL}^{pres}}_{\Phi(\Sigma)} \eta_{(v_1,\ldots,v_n)}(\pi)$$

$\Leftrightarrow$ $\{\eta$ defn. $\}$

$$M' \models^{\mathrm{SOL}^{pres}}_{\Phi(\Sigma)} \pi(v_1, \ldots, v_n)$$

$\Leftrightarrow$ $\{\models$ defn. $\}$

$$(M'_{v_1}, \ldots, M'_{v_n}) \in M'_\pi$$

$\Leftrightarrow$ $\{v_i$ elements are constants $\}$

$$M'_{v_1} \times \cdots \times M'_{v_n} \subseteq M'_\pi$$

$\Leftrightarrow$ {β defn. }

$$\beta_\Sigma(M')_{v_1} \times \cdots \times \beta_\Sigma(M')_{v_n} \subseteq \beta_\Sigma(M')_\pi$$

$\Leftrightarrow$ {Expression evaluation; $\models$ defn. }

$$\beta_\Sigma(M') \models^{\mathcal{A}}_\Sigma (v_1\texttt{->}\ldots\texttt{->}v_n)\ \texttt{in}\ \pi$$

When $e := \pi, \pi \in X_\Sigma$:

$$M' \models^{\mathrm{SOL}^{pres}}_{\Phi(\Sigma)} \eta_v(\pi)$$

$\Leftrightarrow$ {η defn. }

$$M' \models^{\mathrm{SOL}^{pres}}_{\Phi(\Sigma)} v = \pi$$

$\Leftrightarrow$ {$\models$ defn. }

$$M'_v = M'_\pi$$

$\Leftrightarrow$ {v and π are constants }

$$\{M'_v\} \subseteq \{M'_x\}$$

$\Leftrightarrow$ {β defn. }

$$\beta_\Sigma(M')_v \subseteq \beta_\Sigma(M')_\pi$$

$\Leftrightarrow$ {$\models$ defn. }

$$\beta_\Sigma(M') \models^{\mathcal{A}}_\Sigma v\ \texttt{in}\ \pi$$

When $e := e\ .\ e'$:

$$M' \models^{\mathrm{SOL}^{pres}}_{\Phi(\Sigma)} \eta_V(e\ .\ e')$$

$\Leftrightarrow$ {η defn. }

$$M' \models^{\mathrm{SOL}^{pres}}_{\Phi(\Sigma)} (\exists y : U)\ \eta_{(V',y)}(e) \wedge \eta_{(y,V'')}(e')$$

$\Leftrightarrow$ {$\models$ defn. }

There is a y-expansion M'' of M' such that
$M'' \models^{\mathrm{SOL}^{pres}}_{\Phi(\Sigma)^y} \eta_{(V',y)}(e)$ and $M'' \models^{\mathrm{SOL}^{pres}}_{\Phi(\Sigma)^y} \eta_{(y,V'')}(e')$

$\Leftrightarrow$ {I.H., lemma 3, $\models$ defn. }

There is a y–expansion M'' of M' such that

$$\beta_{\Sigma^y}(M'') \models^{\mathcal{A}}_{\Sigma^y} (V'{-}{>}y)\ \mathtt{in}\ e\ \mathtt{and}\ (y{-}{>}V'')\ \mathtt{in}\ e'$$

$\Leftrightarrow$ $\{$lemma 3, $\models$ defn. $\}$

$$\beta_{\Sigma}(M') \models^{\mathcal{A}}_{\Sigma} V\ \mathtt{in}\ e\ .\ e'$$

When $e :=$ ^e:

$$M' \models^{\mathrm{SOL}^{pres}}_{\Phi(\Sigma)} \eta_{(v_1,v_2)}(\hat{}e)$$

$\Leftrightarrow$ $\{\ \eta$ defn. $\}$

$$M' \models^{\mathrm{SOL}^{pres}}_{\Phi(\Sigma)} \exists R.\big(R(v_1,v_2) \wedge \alpha_{\Sigma^R}(e\ \mathtt{in}\ R\ \mathtt{and}\ e.R\ \mathtt{in}\ R) \wedge \\ \forall S.\ \alpha_{\Sigma^{R,S}}((e\ \mathtt{in}\ S\ \mathtt{and}\ e.S\ \mathtt{in}\ S)\ \mathtt{implies}\ R\ \mathtt{in}\ S)\big)$$

$\Leftrightarrow$ $\{\ \models$ defn. $\}$

There is an R-expansion M^R of M' such that
$$M^R \models^{\mathrm{SOL}^{pres}}_{\Phi(\Sigma)^R} R(v_1,v_2) \wedge \alpha_{\Sigma^R}(e\ \mathtt{in}\ R\ \mathtt{and}\ e.R\ \mathtt{in}\ R)$$
and for any S-expansion $M^{R,S}$ of M^R,
$$M^{R,S} \models^{\mathrm{SOL}^{pres}}_{\Phi(\Sigma)^{R,S}} \alpha_{\Sigma^{R,S}}\big((e\ \mathtt{in}\ S\ \mathtt{and}\ e.S\ \mathtt{in}\ S)\ \mathtt{implies}\ R\ \mathtt{in}\ S\big)$$

$\Leftrightarrow$ $\{\models$ defn, α defn. $\}$

There is an R-expansion M^R of M' such that
$$\Big(M^R \models^{\mathrm{SOL}^{pres}}_{\Phi(\Sigma)^R} R(v_1,v_2)\ \text{and}\ M^R \models^{\mathrm{SOL}^{pres}}_{\Phi(\Sigma)^R} (\forall u_1,u_2). \\ \big(\eta_{(u_1,u_2)}(e) \Rightarrow \eta_{(u_1,u_2)}(R)\big) \wedge \big(\eta_{(u_1,u_2)}(e.R) \Rightarrow R(u_1,u_2)\big)\Big)$$
and for any S-expansion $M^{R,S}$ of M^R,
$$M^{R,S} \models^{\mathrm{SOL}^{pres}}_{\Phi(\Sigma)^{R,S}} \big((\forall u_1,u_2).(\eta_{(u_1,u_2)}(e) \Rightarrow \eta_{(u_1,u_2)}(S)) \wedge \\ (\eta_{(u_1,u_2)}(e.S) \Rightarrow S(u_1,u_2))\big) \Rightarrow \big((\forall u_1,u_2).R(u_1,u_2) \Rightarrow S(u_1,u_2)\big)$$

$\Leftrightarrow$ $\{\ \alpha$ defn., $\models$ defn. $\}$

There is an R-expansion M^R of M' such that
$M^R \models^{\mathrm{SOL}^{pres}}_{\Phi(\Sigma)^R} R(v_1,v_2)$ and
$\Big($for any (u_1,u_2)-expansion $(M^R)^{(u_1,u_2)}$ of M^R,
$\big((M^R)^{(u_1,u_2)} \models^{\mathrm{SOL}^{pres}}_{\Phi(\Sigma)^{R,u_1,u_2}} \eta_{(u_1,u_2)}(e)$ implies
$(M^R)^{(u_1,u_2)} \models^{\mathrm{SOL}^{pres}}_{\Phi(\Sigma)^{R,u_1,u_2}} \eta_{(u_1,u_2)}(R)\big)$ and
$\big((M^R)^{(u_1,u_2)} \models^{\mathrm{SOL}^{pres}}_{\Phi(\Sigma)^{R,u_1,u_2}} \eta_{(u_1,u_2)}(e.R)$ implies
$(M^R)^{(u_1,u_2)} \models^{\mathrm{SOL}^{pres}}_{\Phi(\Sigma)^{R,u_1,u_2}} R(u_1,u_2)\big)\Big)$
and for any S-expansion $M^{R,S}$ of M^R,

$$\Big(\text{for any } (u_1, u_2)\text{-expansion } (M^{R,S})^{(u_1,u_2)} \text{ of } M^{R,S}$$
$$\big((M^{R,S})^{(u_1,u_2)} \models^{\text{SOL}^{pres}}_{\Phi(\Sigma)^{R,S,u_1,u_2}} \eta_{(u_1,u_2)}(e) \text{ implies}$$
$$(M^{R,S})^{(u_1,u_2)} \models^{\text{SOL}^{pres}}_{\Phi(\Sigma)^{R,S,u_1,u_2}} \eta_{(u_1,u_2)}(S)\big) \text{ and}$$
$$\big((M^{R,S})^{(u_1,u_2)} \models^{\text{SOL}^{pres}}_{\Phi(\Sigma)^{R,S,u_1,u_2}} \eta_{(u_1,u_2)}(e.S) \text{ implies}$$
$$(M^{R,S})^{(u_1,u_2)} \models^{\text{SOL}^{pres}}_{\Phi(\Sigma)^{R,S,u_1,u_2}} S(u_1, u_2)\big)\Big) \text{ implies}$$
$$\text{that for any } (u_1, u_2)\text{-expansion } (M^{R,S})^{(u_1,u_2)} \text{ of } M^{R,S}$$
$$\Big((M^{R,S})^{(u_1,u_2)} \models^{\text{SOL}^{pres}}_{\Phi(\Sigma)^{R,S,u_1,u_2}} R(u_1, u_2) \text{ implies}$$
$$(M^{R,S})^{(u_1,u_2)} \models^{\text{SOL}^{pres}}_{\Phi(\Sigma)^{R,S,u_1,u_2}} S(u_1, u_2)\Big)$$

$\Leftrightarrow$ {I.H., Lemma 3 }

$$\text{There is an } R\text{-expansion } M^R \text{ of } M' \text{ such that}$$
$$\beta_{\Sigma^R}(M^R) \models^{\mathcal{A}}_{\Sigma^R} (v_1\texttt{->}v_2) \texttt{ in } R \text{ and}$$
$$\Big(\text{for any } (u_1, u_2)\text{-expansion } (M^R)^{(u_1,u_2)} \text{ of } M^R,$$
$$\big(\beta_{\Sigma^{R,u_1,u_2}}((M^R)^{(u_1,u_2)}) \models^{\mathcal{A}}_{\Sigma^{R,u_1,u_2}} (u_1\texttt{->}u_2) \texttt{ in } e \text{ implies}$$
$$\beta_{\Sigma^{R,u_1,u_2}}((M^R)^{(u_1,u_2)}) \models^{\mathcal{A}}_{\Sigma^{R,u_1,u_2}} (u_1\texttt{->}u_2) \texttt{ in } R\big) \text{ and}$$
$$\big(\beta_{\Sigma^{R,u_1,u_2}}((M^R)^{(u_1,u_2)}) \models^{\mathcal{A}}_{\Sigma^{R,u_1,u_2}} (u_1\texttt{->}u_2) \texttt{ in } e.R \text{ implies}$$
$$\beta_{\Sigma^{R,u_1,u_2}}((M^R)^{(u_1,u_2)}) \models^{\mathcal{A}}_{\Sigma^{R,u_1,u_2}} (u_1\texttt{->}u_2) \texttt{ in } R\big)\Big)$$
$$\text{and for any } S\text{-expansion } M^{R,S} \text{ of } M^R,$$
$$\Big(\text{for any } (u_1, u_2)\text{-expansion } (M^{R,S})^{(u_1,u_2)} \text{ of } M^{R,S}$$
$$\big(\beta_{\Sigma^{R,S,u_1,u_2}}((M^{R,S})^{(u_1,u_2)}) \models^{\mathcal{A}}_{\Sigma^{R,S,u_1,u_2}} (u_1\texttt{->}u_2) \texttt{ in } e \text{ implies}$$
$$\beta_{\Sigma^{R,S,u_1,u_2}}((M^{R,S})^{(u_1,u_2)}) \models^{\mathcal{A}}_{\Sigma^{R,S,u_1,u_2}} (u_1\texttt{->}u_2) \texttt{ in } S\big) \text{ and}$$
$$\big(\beta_{\Sigma^{R,S,u_1,u_2}}((M^{R,S})^{(u_1,u_2)}) \models^{\mathcal{A}}_{\Sigma^{R,S,u_1,u_2}} (u_1\texttt{->}u_2) \texttt{ in } e.S \text{ implies}$$
$$\beta_{\Sigma^{R,S,u_1,u_2}}((M^{R,S})^{(u_1,u_2)}) \models^{\mathcal{A}}_{\Sigma^{R,S,u_1,u_2}} (u_1\texttt{->}u_2) \texttt{ in } S\big)\Big) \text{ implies}$$
$$\text{that for any } (u_1, u_2)\text{-expansion } (M^{R,S})^{(u_1,u_2)} \text{ of } M^{R,S}$$
$$\Big(\beta_{\Sigma^{R,S,u_1,u_2}}((M^{R,S})^{(u_1,u_2)}) \models^{\mathcal{A}}_{\Sigma^{R,S,u_1,u_2}} (u_1\texttt{->}u_2) \texttt{ in } R \text{ implies}$$
$$\beta_{\Sigma^{R,S,u_1,u_2}}((M^{R,S})^{(u_1,u_2)}) \models^{\mathcal{A}}_{\Sigma^{R,S,u_1,u_2}} (u_1\texttt{->}u_2) \texttt{ in } S\Big)$$

$\Leftrightarrow$ {inclusion defn., $\models$ defn. }

$$\text{There is an } R\text{–expansion } M^R \text{ of } M' \text{ such that}$$
$$\beta_{\Sigma^R}(M^R) \models^{\mathcal{A}}_{\Sigma^R} (v_1\texttt{->}v_2) \texttt{ in } R \text{ and}$$
$$\beta_{\Sigma^R}(M^R) \models^{\mathcal{A}}_{\Sigma^R} e \texttt{ in } R \texttt{ and } e.R \texttt{ in } R$$
$$\text{and for all } S\text{–expansions } M^{R,S} \text{ of } M^R$$
$$\beta_{\Sigma^{R,S}} \models^{\mathcal{A}}_{\Sigma^{R,S}} (e \texttt{ in } S \texttt{ and } e.S \texttt{ in } S) \texttt{ implies } R \texttt{ in } S$$

$\Leftrightarrow$ {transitive closure defn. }

$$\beta_\Sigma(M') \models^{\mathcal{A}}_\Sigma (v_1\texttt{->}v_2) \texttt{ in } \texttt{^}e$$

When $e := \sim e$:

$$M' \models^{\mathrm{SOL}^{pres}}_{\Phi(\Sigma)} \eta_{(v_1,\ldots,v_n)}(\sim e)$$

$\Leftrightarrow$ {α defn. }

$$M' \models^{\mathrm{SOL}^{pres}}_{\Phi(\Sigma)} \eta_{(v_n,\ldots,v_1)}(e)$$

$\Leftrightarrow$ {I.H }

$$\beta_\Sigma(M') \models^{\mathcal{A}}_\Sigma (v_n\texttt{->}\ldots\texttt{->}v_1) \texttt{ in } e$$

$\Leftrightarrow$ {Galois connection }

$$\beta(M') \models^{\mathcal{A}}_\Sigma (v_1\texttt{->}\ldots\texttt{->}v_n) \texttt{ in } (\sim e)$$

When $e := e + e'$:

$$M' \models^{\mathrm{SOL}^{pres}}_{\Phi(\Sigma)} \eta_V(e + e')$$

$\Leftrightarrow$ {α defn., $\models$ defn. }

$$M' \models^{\mathrm{SOL}^{pres}}_{\Phi(\Sigma)} \eta_V(e) \text{ or } M' \models^{\mathrm{SOL}^{pres}}_{\Phi(\Sigma)} \eta_V(e')$$

$\Leftrightarrow$ {I.H }

$$\beta_\Sigma(M') \models^{\mathcal{A}}_\Sigma V \texttt{ in } e \text{ or } \beta_\Sigma(M') \models^{\mathcal{A}}_\Sigma V \texttt{ in } e'$$

$\Leftrightarrow$ {$\models$ defn., sum defn. }

$$\beta_\Sigma(M') \models^{\mathcal{A}}_\Sigma V \texttt{ in } e + e'$$

Proofs for the remaining cases are analogous. □

A.5 Theorem 2

The satisfaction condition holds in (Φ, α, β) : ALLOY $\to$ SOL^{pres}. I.e., given a signature $\Sigma \in |Sign^{\mathcal{A}}|$, a $\Phi(\Sigma)$-model M', and a Σ-sentence ρ

$$M' \models^{\mathrm{SOL}^{pres}}_{\Phi(\Sigma)} \alpha_\Sigma(\rho) \text{ iff } \beta_\Sigma(M') \models^{\mathcal{A}}_\Sigma \rho$$

Proof. When $\rho := e \texttt{ in } e'$:

$$M' \models^{\text{SOL}^{pres}}_{\Phi(\Sigma)} \alpha_\Sigma(e \; \texttt{in} \; e')$$

$\Leftrightarrow$ $\{\alpha$ defn. $\}$

$$M' \models^{\text{SOL}^{pres}}_{\Phi(\Sigma)} (\forall V : U_1, \ldots, U_n)\, \eta_V(e) \Rightarrow \eta_{V'}(e')$$

$\Leftrightarrow$ $\{$Satisfaction defn. $\}$

For any V-expansion M'' of M',
$M'' \models^{\mathcal{A}}_{\Phi(\Sigma)'} \eta_V(e')$ whenever $M'' \models^{\mathcal{A}}_{\Phi(\Sigma)'} \eta_V(e)$

$\Leftrightarrow$ $\{$Lemma 4 and $\models$ defn. $\}$

For any V-expansion M'' of M',
$\beta_{\Sigma'}(M'') \models^{\mathcal{A}}_{\Sigma'} V \; \texttt{in} \; e \Rightarrow V \; \texttt{in} \; e'$

$\Leftrightarrow$ $\{$Inclusion defn., Lemma 3 $\}$

$$\beta_\Sigma(M') \models^{\mathcal{A}}_\Sigma e \; \texttt{in} \; e'$$

When $\rho := \texttt{not} \; \rho$:

$$M' \models^{\text{SOL}^{pres}}_{\Phi(\Sigma)} \alpha_\Sigma(\texttt{not} \; \rho)$$

$\Leftrightarrow$ $\{\alpha$ defn. $\}$

$$M' \models^{\text{SOL}^{pres}}_{\Phi(\Sigma)} \texttt{not} \; \alpha_\Sigma(\rho)$$

$\Leftrightarrow$ $\{\models$ defn. $\}$

$$M' \not\models^{\text{SOL}^{pres}}_{\Phi(\Sigma)} \alpha_\Sigma(\rho)$$

$\Leftrightarrow$ $\{$I.H. $\}$

$$\beta_\Sigma(M') \not\models^{\mathcal{A}}_\Sigma \rho$$

$\Leftrightarrow$ $\{\models$ defn. $\}$

$$\beta_\Sigma(M') \models^{\mathcal{A}}_\Sigma \texttt{not} \; \rho$$

For implication the proof is analogous
When $\rho := (\texttt{all} \; x : e) \; \rho$:

$$M' \models^{\text{SOL}^{pres}}_{\Phi(\Sigma)} \alpha_\Sigma((\texttt{all} \; x : e) \; \rho)$$

$\Leftrightarrow$ $\{\alpha$ defn. $\}$

$$M' \models^{\text{SOL}^{pres}}_{\Phi(\Sigma)} (\forall x : U)\alpha_\Sigma((x \; \texttt{in} \; e) \; \texttt{implies} \; \rho)$$

$\Leftrightarrow$ $\{\models \text{defn.}\}$

$$\text{For any } x\text{-expansion } M'' \text{ of } M',\\ M'' \models^{\mathrm{SOL}^{pres}}_{\Phi(\Sigma)^x} \alpha_{\Sigma^x}((x \texttt{ in } e) \texttt{ implies } \rho)$$

$\Leftrightarrow$ $\{\text{I.H.}\}$

$$\text{For any } x\text{-expansion } M'' \text{ of } M',\\ \beta_{\Sigma^x}(M'') \models^{\mathcal{A}}_{\Sigma^x} (x \texttt{ in } e) \texttt{ implies } \rho$$

$\Leftrightarrow$ $\{\text{Lemma 3}\}$

$$\text{For any } x\text{-expansion } \beta_{\Sigma^x}(M'') \text{ of } \beta_{\Sigma}(M'),\\ \beta_{\Sigma^x}(M'') \models^{\mathcal{A}}_{\Sigma^x} (x \texttt{ in } e) \texttt{ implies } \rho$$

$\Leftrightarrow$ $\{\models \text{defn.}\}$

$$\beta_{\Sigma}(M') \models^{\mathcal{A}}_{\Sigma} (\texttt{all } x : e)\ \rho$$

□

References

1. Arkoudas, K., Khurshid, S., Marinov, D., Rinard, M.: Integrating model checking and theorem proving for relational reasoning. In: 7th International Seminar on Relational Methods in Computer Science (RelMiCS 2003). Lecture Notes in Computer Science, vol. 3015, pp. 21–33 (2003)
2. Baumgartner, P., Fuchs, A., Tinelli, C.: Implementing the model evolution calculus. Int. J. Artif. Intell. Tools **15**(1), 21–52 (2006)
3. Benzmüller, C., Rabe, F., Sutcliffe, G.: Thf0—the core of the tptp language for higher-order logic. In: Proceedings of the 4th International Joint Conference on Automated Reasoning, IJCAR '08, pp. 491–506. Berlin, Heidelberg, Springer (2008)
4. Benzmüller, C., Theiss, F., Paulson, L., Fietzke, A.: LEO-II—a cooperative automatic theorem prover for higher-order logic. In: Armando A., Baumgartner P., Dowek G. (eds.) Automated Reasoning, 4th International Joint Conference, IJCAR 2008, Sydney, Australia, August 12–15, 2008, Proceedings. LNCS, vol. 5195, pp. 162–170. Springer (2008)
5. Braüner, T.: Proof-theory of propositional hybrid logic. Hybrid Logic and Its Proof-Theory (2011)
6. Diaconescu, R.: Institution-independent Model Theory. Birkhäuser, Basel (2008)
7. Goguen, J.A., Burstall, R.M.: Institutions: abstract model theory for specification and programming. J. ACM **39**, 95–146 (January 1992)
8. Hildebrandt, T.T., Mukkamala, R.R.: Declarative event-based workflow as distributed dynamic condition response graphs. In Proceedings of the 3rd PLACES Workshop, EPTCS, vol. 69, pp. 59–73 (2010)
9. Jackson, D.: Software Abstractions: Logic, Language, and Analysis. The MIT Press, Cambridge (2006)
10. Macedo, N., Cunha, A.: Automatic unbounded verification of Alloy specifications with Prover9. CoRR, abs/1209.5773 (2012)

11. Madeira, A., Faria, J.M., Martins, M.A., Barbosa, L.S.: Hybrid specification of reactive systems: an institutional approach. In: Barthe G., Pardo A., Schneider G. (eds.) Software Engineering and Formal Methods (SEFM 2011, Montevideo, Uruguay, November 14–18, 2011). Lecture Notes in Computer Science, vol. 7041, pp. 269–285. Springer (2011)
12. Manzano, M.: Extensions of First Order Logic. Cambridge Tracts in Theoretical Computer Science. Cambridge University Press, Cambridge (1996)
13. Martins, M.A., Madeira, A., Diaconescu, R., Barbosa, L.S.: Hybridization of institutions. In: Corradini A., Klin B., Cîrstea C. (eds.) Algebra and Coalgebra in Computer Science (CALCO 2011, Winchester, UK, August 30–September 2, 2011). Lecture Notes in Computer Science, vol. 6859, pp. 283–297. Springer (2011)
14. Mossakowski, T., Haxthausen, A., Sannella, D., Tarlecki, A.: CASL: The common algebraic specification language: semantics and proof theory. Comput. Inform. **22**, 285–321 (2003)
15. Mossakowski, T., Maeder, C., Lüttich, K.: The heterogeneous tool set, Hets. In: Grumberg O., Huth M. (eds.) Tools and Algorithms for the Construction and Analysis of Systems (TACAS 2007—Braga, Portugal, March 24 - April 1, 2007). Lecture Notes in Computer Science, vol. 4424, pp. 519–522. Springer (2007)
16. Mukkamala, R.R.: A formal model for declarative workflows: dynamic condition response graphs. PhD thesis, IT University of Copenhagen (2012)
17. Neves, R., Madeira, A., Martins, M.A., Barbosa, L.S.: Giving alloy a family. In: Zhang C., Joshi J., Bertino E., Thuraisingham B. (eds.) Proceedings of 14th IEEE International conference on information reuse and intergration, pp. 512–519. IEEE Press (2013)
18. Nipkow, T., Wenzel, M., Paulson, L.C.: Isabelle/HOL: A Proof Assistant for Higher-Order Logic. Springer, Berlin (2002)
19. Riazanov, A., Voronkov, A.: The design and implementation of vampire. AI Commun. 15(2–3), 91–110 (August 2002)
20. Ulbrich, M., Geilmann, U., El Ghazi, A.A., Taghdiri, M.: A proof assistant for alloy specifications. In: Flanagan C., König B. (eds.) Proceedings of the 18th International Conference on Tools and Algorithms for the Construction and Analysis of Systems (TACAS), Lecture Notes in Computer Science, vol. 7214, pp. 422–436. Springer (2012)
21. Weidenbach, C., Dimova, D., Fietzke, A., Kumar, R., Suda, M., Wischnewski, P.: SPASS version 3.5. In: Schmidt R.A. (ed.) Proceedings of the 22nd International Conference on Automated Deduction, CADE 2009, Lecture Notes in Artificial Intelligence, vol. 5663, pp. 140–145. Springer (2009)

A Framework for Verification of SystemC Designs Using SystemC Waiting State Automata

Nesrine Harrath, Bruno Monsuez and Kamel Barkaoui

Abstract The SystemC waiting-state automaton is a compositional abstract formal model for verifying properties of SystemC at the transaction level within a delta-cycle: the smallest simulation unit time in SystemC. In this chapter, how to extract automata for SystemC components where we distinguish between threads and methods in SystemC. Then, we propose an approach based on a combination of symbolic execution and computing fixed points via predicate abstraction to infer relations between predicates generated during symbolic execution. Finally, we define how to apply model checking to prove the correctness of the abstract analysis.

Keywords SystemC · Automata · Symbolic execution · Predicate abstraction · Formal verification · Model checking

1 Introduction

Traditionally, embedded systems were developed by separating the hardware part from the software part. It takes several iterations in the design process to reach an implementation that is functionally correct and satisfies the performance

N. Harrath (✉) · B. Monsuez
ENSTA ParisTech, Department of Electronics and Computer Engineering,
91762 Palaiseau, France
e-mail: nesrine.harrath@ensta-paristech.fr
http://www.ensta-paristech.fr

B. Monsuez
e-mail: bruno.monsuez@ensta-paristech.fr

N. Harrath · K. Barkaoui
CNAM Paris, VESPA/CEDRIC, 75003 Paris, France

K. Barkaoui
e-mail: kamel.barkaoui@cnam.fr
http://cedric.cnam.fr

T. Bouabana-Tebibel and S. H. Rubin (eds.), *Integration of Reusable Systems*,
Advances in Intelligent Systems and Computing 263, DOI: 10.1007/978-3-319-04717-1_4,

requirements. Those iterations consume large amounts of costly development time, especially since they occur in a phase of the design strategy where there is already a lot of implementation details involved. Yet, this technique is no long appropriate for nowadays embedded systems design due to market pressure that require quick, valid, efficient and safe systems. Thus, due to the design trends mentioned above, new modeling languages that support both hardware and software parts have emerged. Among others, we mention the SystemC language [1], which is a system level design language that supports design space exploration and performance evaluation efficiently throughout the whole design process even for large and complex HW/SW systems. SystemC is a C++ based modeling platform that supports system-level modeling, architectural exploration, performance modeling, software development, functional verification, and high-level synthesis.

SystemC allows the description of both hardware and software, and the designs are executable on different levels of abstraction. As a consequence, co-simulation, i.e., the simultaneous execution of hardware and software, can be used for **validation** and **verification** throughout the whole design process. However, simulation is necessary but *not sufficient*. First, because it is not complete, since it can neither be applied to all possible input scenarios (in particular for real-time and non-terminating systems), nor can it be assured that all possible executions are covered in the case of non-deterministic systems. Besides, it is very difficult to ensure the consistency between different abstraction levels, or to reuse verification results in later development stages, although during co-simulation, models in SystemC can be stepwise refined throughout the conceptual design. Finally, and more precisely, the evaluation of the simulation results should be done manually by the designer, which needs to be computed automatically. This is why we need to exploit new methodologies for program analysis and verification to help designers detect and correct errors in early stages of the conceptual design before proceeding to implementation.

Several attempts have been made to model SystemC designs in a formal way. But each of them has some restrictions and limitations: either the model they propose describe SystemC designs at a low level (RTL or cycle accurate) (e.g. [2] and [3]) and does not treat the transactional level (TLM). Or, they don't support the notion of delta-cycle (e.g. [4]). There is one interesting approach similar to ours: the work of Shyamasundar et al. [5]. They give a compositional approach that incorporates the reactive synchronous features of SystemC succinctly. Our approach also removes the previous constraints and propose a compositional approach for describing SystemC designs at the **transaction level** within a **delta-cycle** using the SystemC waiting-state automaton. It succinctly captures the reactive features of SystemC components communicating through either high level transactions or low-level signal and event communications.

In [6], we define the reactive compositional semantics of SystemC and we introduce the use of abstraction techniques. But, in this chapter, we focus on the validation of the correctness of the previous analysis with respect to the system specifications. We validate but also we illustrate how we can generate an abstract representation of SystemC components without loss of precision.

2 Background on SystemC

SystemC is a System-Level Modeling language based on C++ that is intended to enable system level design in response to the need of a very fast executable specification to validate and verify system concepts. Using the SystemC library, a system can be specified at various levels of abstraction. For hardware implementation, models can be written either in a functional style or in the register-transfer level. The software part of a system can be naturally described in the C or C++ language.

2.1 The SystemC Language

Syntactically, a SystemC program consists of a set of modules, a module contains one or more processes to describe the parallel aspect of the design. A module can also contain other modules, representing the hierarchical nature of the design. Processes inside a module are of three types: METHOD, THREAD and CTHREAD. However, METHOD and CTHREAD processes can be modeled as THREADs without loss of generality. Processes are communicating via signals. Modules communicate via channels. Channels are abstracted and are accessed via their interface methods. The simulation kernel, together with modules, ports, processes, events, channels, and interfaces constitute core language of C++. That is accompanied by a collection of data types. Over this core, SystemC provides many library-defined elementary channels, such as signals, FIFOs, semaphore, and Mutex. On top of this are defined more sophisticated libraries, including master/slave library, and process networks. A transaction-level modeling library (TLM 1.0) was announced in 2005. SystemC has been developed with heavy inter-module communication.

The semantics of SystemC combine the semantics of C++ with the simulation semantics of the kernel. The simulation semantics is event driven rather than cycle driven. But at the same time, SystemC has a discrete model of time, which means that it has also cycle-level semantics. Also time is modeled through macro-time(execution time) in some pre-defined quantifiable unit; a process waits for a given amount of time, expiration of which is notified through an event.

2.2 The Formal Semantics of SystemC

In this Subsection, we briefly present how we define the behavioral semantics of a subset of SystemC language using the Structural Operational Semantics (SOS) of Plotkin [7]. Those semantics are presented in details in [6]: we define small step semantics that capture at the same time formal semantics of SystemC structures as well as synchronous and asynchronous interactions between concurrent threads at both the delta cycle level (low level) and the transaction level (high level). The goal of developing SystemC formal semantics is to provide a complete and unambiguous specification of the language. Besides, giving a general semantics to SystemC is an instance of the problem of formalizing a general purpose programming language that contains concurrency. This problem has already been studied for other languages such as Java [8], with the aim of performing formal verification.

We suppose that each module communicate through local and environment variables. The *local* variables are internal signals, internal variables, and the program counter for the process. The *environment* variables are input/output events, input/output channels, and global variables. As regards to the execution semantics of SystemC scheduler [9], there is at most one process that is reacting to the environment. To describe how a statement changes the configurations of the environment, we write the transitions rules for processes as mentioned below:

$$\langle stmt, \sigma\rangle \xrightarrow[E_o]{E} \langle stmt', \sigma'\rangle$$

where:

- *stmt* is a SystemC statement that corresponds to the location of the program counter, before the reaction, and *stmt'* is the statement with the location of the program counter after the transition,
- σ and σ' are the states before and after the reaction respectively. They represent a function: $\mathcal{V} \cup \mathcal{CH} \mapsto values$, where $\mathcal{V}$ is the set of local and shared variables and $\mathcal{CH}$ is the set of channels.
- E is the environment (set of events and variables that activate the process) while the transition is executed, E_o is the output emitted during the transition. In general, an environment is a 5-tuple $E = (E^I, E^\delta, E^T, \mathcal{V}, \mathcal{RQ})$ where:
  - E^I is the set of immediate events,
  - E^δ is the set of next delta events,
  - E^T is the set of timed events,
  - $\mathcal{V}$ is the set of next delta updates for variable.
  - $\mathcal{RQ}$ is a sequence consisting of pending requests to update channels. A request is a pair $(ch, exp(\sigma))$ where $ch \in \mathcal{CH}$ and $exp(\sigma)$ represents the value assigned to ch.

We resort to the configuration above to be conform to the syntax of the SystemC waiting-state automata (WSA) (see Sect. 3). The SystemC WSA is an abstract compositional model that we use to model and verify SystemC components in a bottom-up approach. We use all the information about the environment variables defined over the transitions to extract the preconditions and the post-conditions for each transition in the WSA. Besides, The operational semantics that we define describe also the simulation semantics of SystemC where we consider a network of synchronous and asynchronous components that are communicating, in a concurrent manner, through either high-level transactions or low-level signal and event communications. Thus, the behavior of the scheduler is incorporated in the definition of the operational semantics of SystemC, so we don't need to separately model the SystemC scheduler.

3 Modeling SystemC with WSA

The SystemC waiting-state automaton, as first presented in [10], is a compositional model that we use to represent the SystemC components in a bottom-up approach. It is based on the analysis of the $wait/notify$ mechanism of SystemC which plays an important role in the SystemC scheduler. Besides, the SystemC WSA provides the ability to model the process either at a high level where we abstract details from system description or at a low level where we consider interactions between processes with respect to the delta cycle semantics: The delta-cycle semantics guarantee that the combinational logic behavior can be simulated even if there are combinational feedback loop within the circuit. The simulation time does not advance during a delta-cycle, and as a result all processes that execute during the delta-cycle appear to be executing simultaneously.

We have chosen to model SystemC designs using automata because it is suitable to model parallelism between different components which is essential for hardware description. This choice will be different if we have distributed systems where a few heterogeneous components communicate in parallel or for sequential processes. As for model checking, we define an internal finite representation with a sufficiently small number of states for SystemC designs using automata. This representation is amenable to verify additional properties of modules: structural properties (liveness and determinism), properties related to the QoS (quality of service) and functional properties.

3.1 Syntax

The SystemC waiting-state automaton (WSA) is defined as a transition system A over a set $\mathcal{V}$ of variables. A SystemC WSA, over a set $\mathcal{V}$ of variables, is a tuple $A = (S; E; \mathcal{T})$, where S is a finite set of states , E a finite set of events and $\mathcal{T}$ a finite set of transitions where every transition is a 6-tuple $(s; e_{in}; p; e_{out}; f; s')$:

- s and s' are two states in S, representing respectively the initial state and the final state;
- e_{in} and e_{out} are two sets of events : $e_{in} \subseteq E$; $e_{out} \subseteq E$;
- p is a predicate defined over variables in $\mathcal{V}$, i.e., $FV(p) \subseteq \mathcal{V}$, where $FV(p)$ denotes the set of free variables in the predicate p;
- f is an effect function that modifies the variables $\mathcal{V}$, it is a first order formulas defined over the set of predicates;

We often write $s \xrightarrow[e_{out},f]{e_{in},p} s'$ for the transition $(s; e_{in}; p; e_{out}; f; s')$. The effect function set $\mathcal{F}(A)$ of the automaton $A(V)$ is the set of all effect functions in $A(\mathcal{V})$: $\mathcal{F}(A) = \{f | \exists t \in \mathcal{T} s.t.\ proj_6^5(t) = f\}$, where $proj_6^5$ denotes the fifth projection of a 6-tuple (or the fifth element of the transition relation). We also use $\mathcal{T}(s)$ to denote the set of transitions from a given state s, i.e., $\mathcal{T}(s) = \{t | t \in \mathcal{T}\ and\ proj_6^1 = s\}$.

Naturally, we expect that a SystemC automaton represents faithfully the process from which it is derived. However, in a SystemC process, the transition from a waiting state to another is only triggered by the events and the predicates determine which state the process will enter after being wake up, which means that transition from the same state must have the same set of incoming events e_{in}. We say that a SystemC waiting-state automaton $A = (S; E; \mathcal{T})$ is faithful if for every two transitions t and t', $proj_6^1(t) = proj_6^1(t') \Rightarrow proj_6^2(t) = proj_6^2(t')$ and for every state $s \in \mathcal{S}$, $\bigvee_{t \in \mathcal{T}(s)} proj_6^3(t)$ always holds.

3.2 Main Properties of SystemC WSA Model

The SystemC WSA [10, 11] is defined first as an abstraction of SystemC semantics, i.e., it is an abstract representation of SystemC designs that only includes the process related information (execution and activation events). Besides, it is a compositional model where each component is developed in a way that the possible interference from its environment is already taken into account, so components are guaranteed to be interference free. In fact, in a concurrent system, components interact with each other, and the correct functioning of different components is often mutually dependent. Therefore, achieving compositionality in the presence of concurrency is much more difficult than in sequential programming. Thereby, the SystemC WSA is efficient to verify functional and non-functional properties (as we prove in [10] and [11]): the model verifies first that each component of the system is locally correct since each component behavior is modeled using a finite automaton and then it verifies that the whole system is correctly operating even in a fully concurrent environment. The model helps to detect if there exists deadlocks or non deterministic behavior. The non-functional properties includes *synchronization*, *sharing*, *interaction* and *time properties*. The model verifies these properties due to the way it is automatically built because the process of building the automata [6] is based on the presence/absence of events in the environment which activate the component. Besides, the model is annotated with information about the continuous time [11] that a transition may take to move from one state to another.

4 Mapping SystemC Designs to SystemC WSA

In our approach, we get the description of a system in SystemC. Then we distinguish the constituent components of the system and their communication relation. This section presents another extension of our previous work [6], where we define how to extract the abstract automaton for each SystemC component although we previously mentioned that different SystemC components are approximately similar if we consider the semantic point of view.

The processes in a SystemC design, either SC_METHOD or SC_THREAD, build up the components of the system. Each process is modeled as a SystemC waiting state automaton. The communication and coordination between processes is also mentioned in the operational semantics of SystemC (Sect. 2.2). The behavior of the whole system is compositionally obtained by joining the automata of the processes and their communication scenarios. But first we need to determine the constituent components of the system. Then, we generate the abstract model for each component separately.

4.1 Determining the Constituent Components

Each process of a SystemC design (SC_METHOD or SC_THREAD) is considered as a component of the system. The SC_METHOD is an uninterruptible process and has no wait statements. Its automaton must have only one state which is the initial state and the transitions are triggered iff one or more event in the sensitivity list of the process occur or change value. However, the SC_THREAD synchronizes with the environment only through the wait statements and each wait statement presents a state in the abstract model of the process. Transitions are built from the control flow graph of the process. After determining the components of the system, an abstract automaton is derived for each component. These automata are captured through the *wait* statements in the control flow graph of the related processes.

But before, we need to symbolically execute the program in order to generate the control flow graph of the components. Therefore, we use a conjoint symbolic execution (SE) [12] that executes the program using symbolic values of the variables instead of real ones. We call it conjoint symbolic execution because it includes both symbolic execution together with the formal semantics of SystemC (for more details and due to the space limitation, you may refer to [6]).

Thus, the program is first visualized as a *control flow graph* (CFG). The nodes of this graph represent the basic commands and guard expressions of the process, and the edges stand for flow of control between the nodes. We annotate the (CFG) with exemplary of logical expressions defined over variables: the assignment statement is transformed into equality and the *path condition* (PC) that we define over conditional instructions. The *PC* is a (*quantifier-free*) boolean formula defined over the symbolic inputs, it accumulates constraints which the inputs must satisfy so that the execution follows the particular associated execution path. We suppose here that the *PC* is also a first-order formulas which always hold when control flow reaches a specific program point such as a loop entry. Therefore a path condition (PC) is also included in the state that will keep track of all the decisions made along the execution, working as an accumulation of assertions made on that symbolic variables, refining their values domains and helping decide which of the **then** or the **else** branches should be taken. We can see that, by construction, the SE only generates feasible paths.

In the following, we show how the automata of the SC_METHOD processes and the SC_THREAD processes are derived respectively.

```
Begin
create initial state: S0
For each path in the control flow graph
  For each combination in the sensitivity list
      Add a transition: S0 --> S0
      Condition= predicates of the path
      event= set of input and output events
      Action= effect functions of the path
   EndFor
EndFor
END
```

Fig. 1 Algorithm to construct the SystemC waiting-state automata of SC_METHOD processes

4.2 WSA for SC_METHOD: Algorithm to Extract Automata for SC_METHODS

Figure 1 shows the algorithm to construct the waiting-state automata of SC_METHOD processes. The waiting-state automaton of an SC_METHOD process has only one state which is the initial state. The transitions are added to the abstract automaton of the process using the paths from and to the initial state of the control flow graph. The occurrence of events on each combination of the signals in the sensitivity list of the process, can activate the process independently. Therefore, for each path, at most $2^N - 1$ transitions will be added; where N is the number of the signals in the sensitivity list of the process. The entry-conditions and the exit-conditions sets of the transitions of the waiting state automaton are equal to the condition set and the action set of the path, respectively.

4.3 WSA for SC_THREAD: Algorithm to Extract Automata for SC_THREADS

Figure 2 shows the algorithm to construct the waiting-state automata of SC_THREAD processes. For each path from one waiting state to the next waiting state, there exists a transition. The entry-conditions and the exit-conditions sets of the these transition are equal to the condition and action sets of the corresponding path, respectively. The first transition of the waiting state automaton starts from the first waiting state. For each subsequent waiting state, a state is added from which the second transition in the waiting-state automaton starts. Here, loops are treated like the loops in SC_METHOD processes with a little bit difference, considering wait statements in them.

```
BEGIN
create initial state: S0
State: current_state
FOR each path in the control flow graph
 current_state=S0
 FOR each path from current_state to next wait stmt
  IF  there exists wait stmt after that first wait stmt
   Create a new state: S
   Add a transition: current_state --> S
   Condition= predicates of the path from
   current_state to S
   Event= set of input and output events
   Action= effect functions of the path from
   current_state to S
  ELSE
   Add a transition: current_state --> S0
   Condition= predicates of the path from
   current_state to S0
   event= set of input and output events
   Action= effect hunctions of the path from
   current_state to S0
   EndIF
  EndFor
EndFor
END
```

Fig. 2 Algorithm to construct the SystemC waiting-state automata of SC_THREAD processes

5 Applying Predicate Abstraction to SystemC Programs: Overview of the Automation Chain for Predicate Inference

Abstraction techniques like *predicate abstraction* [13] which is a special variant of *abstract interpretation* are widely used for semantics based static analysis of software. These techniques are based on two main key-concepts: the correspondence between the concrete and the abstract semantics through the *Galois connections*, and the feasibility of a fixed point computation of the abstract semantics, through the fast convergence of *widening* operators.

In the previous section, we briefly enumerate the usefulness of symbolic execution to generate the set of the execution traces by generating the control flow graph of the SystemC program. However, the symbolic execution is itself not approximative, but as precise as possible (which corresponds to generating abstract formulas instead of real ones). Instead, the necessary approximation is performed by explicit *abstraction* operations, which make use of an arbitrary finite set of predicates over

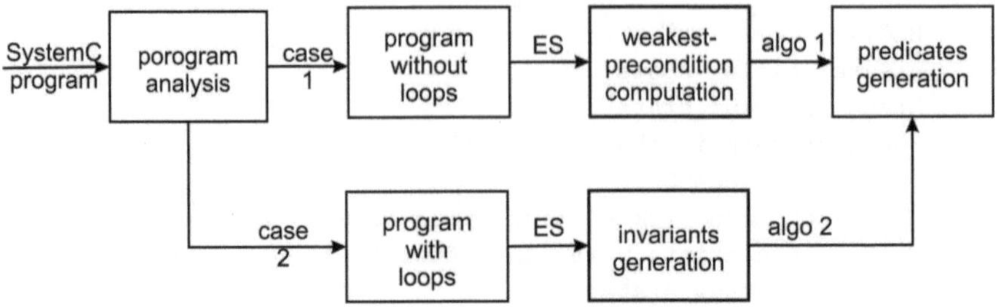

Fig. 3 An overview about the automation approach for predicate inference

the variables of the program. A similar approach that combines the use of symbolic execution together with predicate abstraction is already presented in [14]. To apply abstraction techniques to SystemC programs, we need to distinguish between two cases (Fig. 3): First, we study programs without loops where we compute the set of weakest preconditions to merge the set of transitions between each two waiting states in the control flow graph. Second, we study a special case of programs with loops where we use another technique for abstract analysis which is invariants generation for loops. In this section, we illustrate the use of predicate abstraction on some trivial examples.

5.1 Background

Predicate abstraction [13], in particular, is one of the successful abstraction techniques. In predicate abstraction, the concrete system is approximated by only keeping track of certain predicates over the concrete state variables. Each predicate corresponds to an abstract boolean variable. Any concrete transition corresponds to a change of values for the set of predicates and it is subsequently translated into an abstract transition. Using this technique, it is possible to not only reduce the complexity of the system under verification, but also, for software systems, to extract finite models that are amenable to model checking algorithms. The technique of predicate abstraction was first used for verifying low level languages such as C. But, later with the emergence of new languages like SystemC, this technique was also applied on them [15]. Let us consider the following definitions:

Definition 1 Let $A = (S, T, I)$ be the state graph of a program P where S is the set of states, T is the set of transitions and I the set of initial states. Let $\widetilde{S}$ a lattice of abstract states and $(\alpha : P(S) \mapsto \widetilde{S}, \gamma : \widetilde{S} \mapsto P(S))$ a *Galois connection* [1], where the abstraction function α associates with any set of *concrete* states a corresponding *abstract* state (the abstract state space is a lattice where larger abstract states represent

[1] a Galois connection is a pair of functions $(\underset{\sim}{\alpha}, \gamma)$ satisfying $\alpha(\gamma(\widetilde{s})) = \widetilde{s}$ and $\varphi \Rightarrow \gamma(\alpha(\varphi))$. Given γ, α is implicitly defined by $\alpha(\varphi) = \cap\{\widetilde{s} \in \widetilde{S} | \varphi \Rightarrow \gamma(\widetilde{s})\}$.

larger sets of concrete states). The concretization function γ associates with every abstract state the set of concrete states that it represents. We assume that the abstract model can make a transition from a state $\widetilde{s}$ to a state $\widetilde{s}'$ iff there is a transition from s to s' in the concrete model, where $\widetilde{s}$ is the abstract state of s and $\widetilde{s}'$ is the abstract state of s' . We denote the transition relation:

$$R := \{(\widetilde{s}, \widetilde{s}') | \exists s, s' \in S : R(s, s') \wedge \alpha(s) = \widetilde{s} \wedge \alpha(s') = \widetilde{s}'\}$$

Definition 2 Formally, we assume that the program maintains a set of n predicates $\{p_1, .., p_n\}$ ordered by implication. A predicate p_i denotes the subset of states that satisfy the predicate $\{s \in S | s \models p_i\}$. The range of the abstraction consists of boolean formulas constructed using a set of boolean variables $\{B_1, .., B_n\}$ defined over the set of predicates and ordered by implication. When applying all predicates to a specific concrete state, we obtain a vector of n boolean values, which represents an abstract state $\widetilde{s}$. If X ranges over sets of concrete states and Y ranges over boolean formulas in $\{B_1, .., B_n\}$ then the abstraction and the concretization function α and γ have the following properties:

$$\alpha(X) = \bigwedge\{Y | X \Rightarrow \gamma(Y)\}$$

$$\gamma(X) = \bigvee\{X | \alpha(X) \Rightarrow Y\}$$

The main challenge in predicate abstraction is to identify the predicates that are necessary for proving the given property. In [16], the predicate abstraction was applied to C programs, the authors have defined an algorithm for inferring predicates based on branch statements and using *weakest precondition* (WP).

In this present work we suppose that the abstraction is applied on transitions instead of states like in [17].

Definition 3 A predicate transformer [18] is a total function between two predicates on the state space of a statement. We distinguish two kinds of predicate transformers: the **weakest-precondition** and the **strongest-postcondition**. Technically, predicate transformer semantics perform a kind of symbolic execution of statements into predicates: execution runs backward in the case of weakest-preconditions, or runs forward in the case of strongest-post-conditions. We focus here only on the weakest-precondition transformer.

Given S a statement, the weakest-precondition of S is a function mapping any postcondition Q to a precondition. Actually, the result of this function, denoted $wp(S, Q)$, is the *weakest* precondition on the initial state ensuring that execution of S terminates in a final state satisfying p. We show in Fig. 4 the definition of weakest-precondition for some examples of sequential statements.

$$\begin{array}{|l|}\hline wp(skip, Q) \Leftrightarrow Q \\ wp(abort, Q) \Leftrightarrow true \\ wp(x := e, Q) \Leftrightarrow Q[x \leftarrow e] \\ wp(c_1; c_2, Q) \Leftrightarrow wp(c_1, wp(c_2, Q)) \\ wp(if\ b\ then\ c_1\ else\ c_2, Q) \Leftrightarrow (b \Rightarrow wp(c_1, Q)) \wedge (\neg b \Rightarrow wp(c_2, Q)) \\ wp(if\ b\ then\ c, Q) \Leftrightarrow (b \Rightarrow wp(c, Q)) \wedge (\neg b \Rightarrow Q) \\ \hline \end{array}$$

Fig. 4 Rules for weakest-precondition

5.2 Handling Programs Without Loops

We define each execution trace generated during symbolic execution as follows: it starts from a state that represents a *wait* statement and then we consider all the consecutive transitions that lead to the next *wait* statement in the control flow graph. Otherwise, σ_0 and σ_n represent wait states and all the intermediate states from σ_1 to σ_{n-1} represent the regular sequential constructs (including assignments, channel statements, event statements, guarded statements).

$$\langle stmt_0, \sigma_0 \rangle \xrightarrow[e^1_{out}, f^1]{e^1_{in}, p^1} \langle stmt_1, \sigma_1 \rangle \ldots \xrightarrow[e^n_{out}, f^n]{e^n_{in}, p^n} \langle stmt_n, \sigma_n \rangle$$

The goal of this analysis is to explain how to generate one-transition system from a set of consecutive transitions (an execution trace). To do so, we define an abstraction rule that starts from an initial subset of predicates and defines different transformations applied to that subset. The purpose of this transformation is to build a candidate predicate for each transition in the SystemC waiting-state automata.

Algorithm We consider P the set of predicates and F the set of functions. We use the standard definitions for constructively computing weakest precondition (wp). For all the pair of wait states (paths from a wait statement to the next wait statement in the control flow graph (CFG)) we compute the weakest precondition between the points, and add it to the SystemC waiting-state automaton if the weakest precondition is satisfied.

But first, we define the function F_E that describes the changes in the set of output events when we merge consecutive transitions. $F_{E_{out}}$ eliminates an output event e_{out} from the set of output events when it figures in the following input events in the forward transitions, otherwise it adds e_{out} to the set of output events.

$$F_E(e_{out}, E^{out}) =_{df} \begin{cases} E^{out} \backslash \{e_{out}\}\ if\ e_{out} \in E^{in} \\ E^{out} \cup e_{out}\ otherwise \end{cases}$$

We formally define the following abstraction rule that transforms a series of transitions in an execution trace into a one-transition trace with only one entry state and one exit state:

$$\frac{\langle stmt_1, \sigma_1 \rangle \xrightarrow[e^*_{out}, f^*]{e^*_{in}, p^*} \langle stmt_n, \sigma_n \rangle}{\langle stmt_1, \sigma_1 \rangle \xrightarrow[e^1_{out}, f^1]{e^1_{in}, p^1} \dots \xrightarrow[e^n_{out}, f^n]{e^n_{in}, p^n} \langle stmt_n, \sigma_n \rangle} \tag{1}$$

where $e^*_{in} = \bigcup_{i=1}^{i=n} e^i_{in}$ and $e^*_{out} = \bigcup_i F_E(e^i_{out}, E^{out})$.

Now, we use predicate abstraction to infer the relation between the set of predicates p^i and the functions f^i in order to define how we generate p^* and f^*: For each predicate p^i, we select the subset of functions $F^i \subset F$ that modifies p^i in the transitions that are triggered before p^i. More precisely, any free variable of F^i is incorporated as terms of the predicate p^i, i.e., the predicate p^i is modified by F^i during the execution of the trace. The goal is then to compute for each p^i the set of weakest preconditions of the last function from the subset F^i with respect to p^i. Besides, we consider the same order of the functions f^i as in the initial execution trace, because in our study we will consider only the last function that modifies each predicate since the intermediate transitions are simultaneous. This is why, we call f^i_F: the last function in F^i that modifies p^i. We consider the following execution trace where we consider only parameters predicate p and the effect function f:

$$\xrightarrow[f^1]{p^1} \xrightarrow[f^2]{p^2} \xrightarrow[f^3]{p^3} \dots \xrightarrow[f^{n-1}]{p^{n-1}} \xrightarrow[f^n]{p^n}$$

For each predicate p^i, each subset F^i of functions that modifies p^i and each function $f^i_F \in F^i$ that modifies the predicate p^i the last, we define the new set of predicates as follows:

$$\begin{cases} p'^1 = p^1 \\ p'^2 = p^2 \wedge \mathcal{WP}(f^2_F, p^2) \\ p'^3 = p^3 \wedge \mathcal{WP}(f^3_F, p^3) \\ .. \\ p'^n = p^n \wedge \mathcal{WP}(f^n_F, p^n) \end{cases}$$

Where $\mathcal{WP}$ is the weakest precondition for the function f^i_F that verifies the predicate p^i. The previous formulas are valid only with one condition: when the free variables $\mathcal{FV}$ of the predicate p^i (where p^i is true in the present environment) are included in the modified variables $\mathcal{MV}$ of f^i_F (where f^i_F represents the previous environment in which the transition is taking place); i.e., $\mathcal{FV}(p^i) \subset \mathcal{MV}(f^i_F)$. We use the conjunction $\wedge$ to accumulate predicates and the order of predicates is preserved. We use also the symbol $\circ$ to compose functions.

Then, we define the predicate p^* for the equation 1 as follows:

$$p^* = \bigwedge_i p'^i$$

f^* is the composition of all the functions f_F^i, i.e:

$$f^* = \underbrace{f_F^1 \circ .. \circ f_F^n}_{n\ times}.$$

We get as a result the configuration below which conforms to the SystemC waiting-state automaton definition:

$$\langle stmt_0, \sigma_0 \rangle \xrightarrow[e^*_{out}, f^*]{e^*_{in}, p^*} \langle stmt_n, \sigma_n \rangle$$

We do the same for all execution traces. The abstraction rule as defined in Eq. (1) is only valid for *program without loops*.

Example Consider the following program with two variables x and y, this program executes two tests on x and y and modifies both variables. We illustrate the previous results on this example.

```
1  if (x = 1){
2  x = x + 1;
3  y := y + 1;
4  } else {
5  if (y > 0) {
6  y := y − 1;
7  }}
```

We have two execution traces in this example:

$$\begin{cases} \xrightarrow{(x=1)} \xrightarrow[x=x+1]{} \xrightarrow[y=y+1]{} \\ \xrightarrow{(x \neq 1)} \xrightarrow{(y>0)} \xrightarrow[y=y-1]{} \end{cases}$$

For each trace, we determine the p^* then the f^*. But first, we fix the set of candidate predicates and functions. For lines 1, 4 and 5, we associate respectively the set of predicates p^1, p^4 and p^5. For lines 2, 3 and 6, we associate respectively the set of the following functions: f^2, f^3 and f^6. They are defined as follows:

$$\begin{cases} p^1 = (x = 1),\ p^4 = (x \neq 1)\ and\ p^5 = (y > 0) \\ f^2 = (x = x + 1),\ f^3 = (y = y + 1)\ and\ f^6 = (y = y - 1) \end{cases}$$

Let us consider the set of pairs (f^2, p^1) and (f^6, p^5). we define now: $p'^1 = \mathcal{WP}(f^2, p^1)$ and $p'^2 = \mathcal{WP}(f^6, p^5)$, the goal is to infer the new predicates from the initial set of predicates and consider functions (actions) that modify each predicate. This is the case of (p^1, f^2) and (p^5, f^6) since f^2 is the last and the only function that modifies p^1 and f^6 is the last and the only function that modifies p^5. We get as a result two predicates:

$$\begin{cases} p'^1 = \mathcal{WP}(x = x + 1, (x = 1)) = (x = 2) \\ p'^2 = \mathcal{WP}(y = y - 1, (y > 0)) = (y \geq 0). \end{cases}$$

Now we determine the function f^*: let us consider first $F = \{x = x + 1, y = y + 1, y = y - 1\}$, we consider the same order of the functions as in the execution trace and as we previously explain. We extract from F the subsets of functions that modify the same predicate from the initial set of predicates. Here, we have three subsets: $F^1 = \{x = x + 1\}$, $F^2 = \{y = y + 1\}$ and $F^3 = \{y = y - 1\}$. From F^1 we extract the function $f'^1 = f^2$ since we have just one element in this subset. From F^2 we extract the last and the only function that modifies the variable y, we get then the following funtion: $f'^2 = f^3$. As a result $f^* = f'^1 \circ f'^2 = f^2 \circ f^3$. Besides, we consider for the second path the function $f'^* = f^6$.

To conclude, the first trace is transformed into one transition trace of the form: $s_0 \xrightarrow[f^*]{p^*} s_1$, such that $p^* = (x = 1)$ and $f^* = (x = x + 1) \circ (y = y + 1)$.

The second trace is also transformed into the following transition: $s'_0 \xrightarrow[f'^*]{p'^*} s'_1$ such that $p'^* = p'^2 \wedge p^4$ and $f'^* = (y = y - 1)$.

As a result, we obtain the following transitions:

$$\begin{cases} s_0 \xrightarrow[(x=x+1)\circ(y=y+1)]{(x=1)} s_1 \\ s'_0 \xrightarrow[y=y-1]{(x\neq1)\wedge(y\geq0)} s'_1 \end{cases}$$

5.3 Handling Programs with Loops

As a simple example of SystemC threads including loops, we consider the following program that computes the maximal element of a table of positive integers T. The ultimate goal of this analysis is to prove that after executing this program, all elements of the array are less than or equal to **max**.

```
max=0;
i=0;
while(i < T.length){
if(T[i] > max) max= T[i];
i++;
}
```

Since loops constructs are of a notorious difficulty in the formal verification of programs, we will focus here on loops and how to use predicate abstraction to automatically infer invariants for loops. Several attempts have been made to automatically infer invariants for loops using predicate abstraction, it was first introduced by [13] and later used in [19]. Our method is based on predicate abstraction, a novel feature of our approach is that it infers predicates by iteration and in a simple way. Thus, we start first by symbolically executing the program in order to generate a set of candidate predicates ($\mathcal{CP}$) from the set of the path conditions ($\mathcal{PC}$) we generate during symbolic execution.

The symbolic execution of the example is as follows (see Appendix A): at the **entry** node, we assume nothing about the program state, so the path condition is set to **true**. Then, the first assignment statement is executed {max:=0;}, we distinguish between conditions and assignments: assignments are written between accolades. max=0 always holds-this is the postcondition of **true** under the assignment statement. Next is the loop entry, we distinguish between two cases: the loop is not entered if the condition ($i < T.length$) is false or it is unfolded at least once if ($i < T.length$). We generate two additional paths conditions: $PC_1 = i < T.length$ and $PC_2 = i >= T.length$. Once PC_1 is verified, we enter the loop body and we execute the **if** statement. The if statement is of the form **if** b **then** p **else** q, it results in two branches. When the condition *b* is true we execute *p* else we execute *q*. Therefore, we have two path conditions b = true, b = false, each PC represents a branch. Then, we resume the execution of *p* and *q* and for each statement we present its symbolic execution. In our example, we generate two PCs: $PC_3 = T[i] > max$ and $PC_4 = T[i] <= max$. Finally, we execute the assignment statement i:=i+1 that increments i by 1, this statement has just one node and can be reached via several paths. Then, we execute first the assignment statements $inst.1-2$ (line 1 and 2) of the program. The effect of the two assignments manifests itself in the two additional assumptions $max = 0$ and $i = 0$. Now, the active statement of the program is the while loop.

If we consider the formula we build before entering the loop as an invariant for the loop, we can then consider the formula φ_0 (Appendix A) as a first candidate for the loop invariant. Our technique is similar to the work of [20] where authors infer invariants for loops in Java programs. Their method is based on a combination of symbolic execution and computing fixed points via predicate abstraction. Next step in our execution process is to enter the loop and to proceed to symbolic execution. We execute $inst.3$ (line 3), here we have two cases: the loop is entered when the condition $i < T.length$ is true and the loop is not entered when the condition is not true. Thus, we build two additional formulas: $(max = 0) \wedge (i = 0) \wedge (i < T.length)$ and $(max = 0) \wedge (i = 0) \wedge (i \geq T.length)$. Next step is to execute the if statement $inst.4$ (line 4), here we have two additional branches and so two additional formulas where each formulas represent the abstract execution of each branch. The idea through this technique is to accumulate the conditions during symbolic execution and each time we enter the loop we add a new invariant. In the example above, we generate a new invariant candidate when we enter a second time the loop: the invariant φ_1 (Appendix A). Naturally, we consider the disjunction of φ_0 and φ_1 as our new invariant candidate ($\varphi_0 \vee \varphi_1$). We resume the symbolic execution of the program since φ_0 and $\varphi_0 \vee \varphi_1$ are not equivalent, we may generate a new invariant φ_2. This technique using only symbolic execution may not terminate. Thus, we resort to predicate abstraction. We can proceed again to symbolic execution to generate a new formula φ_2 for the loop, and then stop or go on accordingly. The problem with this plan is that it may not terminate this is why we resort to predicate abstraction to over-approximate the computing of the fixed point for the loop and generate a set of candidate predicates that satisfies each formula generated during symbolic execution and using the previous steps. We need first to fix a set of predicates so we consider the following set of formulas $\{\varphi_0, \varphi_1\}$. Now, we generate a set of candidate predicates CP that satisfies both φ_0 and φ_1. Each predicate p in CP must satisfy $(\varphi_0 \vee \varphi_1) \rightarrow p$.

We consider the following set of predicates for this example:
$\mathcal{CP} =$

$$\{\underbrace{i = 0}_{p_1}, \underbrace{0 \leq i}_{p_2}, \underbrace{i \leq T.length}_{p_3}, \underbrace{\forall j (0 \leq j < i \rightarrow T[j] \leq max)}_{p_4}\}$$

Algorithm In general the candidate predicates $\mathcal{CP}$ might be chosen by following heuristics, e.g. include all parts of the invariant candidate accumulated before the first unfolding of the loop, the loop guard, the weakest precondition computation and parts of the kth iteration of the loop.

For this example the set of all invariants must verify the formula below, this formula is the conjunction of predicates p_2, p_3 and p_4 in $\mathcal{CP}$.

$$\forall k, \forall j$$

$$\begin{cases} 0 \leq k < T.length, 0 \leq j < T.length \\ \forall k.(k < j \wedge T[k] \leq T[j]) \rightarrow max = T[j] \end{cases}$$

5.4 The Correctness of the WSA Model with Respect to the Concrete Semantics

The SystemC WSA is an abstract representation of the concrete semantics of SystemC programs. It transforms a set of consecutive states into only a pair of states under certain conditions. This transformation is ensured through the definition of an abstract function α that approximates the concrete semantics by keeping track of certain predicates over the concrete state variables.

To prove that the abstract model is correct, we show that there exists an inverse function (concretization function γ) that transforms the abstract semantics into the concrete ones without loss of precision. α and γ are defined over the control flow graph (CFG) and the WSA as follows:

$$\text{CFG} \underset{\gamma}{\overset{\alpha}{\rightleftharpoons}} \text{WSA}$$

Definition For each pair of states $\tilde{s} \in WSA$ and $\tilde{s'} \in WSA$, there exists at least a pair of states $s \in CFG$ and $s' \in CFG$ such that: $\gamma(\tilde{s}) = s$ and $\gamma(\tilde{s'}) = s'$. For each transition relation R in the WSA model, there exists at least one concrete

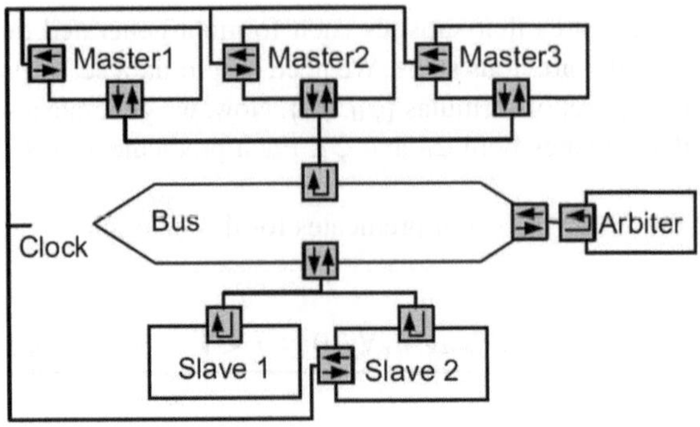

Fig. 5 Simple bus structure

trace T in the CFG such that: $\gamma(R(\tilde{s}, \tilde{s'})) = T(s, s')$ including unfaisable traces, i.e., $\gamma(WSA) = CFG \cup \{unfaisable\ \ traces\}$. Otherwise, $\gamma(WSA) \subset CFG$, i.e., γ is an increasing function over the abstract semantics defined using the WSA model.

Conclusion The SystemC WSA is faithful to the concrete semantics of the program, it abstracts only faisable paths from the concrete semantics. We explain in the previous paragraph that there exists an increasing concretization function that transforms the WSA model into a new graph that contains all possible execution cases including unfaisable paths.

5.5 Simple Bus Case Study

We illustrate our results with predicate abstraction presented in [6] on an example more intricate: The simple bus case study. The Simple Bus is a well-known transaction level example, designed to perform also cycle-accurate simulation. It is made of about 1200 lines of code that implement a high performance, abstract bus model. The complete code is available at the SystemC web site [1].

Figure 5 shows the bus structure. It uses an overall form of synchronization, where modules attached to the *bus* execute on the rising *clock* edge, and the bus itself executes on a falling clock edge. Multiple *masters* can be connected to the bus. Each master is identified by a unique priority, that is represented by an unsigned integer number. The lower this priority number is, the more important the master is. Each master communicates with the bus via an interface, which describes the communication between masters and the bus. Three modes are possible: (1) *Blocking Mode* where data is moved through the bus in a burst mode. Hence, the transaction cannot be interrupted by a request with a higher priority. (2) *Non-Blocking* Mode where the master read or write a single data word. After the transaction is completed, the caller must take care of checking the status of the last request, which can be issued and placed on the queue (BUS, REQUEST), served but is not completed (BUS, WAIT), completed without errors (BUS, OK), or finally did not complete due to an error (BUS, ERROR). (3) *Direct Mode*, where the direct interface functions perform the data transfer through the bus, but without using the bus protocol. They are usually used to debug the state of the memory. The *slave* interface describes the communication between the bus and the slaves. Multiple slaves can be connected to the bus. Each slave models some kind of memory that can be accessed through the slave interface. Two modes are possible: (i) Direct interface where it can perform immediate read or write of data without using the bus protocol. (ii) Indirect interface where the slave can read or write a single data element. The functions return instantaneously and the caller must check the status of the transfer.

The arbiter is responsible for choosing the appropriate master when there is more than one master connected to the bus. The arbiter performs the selection according to the following rules: (1) if the current request is a locked burst request, then it is always selected, (2) if the last request had its lock flag set and is again *requested*, then it is selected from the collection queue and returned, otherwise (3) the request with the highest priority is selected from the collection queue and returned.

This structure includes several SystemC components and nicely makes use of the principles of using SystemC at the transaction level. Besides some of the sample properties, e.g. liveness and safety, cannot be verified using simulation. They require the usage of formal techniques such as model checking (Sect. 6).

To illustrate our method for predicate inference, we take as an example the code of the bus arbiter (Fig. 6). The arbiter that manage priorities between the masters each time they want access to the bus. The code, as mentioned in Fig. 6, includes three independent loops, our goal is to analyze each loop independently and generate the set of abstract formulas for each loop. We define first the set of following tests generated from the loops conditions:

- $test_1(req) : (req \rightarrow status = SB\_WAIT) \wedge (req \rightarrow lock = SB\_LOCK\_SET)$
  $test_1$: `verifies if the request is successful and that it asks for a lock. It verifies as well, whether it was in the WAIT state.`
- $test_2(req) : (req \rightarrow lock = SB\_LOCK\_GRANTED)$
  $test_2$: `verifies if the request for a lock is guaranteed.`

Interpretation of the analysis From the previous analysis, we generate the abstract formulas that describes the abstract behavior of the second loop in the arbiter example. As we previously explained the second loop extract from the table of requests the first request that verifies $test_2$. Due to space limitations in this paper, we won't present the steps how to analyze the behavior of loops and how to use *the passage to limit* to generalize the final result. As a result, we propose the following abstract formulas:

$$P_2 : \exists\, i, \forall j, 0 \leq i < requests.size() \wedge test_2(requests[i]) \wedge 0 \leq j \leq i \wedge \neg test_2(requests[j])$$

```
#include "simple_bus_arbiter.h"
..
simple_bus_arbiter<T>::
arbitrate(const sc_pvector simple_bus_request<T>*> &requests)
{
  int i;
  simple_bus_request<T> *best_request = requests[0];
  // highest priority: status==SIMPLE_BUS_WAIT and lock is set:
  for (i = 0; i < requests.size(); ++i)
    {  simple_bus_request<T> *request = requests[i];
       if ((request->status == SIMPLE_BUS_WAIT) &&
  (request->lock == SIMPLE_BUS_LOCK_SET))
{      return request;
} }
  // second priority: lock is set at previous call,
   for (i = 0; i < requests.size(); ++i){
    if (requests[i]->lock == SIMPLE_BUS_LOCK_GRANTED)
         {   return requests[i];
         }
      }
  // third priority: priority
  for (i = 1; i < requests.size(); ++i)
    { sc_assert(requests[i]->priority != best_request->priority);
       if (requests[i]->priority < best_request->priority)
best_request = requests[i];
    }
  if (best_request->lock != SIMPLE_BUS_LOCK_NO)
    best_request->lock = SIMPLE_BUS_LOCK_GRANTED;
  return best_request;
}
```

Fig. 6 Simple bus arbiter code

We use the same analysis to extract the abstract formulas for the first loop and the third one, we generate the formulas P_1 and P_3 defined as follows:

$$P_1 : \exists\, i, \forall j, 0 \leq i < requests.size() \wedge test_1(requests[i]) \wedge 0 \leq j \leq i \wedge \neg test_1(requests[j])$$

$$P_1 : \exists\, i, \forall j, 0 \leq i < requests.size() \wedge test_2(requests[i]) \wedge 0 \leq j \leq i \wedge \neg test_2(requests[j]).$$

$P_3 : \exists i, \forall j, 0 \leq i < requests.size() \wedge 0 \leq j < requests.size() \wedge$ $(i == j) \vee requests[i] < requests[j] \rightarrow priority \wedge requests[i] \rightarrow lock \neq$ $SB\_LOCK\_NO \wedge request[i] \rightarrow lock = SB\_LOCK\_GRANTED.$

The analysis of the *Simple Bus* code shows then that we browse at most three times the list of queries in order to select what is the next request to be transmitted. The previous result is represented using the three logical formulas P_1, P_2 and P_3. Each formula represents a loop in the arbiter code.

6 Applying Model Checking Techniques on SystemC

Model checking [21] is a technique to automatically verify finite state concurrent systems. It consists in proving if an abstract finite model M defined in a certain logic verifies a property p expressed in the same logic.

Model checking has been successfully adopted for both hardware and software verification. Without loss of generality, the core techniques of model checking rely on the analysis of reachability property of the set of states. Therefore, it is required that the states and the corresponding transitions of the design under verification should be clearly defined. For hardware, the states are the valuation of the flip-flops and the transitions are the combination logic in the circuit; for software, they are the valuations of variables and the statements in the program, respectively.

As shown in Fig. 7, we apply model checking on SystemC using the SystemC waiting-state automata as follows: First, we need to translate the SystemC WSA with timed language constructs into an intermediate model so that we can easily apply model checking techniques. We can use either **timed automata** [22, 23], a transition system annotated with a set of real-valued variables called clocks that increase synchronously with time and associates guards and update operations with every transition, or existing abstract models like **Kripke structures**. Then, we use temporal logics to express the property we want to verify on the abstract model.

Many approaches apply model checking techniques to verify SystemC. These approaches differ on the models they use to interpret the SystemC semantics. Nevertheless, they either fail to not handle all SystemC constructs like [24], or are bound not to scale up specially when the system require non-deterministic behavior [25].

To deal with all these limitations, we propose an efficient model checking approach based on the SystemC waiting-state automta because:

1. the state explosion problem is already reduced in the waiting-state automaton model since we consider only specific states to extract the automata,
2. we don't need to model separately the SystemC scheduler since it is already included in our formal semantics for SystemC [6]. Thus, the scheduling of the concurrent behavior of the system can not influence the execution paths of the design and so the waiting-state automata,
3. the number of states in the waiting-state automata to explore is enormously reduced,
4. our predefined semantics supports all SystemC constructs and communication mechanisms (channels, signals, etc.),

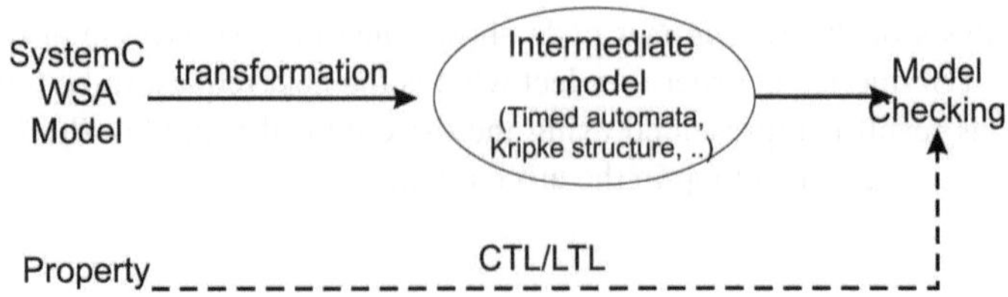

Fig. 7 Applying MC to SystemC WSA

5. signals and variables with large domain, e.g. integers, are already taken into account and present no problem in our modeling approach since they are **symbolically** modeled,
6. to deal with unbounded loops that are not supported in some model checking techniques like the approach of [26], we used predicate abstraction as shown in Sect. 5.

Checked properties In the following, we enumerate the main properties to verify on the SystemC waiting-state automata.

- Safety property: it concerns variables values which have to satisfy certain constraints. This is already reflected during the symbolic execution of SystemC designs, since we use symbolic values of variables instead of real ones. But, we need to prove the previous assumption using model checking.
  We express this property as a set of assertions defined over the set of predicates used in transitions. the failure of those assertions involves refinement of the initial model. If we take the example in Sect. 5.3, we need to verify after symbolically executing the code that all the element of the table are sorted.
- Transaction properties (TLM): check whether a request or a response is (in) valid or whether a transaction is successful. If we take the case of the simple bus (Sect. 5.5), we prove that each data written into the bus arrives to its destination without loss of information.
- System level properties: check on the order of occurrence of event notifications and the order of transactions. This property concerns the order of notification of input events in the abstraction rule (Sect. 5.2) and how to manage the set of requests in the simple bus case study (Sect. 5.5).

We express the previous properties as follows:
`Safety property:`

A transition from a state σ_i to σ_{i+1} is called **safe** when it has no assertion failure. It is written $safe(s_i, s_{i+1})$. Thus, we need to verify that each execution trace defined in Sect. 5.2 satisfy the property defined as follows:

$$allSafe(\sigma_0, \sigma_n) = \bigwedge_{0 \leq i \leq n} safe(\sigma_i, \sigma_{i+1})$$

The relation **allSafe** is used to express that all the consecutive states from σ_0 to σ_n are safe. Thus, we say that a state in the SystemC WSA is `reachable` iff all the execution traces that lead to that state are safe. We prove this by induction over each execution trace.

`Transaction property:`

We label each write transaction into the bus as M_WRITE_DATA and each read transaction from the bus as S_READ_DATA. The checked property consists in verifying whether the number of data written into the bus is equal to the number of requets read from the bus. We express it as follows:

$$assume\ number\_of(M\_WRITE\_DATA) \leq number\_of(S\_READ\_DATA)$$

`System-level property:`

The first property verifies whether the abstraction rule respects the order of notification of the input events. The second property verifies that each request in the table of requests verifies at least one property in $\{P_1, P_2, P_3\}$.

for each input event (Sect. 5.2) $e_{in} \in E \Rightarrow F_E = true$

for each request (Sect. 5.5) $Rq \in requests[n] \Rightarrow P_1 \vee P_2 \vee P_3 = true$

7 Related Works

Over the last years, research activities were mainly focused on exploiting modeling flexibility and exploring different levels of communication and behavior abstraction. More recent work concentrates on formalization and verification. The aim of our work is to propose an effective framework to model, verify and validate SystemC designs using the SystemC waiting-state automata (WSA). Although, the WSA presents many advantages like its refinability and modularity, there is a necessity to automatically build the SystemC waiting-state automata from the control flow graph as mentioned by [27]. In [6], we present a complete and detailed analysis of the compositional semantics of SystemC programs. This analysis is conform to the semantics of both SystemC language and the SystemC WSA, we also introduce the use of abstraction techniques to build the WSA from SystemC programs. But, in this chapter we first distinguish between WSA for threads and methods and we define algorithms for each of them. Second, we illustrate the use of the abstract analysis on real examples of SystemC programs and finally, we apply the model checking to validate the previous analysis.

Now, we briefly present the important works that study SystemC semantics: Große and Drechsler in [2, 3], focused on the verification at the gate level. Their work consists in verifying properties of synchronous sequential circuits using the LTL (*Linear Temporal Logic*). The main drawback of this approach is that it was somehow limited to the gate level and doesn't support the transaction level. Mueller et al. [9] translate SystemC program simulation using Abstract State Machines (ASMs). The ASMs have been extensively used in the definition of different modeling and hardware description languages, but it still be not efficient since it does not capture the synchronization between processes at the waiting states. Later Gawanmeh et al. [28], use also ASM to model SystemC designs.

Habibi et Tahar [29], translate the logic properties and the UML behavior of SystemC codes into Abstract State Machines using the AsmL language (a specification executable language developed by Microsoft). This model is then used to generate Finite State Machines (FSMs), their study focus on solving the state exploration problem using the grouping technique. Instead, our approach has the component assembly nature, where predefined abstractions are done during composition. State exploration is of a less difficulty in our model, at least at the symbolic level since we are using traces abstraction to build the automata.

Kroening et al. [4], represent SystemC models using the Labeled Kripke Structures (LKSs). In fact, the LKS model provides a syntactic way of partitioning a SystemC model into a hardware and a software part. But states in their model include all possible intermediate states within a process, not just those waiting-states as in our model. They also make a classification of processes as runnable processes, waiting processes, etc., which is basically the implementation idea of SystemC scheduler and is avoided in our model.

Karlsson et al. [30], propose a formal representation of SystemC models at a high level of abstraction using Petri-nets. Although this approach is efficient to represent SystemC designs in a formal model but it still be inadequate for complex systems where interactions are intricate. Another disadvantage of this approach is that it does not support verification of properties like concurrency and interactions between processes at the delta-cycle level.

Moy et al. [31], use the Lussy tool to extract information about the system architecture and behavior in the transaction level. They also use abstraction on their design to build an intermediate model, that they call HPIOM (*Heterogeneous Parallel Input/Output Machines*). Their global approach consists in extracting an automaton for each process and an automaton for each TLM SystemC component. They also use abstraction techniques to avoid state explosion on HPIOM automata.

For model checking, several works have recently emerged. We mention for example work of [32, 33] and [34]. In [32], authors translate a SystemC design into Petri nets and then apply CTL model checking. However, the resulting Petri nets become very large even for small SystemC descriptions which is avoided in our model.

Work of [33] translates a SystemC TLM design into Promela. The Promela model is then checked by the model checker SPIN. The translation is entirely manual and properties related to events and transactions are not considered. Finally, author in [34] maps SystemC designs into UPPAAL timed automata. But we can do more in term of expressiveness of the properties.

8 Conclusion and Prospects

Verification of reactive systems, critical systems or embedded systems is a very important issue today. In this chapter, we have presented a new semi-automatic approach for verifying SystemC designs based on the SystemC waiting-state automata model (WSA). We show how to generate the automaton for each component where we distinguish between threads and methods and we define the algorithm for each automaton. Then, we explore predicate abstraction techniques to build automatically the SystemC WSA. Thus, we distinguish between two cases for program analysis, first we consider programs without loops where we define our abstract formulas using the computation of weakest precondition to merge transitions and second we take a special case of programs with loops for which we define how to symbolically infer invariants using symbolic execution.

Future work include case studies of large examples besides the long-term compilation work and using further abstract techniques for programs analysis. Furthermore, we intend to make the process of building waiting-state automata fully automatic. Another line of future work, which is more speculative, concerns different techniques for validation, one could investigate the use of a specific model checker to prove the correctness of the previous results in properties defined for model checking purposes.

Appendix A: A Symbolic execution for the example program

- inst. $1-2$

$$\underbrace{(max = 0) \wedge (i = 0)}_{\varphi_0} \rightarrow$$

```
while(i < T.length){
if(T[i] > max) max= T[i];
i++;
}
```

- inst.3

$$(max = 0) \wedge (i = 0) \wedge (i < T.length) \rightarrow$$

```
if(T[i] > max) max= T[i];
i++;
while(i < T.length){
if(T[i] > max) max= T[i];
i++;
}
```

$$\wedge(max = 0) \wedge (i = 0) \wedge (i \geq T.length) \rightarrow$$

```
exit
```

- inst.4

$(max' = 0) \wedge (i = 0) \wedge (i < T.length) \wedge (T[i] > max') \wedge (max = T[i]) \rightarrow$

```
i++;
while(i < T.length){
if(T[i] > max) max= T[i];
i++;
}
```

$\wedge(max = 0) \wedge (i = 0) \wedge (i < T.length) \wedge (T[i] \leq max) \rightarrow$

```
i++;
while(i < T.length){
if(T[i] > max) max= T[i];
i++;
}
```

$$\wedge(max = 0) \wedge (i = 0) \wedge (i \geq T.length) \rightarrow$$

```
exit
```

$$\Longleftrightarrow \begin{cases} (max' = 0) \wedge (i = 0) \wedge (i < T.length) \wedge \\ (T[i] > max') \wedge (max = T[i]) \\ \bigvee \\ (max = 0) \wedge (i = 0) \wedge (i < T.length) \wedge \\ (T[i] \leq max) \end{cases} \rightarrow$$

```
i++;
while(i < T.length){
if(T[i] > max) max= T[i];
i++;
}
```

- inst.5

$$\underbrace{\begin{matrix} (max' = 0) \wedge (i' = 0) \wedge (i' < T.length) \wedge (T[i'] > max') \wedge (max = T[i']) \\ \bigvee (max = 0) \wedge (i' = 0) \wedge (i' < T.length) \wedge (T[i'] \leq max) \wedge (i = i' + 1) \end{matrix}}_{\varphi_1} \rightarrow$$

```
while(i < T.length){
if(T[i] > max) max= T[i];
i++;
}
```

- $\varphi_0 \vee \varphi_1 \rightarrow$

```
while(i < T.length){
if(T[i] > max) max= T[i];
i++;
}
```

- $0 \leq i \wedge i \leq T.length \wedge \forall j.(0 \leq j < i \rightarrow T[j] \leq max) \rightarrow$

```
while(i < T.length){
if(T[i] > max) max= T[i];
i++;
}
```

- $0 \leq i \wedge \forall j.(0 \leq j < i \rightarrow T[j] \leq max) \wedge i \geq T.length \rightarrow$

```
{}
```

References

1. Main page of the SystemC Initiative. http://www.systemc.org
2. Drechsler, R., Große, D.: Reachability analysis for formal verification of SystemC. In: Euromicro Symposium on Digital Systems Design, pp. 337–340 (2002)
3. Drechsler, R., Große, D.: Formal verification of LTL formulas for SystemC designs. In: IEEE International Symposium on Circuits and Systems, vol. 25, pp. 45–248 (2003)
4. Kroening, D., Sharygina, N.: Formal verification of SystemC by automatic hardware/software partitioning. In: the Third ACM and IEEE International Conference on Formal Methods and Models for Co-Design, pp. 101–110 (2005)
5. Shyamasundar, R.K., Doucet, F., Gupta, R., Kruger, I.H.: Compositional Reactive Semantics of SystemC and Verification in RuleBase. In: Proceedings of the GM R&D Workshop, pp. 227–243. Bangalore, India (2007)
6. Harrath, N., Monsuez, B.: Compositional Reactive Semantics of System-Level Designs Written in SystemC and Formal Verification with Predicate Abstraction. accepted in the International Journal of Critical Computer-Based Systems (IJCCBS) (2013)
7. Plotkin, G.D.: A structural approach to operational semantics. Logic Algebraic Program. **60–61**, pp. 17–139 (2004)
8. Havelund, K., Pressburger, T.: Model checking Java programs using Java pathfinder. Int. J. Softw. Tools Technol. Transfer (STTT) **2**(4), 366–381 (2000)
9. Mueller, W., Ruf, J., Rosenstiel, W.: SystemC Methodologies and Applications. Kluwer Academic Publishers, Boston (2003)
10. Zhang, Y., Védrine, F., Monsuez, B.: SystemC waiting-state automata. In: On First International Workshop on Verification and Evaluation of Computer and Communication Systems, pp. 5–6. eWiC, BCS (2007)
11. Harrath, N., Monsuez, B.: Timed SystemC waiting-state automata. In: On Third International Workshop on Verification and Evaluation of Computer and Communication Systems. eWiC, BCS (2009)
12. King, J.C.: Symbolic execution and program testing. Commun. ACM (Assoc. Comput. Mach.) **19**(7), 385–394 (1976)
13. Flanagan, C., Qadeer, S.: Predicate abstraction for software verification. In: Proceedings of the 29th Annual ACM Symposium on Principles and Programming Languages (POPL), pp. 191–202 (2002)

14. Bubel, R., *Hähnle*, R., Weiße, B.: Abstract interpretation of symbolic execution with explit state updates. In: On the International Symposia on Formal Methods for Components and Objects, pp. 247–277 (2008)
15. Clarke, E., Grumberg, O., Talupur, M., Wang, D.: High level verification of control intensive systems using predicate abstraction. In: Proceedings of First ACM and IEEE International Conference on Formal Methods and Models for Co-Design, IEEE Computer, Society, 25 Sept (2004)
16. Chaki, S., Clarke, E., Große, A., Strichman, O.: Abstraction with Minimum Predicates. Springer, Berlin/Heidelberg (2003)
17. Podelski, A., Rybalchenko, A.: Transition predicate abstraction and fair termination. In: Symposium on Principles of Programming Languages (POPL) (2005)
18. Dijkstra, E.W.: A Discipline of Programming. Prentice Hall PTR, Upper Saddle River (1997)
19. Leino, K.R.M., Logozzo, F.: Loop invariants on demand. In: Yi, K. (ed.) Proceedings, 3rd Asian Symposium on Programming languages and Systems (APLAS). 3780 of LNCS, pp. 119–134 (2002)
20. Schmitt, P.H., Weiß, B.: Inferring invariants by symbolic execution. In: Proceedings of the 4th International Verification, Workshop (VERIFY'07), pp. 195–210 (2007)
21. Clarke, E., Grumberg, I., Peled, D.: Model Checking. The MIT Press, Cambridge (1999)
22. ALUR, R., DILL, D.: Automata for modeling real-time systems. In: Proceedings of 17th International Colloquium on Automata, Languages and Programming (ICALP'90). Lecture Notes in Computer Science, vol. 443, pp. 322–335. Springer, Berlin (1990)
23. ALUR, R., DILL, D.: A theory of timed automata. Theoret. Comput. Sci. **126**(2), 183–235 (1994)
24. Große, D., Dreschsler, R.: CheckSyC: An efficient property checker for RTL SystemC designs. In: IEEE International Symposium on Circuits and Systems, pp. 4167–4170 (2005)
25. Moy, M., Maraninchi, F., Maillet-Contoz, L.: LusSy: A Toolbox for the analysis of systems-on-a-chip at the transactional level. In: IEEE ACSD, pp. 26–35 (2005)
26. Drechsler, R., Große, D.: CheckSyC: An Efficient Property Checker for RTL SystemC Designs. In: Proceedings of the IEEE International Symposium on Circuits and Systems (ISCAS), vol. 4, pp. 4167–4170 (2005)
27. Herber, P.: A Framework for Automated HW/SW Co-Verification of SystemC Designs using Timed Automata. Berlin, 145 (2010)
28. Gawanmeh, A., Habibi, A., Tahar, S.: An executable operational semantics for SystemC using abstract state machines. Technical Report, Concordia University, Department of Electrical and Computer Engineering, pp. 24 (2004)
29. Habibi, A., Moinudeen, H., Tahar, S.: Generating finite state machines from systemc. In: Gielen, G.G.E. (ed.) DATE Designers' Forum. European Design and Automation Association, Leuven, Belgium, pp. 6–81 (2006)
30. Karlsson, D., Eles, P., Peng, Z.: Formal verification of SystemC designs using a petri-net based representation. In: Proceeding on the Conference on Design, Automation and Test in Europe, pp. 1228–1233 (2005)
31. Maillet-Contoz, L., Moy, M., Maraninchi, F.: Lussy: a toolbox for the analysis of systems on-a-chip at the transactional level. In: Proceedings of Fifth International Conference on Application of Concurrency to System Design, pp. 26–35 (2005)
32. Karlsson, D., Eles, P., Peng, Z.: Formal verification of SystemC designs using a petrinet based representation. In: Proceedings of Design, Automation and Test in Europe, pp. 1228–1233 (2006)
33. Traulsen, C, Cornet, J., Moy, M., Maraninchi, F.: A SystemC/TLM semantics in promela and its possible applications. In: SPIN, pp. 204–222 (2007)
34. Herber, P., Fellmuth, J., Glesner, S.: Model checking SystemC designs using timed automata. In: IEEE/ACM/IFIP International Conference on Hardware/Software Codesign and System, Synthesis, pp. 131–136 (2008)

Formal MDE-Based Tool Development

Robson Silva, Alexandre Mota and Rodrigo Rizzi Starr

Abstract Model-driven engineering (MDE) focuses on creating and exploiting (specific) domain models. It is common to use domain-specific languages (DSL) to describe the concrete elements of such models. MDE tools can easily build DSLs, although it is not clear how to capture dynamic semantics as well as formally verify properties. Formal methods are a well-known solution for providing correct software, but human-machine interaction is usually not addressed. Several industries, particularly the safety-critical industries, use mathematical representations to deal with their problem domains. Such DSLs are difficult to capture, using just MDE tools for instance, because they have specific semantics to provide the desired (core) expected behavior. Thus, we propose a rigorous methodology to create GUI (Graphical User Interface) based DSLs formal tools. We aim at providing a productive and trustworthy development methodology to safety critical industries.

Keywords MDE · Formal Methods · DSLs · GUI-based formal tools

1 Introduction

Model Driven Engineering (MDE) is a software development methodology whose goal is on creating and exploiting domain models. In MDE, metamodeling is essential because it provides an abstract notation suitable to describe problem domains in terms of Domain Specific Languages (DSLs).

R. Silva (✉) · A. Mota (✉)
Informatics Center—Federal University of Pernambuco, Recife, PE, Brazil
e-mail: robson.rss@gmail.com

A. Mota
e-mail: acm@cin.ufpe.br

R. R. Starr
Embraer S.A., São José dos Campos, SP, Brazil
e-mail: rodrigo.starr@embraer.com.br

T. Bouabana-Tebibel and S. H. Rubin (eds.), *Integration of Reusable Systems*,
Advances in Intelligent Systems and Computing 263, DOI: 10.1007/978-3-319-04717-1_5,

DSLs [17] are languages designed to fill the lack of expressivity of general purpose languages, in their application domain, focusing on the key concepts and features of that domain.

There are several ways of creating DSLs based on metamodels. The Eclipse Modeling Framework (EMF) is the most well-known solution [24]. Within EMF, the definition of a DSL syntax is usually given using meta-languages such as ECore [24] (used to specify metamodels) and OCL (Object Constraint Language) to handle static semantics [20]; OCL can also describe basic dynamic behavior of a language through pre/post-conditions on operations [7]. But OCL is not used to prove properties about the DSL; only runtime checking (Testing).

Metamodeling frameworks still do not have a standard way to provide static and/or dynamic semantics in a rigorous way. Furthermore, it is rarely used to formally verify, by means of an automatic inference engine and theorem prover, certain properties about the DSL being specified. So, instead of using only metamodeling frameworks to define DSLs we need to rely on the support of formal methods.

Formal methods are mathematically based languages, tools and techniques used to specify, develop and verify systems. Nowadays, we have several of them, where from JML [1], ***Perfect*** Developer [3] (PD for short) and SCADE [4] we can even synthesize code automatically. But formal methods usually do not address graphical user interfaces (GUIs).

This work proposes a rigorous methodology to create a GUI-based DSL formal tool from a formal specification of a DSL (L) capturing the: syntax and static semantics of L (SS_L), as well as its dynamic semantics (DS_L). From L we can check desirable properties, adjust if necessary, and translate part of it (SS_L) in terms of modeling artifacts. Particularly, from SS_L we show how a metamodel MM and a set of constraints SC_{MM} over MM can be extracted automatically by using systematic translation rules that we have developed.

From MM and SC_{MM} we generate a user-friendly constraint preserving front-end (interface between the user and back-end) for the DSL. Additionally, from the formal specification L we create an executable back-end (which concentrates on business-like functionalities) using a code synthesizer. Finally, we systematically link the front and back-ends. We investigate the soundness of our translations indirectly by exercising invariants and operation contracts through the constraint-preserving generated GUI.

To illustrate the methodology, we create a Fault Tree (FT) Analysis tool [22]. We specify an FT using the *Perfect* language [3] (*P*L for short) and prove some properties using the PD Tool [3]. We extract a metamodel and constraints from this specification using a tool we have created, called *PL2EMF* [22].

We aim at providing a productive and trustworthy development methodology to safety critical industries. Although we base our work on a formal specification, that the MDE community can consider too difficult and of great effort to work in practice, we argue that for the problem domain of critical systems, where there are several mature Mathematical models available, such a formal specification would be simply such Mathematical models rewritten in terms of a formal language with tool support.

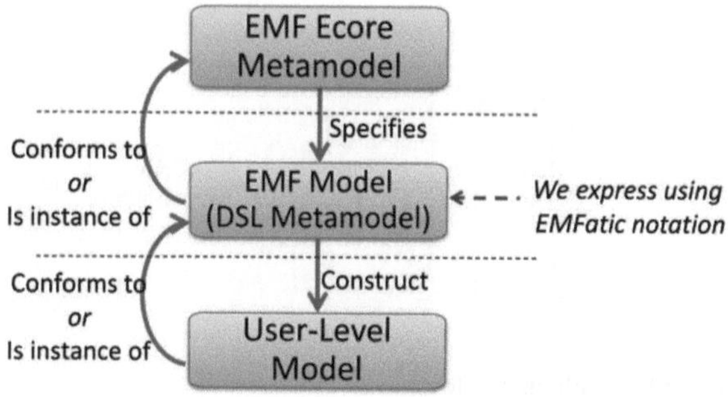

Fig. 1 EMF models relationships

This work is organized as follows. Section 2 gives an overview of the underlying concepts used in our proposed methodology, which is presented in Sect. 3. Sections 4 and 5 detail the steps of our methodology. Section 6 shows our case study. Section 7 presents related works and Sect. 8 our conclusions and future work.

2 Background

Model Driven Engineering. MDE is a software development philosophy in which models, constructed using concepts related to the domain in question, play a key role in the software development process. The idea is on building high-level technology-independent models focused exclusively on the problem domain. These models can be transformed into code for various platforms and technologies. Unlike other approaches, with mechanisms for automatic transformations, MDE becomes very productive. One can even reuse the knowledge about the domain besides the reuse of software components.

Eclipse Modeling. The Eclipse Modeling Project follows the MDE approach. It is structured into projects that provide several capabilities, such as: abstract and concrete syntax development, model-to-model and model-to-text transformations.

The EMF framework includes a metamodel (EMF's Ecore metamodel) for describing the structure of EMF models that is the basis for constructing user-level models. Figure 1 shows how these concepts are related. The Graphical Modeling Framework (GMF) [8] is an MDE framework that provides the graphical concrete syntax for a DSL and maps it to its abstract syntax (metamodel). GMF is widely used to develop graphical Eclipse-based editors for EMF-based DSL languages.

Figure 2 gives the basic usage flow for developing a graphical editor using GMF. The starting point is the definition of an Ecore metamodel. From this metamodel, GMF provides wizards to create additional models related to the graphical concrete syntax. The graphical model specifies the shapes of the aimed editor. The tooling model states the available tools. The mapping model binds the information from the

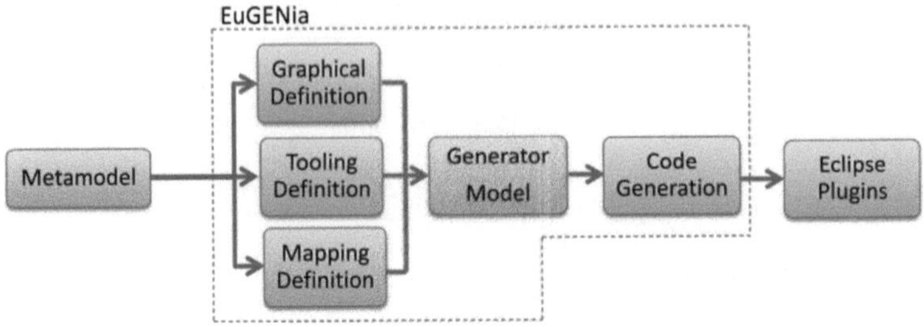

Fig. 2 Overview of GMF development flow

domain model, graphical model and tooling model. The generator model is used as input for the GUI code generator.

For several technical reasons, implementing a graphical editor with GMF is a hard-working and error prone task even for experienced users. Thus we use EuGENia [13] that assumes some high-level annotations in the Ecore metamodel [13]. From this annotated metamodel, EuGENia automatically produces the required GMF intermediate models (see dotted lines in Fig. 2) to generate a fully-functional GUI editor.

Spoofax Language Workbench. Spoofax is a language workbench for developing textual DSLs with full-featured Eclipse editor plugins [23]. It integrates language processing techniques for parser generation, meta-programming, and IDE development in a single environment [10].

With Spoofax, a DSL grammar can be written, in a declarative and modular way, using the SDF. Using this single formalism the complete syntax (lexical and context-free) of a language can be defined and integrated [9].

Based on a grammar expressed in SDF grammar, the Spoofax language workbench automatically provides basic editor facilities such as syntax highlighting and code folding, which can be customized using high-level descriptor languages. Using its parser generator tool, a parser can be created from this grammar. This generated parser can be used in an interactive environment, supporting error recovery in case of incorrect syntax or incomplete programs [10].

We express the semantic definitions, using Stratego language, by means of rewrite rules that provide an integrated solution for analysis, program transformation, code generation rules and more sophisticated editor services such as error marking, reference resolving, and content completion [10].

Defining transformations rules by using the Stratego Language, the Spoofax Workbench provides the automatic generation of a functional program transformation infrastructure that is able to perform the transformations we defined.

All of these generated services and infrastructure are integrated with Eclipse. This allows the application to be delivered as a stand-alone plugin.

Constraints. We use the Epsilon platform to handle instances of the metamodel. We indeed use its validation language (EVL) [12].

The Epsilon Object Language (EOL) [14] is the core of the platform. Its model navigation and modification facilities are based on OCL. As we use EVL, EOL is used as well. EVL has been designed atop the Epsilon platform, and therefore instead of pure OCL, it uses the OCL-based EOL as a query and navigation language [12].

EVL enhances OCL by modeling behavioral aspects as well, bringing several improvements such as: support for detailed user feedback, for warnings/critiques, for dependent constraints, and for semi-automatically repairing inconsistent model elements and so on [12]. On the other hand, it lacks static and/or dynamic formal semantics and the capability to verify formally, by means of (automatic) theorem prover, DSL properties. So, instead of using only EVL/EOL we rely on formal specification language.

The Perfect Developer Tool. (PD for short) is part of the *Escher Verification Studio* [3]. PD provides a theorem prover to reason about the requirements, contracts and code. It is also able to synthesize code.

A specification in *PL* has a set of related classes, each one with their variables, invariants, schemas, functions, pre and post-conditions, constructors, (loop) invariants, variants, properties, axioms, assertions, and so on. *Schemas* are operations that change the state and *functions* are side-effect free operations. *Schema* names are prefixed with the symbol '!' to denote that they change the system state. A class definition in *PL* is divided into sections (abstract, internal, confined and interface), each one with its specific elements.

Variables declared in the ***abstract*** section represent the abstract data model of the class (its abstract state-space). Inside this section we can have, for example, invariants that define what must be true regardless of the current model state. We can have several levels of data refinements of the abstract class data model. The ***internal*** section (not shown here), is used to declare the data refined. For each level of refinement, it is required to define a *retrieve relation* between the previous level and the current level of refinement. The ***confined*** (not shown here) and ***interface*** sections are used to declare the public interface of a class. The main difference between the elements declared in these sections is that the elements of ***confined*** are only accessible by derived classes and the elements of ***interface*** are also visible by non-derived classes [3].

To illustrate some *PL* elements we show part of the specification of the Fault Tree (FT for short) formalism (Fig. 3). Basically, an FT [19] is a kind of combinatorial model commonly used to find how an undesired event of interest (called the top event) might be caused by some combination of other undesired events (failures). The most common elements of an FT are: *AND*- and *OR-Gates*, Basic Events. Further details are presented in, see Sect. 6.

We start by defining the class *FTEdge*, left-hand side of Fig. 3, we have one attribute with type ***from*** *Comp* and other with type ***from*** *Gate*. This means these variables can be assigned to any descendants of *Comp* and *Gate*, respectively. The same is valid for *nodes:**set of from*** *Comp*, in the *FT* class. This says that *nodes* can contain an arbitrary number of elements (descendants of) *Comp*. In Fig. 3 (FT class), we can see the schema *addFTNode*. It changes the state of the system by

```
class FTEdge ^=
abstract
  var src: from Comp;
  var tgt: from Gate;
interface
  function src;
  function tgt;
  build{!src: from Comp,
      !tgt: from Gate};
end;
```

```
class FT^=
abstract
  var nodes: set of from Comp;
  var edges: set of FTEdge;
invariant
  //list of invariants (shown below)...
interface
  schema !addFTNode (n: from Comp)
    pre ~checkFTNode(n),
    (n.idc = "TOP_EVENT"
      ==> ~existsTOPEventId &
      (n within from Gate))
    post nodes! = nodes.append(n);
...
end;
```

Fig. 3 *FTEdge* and a fragment of the main class, *FT*, of the formal specification

adding a new node *n*, passed as parameter, if it was not already in the set of *nodes* (*~checkFTNode(n)*). Additionally, if node *n* is the top event (*n.idc* = *"TOP_EVENT"*), it must be a gate (*n* ***within from*** *Gate*) and there must not be another event labeled as top event (function *existsTOPEventId*, without parameters). If this precondition holds, the postcondition of this operation states that the attribute *nodes!* (after the operation) is equal to itself before the operation, *nodes*, appended with the new node *n*.

Invariants must always hold. Any instance of the FT model must satisfy the following (the list is not exhaustive):

inv1 There is a unique top event:
*#(**those** n::nodes :- n.idc = "TOP_EVENT") <= 1*

inv2 Source and target of an edge must be different:
*~(**exists** e::edges :- e.src.idc = e.tgt.idc)*

inv3 The id's are unique:
***forall** n1::nodes :-~(**exists** n2::nodes.remove(n1) :- n1.idc = n2.idc)*

inv4 One source component must have only one target:
***forall** f::edges :-~(**exists** e::edges.remove(f) :- f.src.idc = e.src.idc)*

3 Proposed Methodology

Figure 4 shows the general idea of the methodology. In Step A we create a DSL formal specification (L) composed of a syntax and static semantics (SS_L), and dynamic semantics (DS_L) parts. Therefore, L is a formal specification of a DSL. This allows us to formally check desired properties.

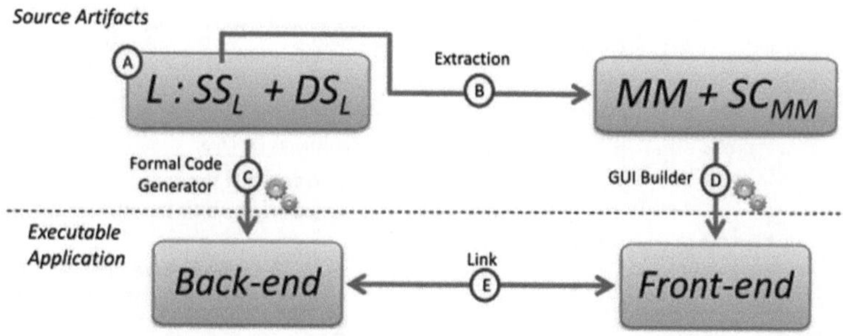

Fig. 4 Methodology high-level view

When the DSL semantics, in particular, is too abstract, one can refine it into more concrete descriptions (by refinement). The *refinement* is considered optional, inside the Step A, because: (i) the chosen formalism may not be able to support it; (ii) depending on the nature of the specification being designed, *refinement* may not be necessary.

The syntax and static semantics (SS_L) parts of L are used to create a corresponding metamodel MM and a set of constraints SC_{MM} over MM (Step B). See Sect. 4 for details. With MM and SC_{MM} we create the constraint preserving GUI (Step D). From the complete DSL formal specification, we propose using a verifiable formal code generator to create a back-end for this DSL (Step C). Finally, we integrate front and back-end (Step E). Before applying the previous steps, we need to choose the right techniques and tools to provide the appropriate technological support. Such choices must follow the requirements:

Req–1: The formalism chosen must have tool support able to prove properties and synthesize code (back-end) from a formal specification. Optionally, it can support refinement.
Req–2: Automatically extract a metamodel and a set of constraints over this metamodel from a formal specification.
Req–3: The target metamodel notation should be supported by modeling tools able to generate graphical editors (front-end) that manipulate instances of this metamodel.
Req–4: The constraint language should be able to specify and evaluate constraints on models of the chosen metamodel notation.
Req–5: The link between back-end and front-end should be designed in a way that it is transparent for the user of the final application.

□

In this chapter, we show one of many possible technological instantiations:

Req–1: it suffices to use the *Perfect* Language and the PD Tool [3];
Req–2: we developed an extractor as an Eclipse Plugin Tool, called *PL2EMF*, using the Spoofax Language Workbench [23], more details in Sect. 4;
Req–3: we use the EMFatic language, supported by EMF and GMF;
Req–4: we use the EVL language;

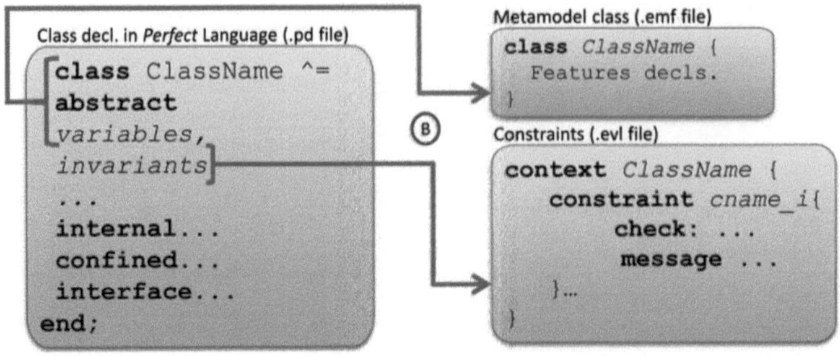

Fig. 5 Detailed metamodel extraction

Req–5: as we chose to use Eclipse Modeling, we establish the link between back and front-end by creating an integration plugin, presented in Sect. 5.

4 Metamodel Extractor

In this section, we show how we designed our metamodel and constraints extraction strategy and implemented this in a tool called *PL2EMF*.

4.1 Overview

We implemented a Metamodel Extractor named *PL2EMF* to obtain a metamodel, expressed in EMFatic, and constraints, in EVL, from a formal specification written in the *Perfect* Language. The extraction is performed by a set of translation rules (Sect. 4.3) based on a subset of the *P*L grammar (Sect. 4.2). These are the main artifacts used to generate our tool by means of the Spoofax Language Workbench.

There are several other Language Workbench, such as XText [5], JetBrains MPS [26]. However, we chose to use Spoofax due to its support for the Stratego Language (a language for program transformations). It is very productive to implement translation rules and for manipulating Abstract Syntax Trees. Figure 5 shows, more detailed, how this translation strategy works. Classes in *P*L become classes of the metamodel and their variables, in the ***abstract*** section, become class features. The set of class invariants are translated to a set of constraints. The remaining parts of the *Perfect* specification are ignored.

```
01: module PerfLang
02:  imports Common
03: exports
04:  context-free start-symbols
05:   Classes
06:  context-free syntax
07:   ClDeclaration* -> Classes {cons("Classes")}
08:   ClassModifier ClassDeclName "^=" ClassBody "end" ";"
09:                   -> ClassDeclaration {cons("ClassDecl")}
10:   "final" -> ClassModifier {cons("ClModifFinal")}
11:   ...
12:   "class" ID -> ClassDeclName {cons("ClassDeclName")}
13:   AbstractDeclarations -> ClassBody {cons("ClBodyA")}
14:   "inherits" ID AbstractDeclarations -> ClassBody {cons("ClBodyB")}
15:   "abstract" AbsMbrDecl* -> AbstractDeclarations {cons("AbsMbrsDecls")}
16:   "var" ID ":" TypeExpr ";" -> AbsMbrDecl {cons("AbsVarDecl")}
17:   "invariant" ListOfPredicates ";" -> AbsMbrDecl {cons("ClInv")}
18:   BasType -> PredefType {cons("PredefType")}
19:   PredefType  -> Type
20:   ClassNameAsType -> Type
21:   "set" "of" Type -> SetOfType {cons("SetOfType")}
22:   ...
23:   Type -> TypeExpr
24:   SetOfType -> TypeExpr
25:   ...
26:   {Predicate ","}+ -> ListOfPredicates {cons("ListOfPredicates")}
27:   Expression -> Predicate {cons("Predicate")}
28:   ForAllExpr -> Expression {cons("Expression")}
29:   "forall" BndVarDecl  ":-" Expression -> ForAllExpr {cons("ForAllExpr")}
30:   ...
31:   ID "::" CollExpr -> BndVarDecl {cons("BoundVarDecl")}
32:   ...
```

Fig. 6 *PL* grammar expressed in SDF: `main` module

4.2 Syntax

In Fig. 6 we show the *PL* context-free grammar, written in the SDF language, tailored to metamodeling needs. Additionally, we define the *PL* lexical syntax (not shown) also using SDF. For the complete grammar, please see [3].

In line `01`, of Fig. 6, we can see the name (`PerfLang`) of our main module. A module can import other modules. We import the `Common` module (line 02).

A module can contain a number of sections. The `exports` section in line `03`, is used to define the syntactic aspects (visible to other modules that import it). In this case, we have: start symbols (line `04`) and context-free syntax (line `06`).

Like in a traditional BNF, here syntax is defined by means of productions. We have a `lexical syntax` section (to declare tokens. We omit this) and a `context-free syntax` section (for operators, statements and so on). In module `PerfLang` (Fig. 6) we declare the context-free syntax (starting in line `06`).

Productions in SDF, differently from a BNF, have the form $a_1 ... a_n \rightarrow a_0$, where $a_1 ... a_n$ are n strings. When these strings match with an input text, they trigger the production rule and create the symbol a_0. That is, productions take a list of symbols and produce another symbol. The terminology of terminal and non-terminal is not very suitable for SDF, since only single characters are terminals and almost everything else is a non-terminal. For this reason, every element of a production is called a symbol [21].

SDF includes a declarative disambiguation construct to uniquely identify a symbol in the abstract syntax: the `cons(n)` annotation, where `n` uniquely identifies a symbol [11]. It also provides several regular expression operators to simplify common patterns that appear in defining productions in order to reduce the effort and enhance expressivity in defining grammars. In line `07` of Fig. 6 the declaration `ClDeclaration*` means zero or more symbols `ClDeclaration`. That is, it is allowed to declare zero or more *PL* classes in one file. The declaration `{Predicate","}+`, line `26`, means one or more symbols `Predicate` separated by `","` are allowed.

4.3 Translation Rules

This section presents how our translation rules are written in Stratego transformation language. The translation rules take a *PL* specification (according to our grammar) and produce a corresponding metamodel and constraints as output.

Basic transformations are defined using conditional term rewrite rules that are combined with strategies to control the application of rules. With Stratego, basic rules can be defined separately from the strategy that applies them. This allows a modular understanding [25]. The rules defined using Stratego act over the abstract syntax tree of a *PL* specification.

The abstract syntax tree is expressed in terms of an Annotated Term Format (ATerms, for short). An ATerm is a structured representation generated after a parser reads the input text (a PL specification). Given the grammar using SDF, this parser is generated automatically by the Spoofax Workbench.

A basic unconditional rewrite rule, using Stratego language, has the following form: `r: t1 -> t2`, where `r` is the rule name, `t1` is the left-hand side and `t2` the right-hand side term pattern. The rule `r` applies to a term `t` when the pattern `t1` matches `t`, resulting in the instantiation of `t2`. Conditional basic rules have the form: `r: t1 -> t2 where s` (`s` is a condition). We also use rules of the form:

```
r:
  t1 -> t2
  with
    x := y;
    ...
```

Using the `with` clause, instead of `where`, we can set values of new local variables as well as call other rules.

Our rules produce its output by string interpolation, using the `$[ ... ]` brackets. This constructs the metamodel as text fragments. Variables can be inserted using brackets without a dollar: `[ ... ]`. String interpolation allows the combination of text with variables. Any indentation used is preserved in the end result, except the indentation leading up to the quotation.

The rules are divided in four groups: *Class declaration*, *Abstract Declarations*, *Corresponding Types* (not shown) and *Invariants*. For conciseness, we show only one (or two) rule(s) of each group. See [22] for all the rules.

Class Declaration. Rule 1 (line `01`) starts the translation and it triggers all the other rules. It represents a complete class declaration, whose main elements are its modifier (`modif`), its name (`name`), and its body (`body`).

Rule 1 triggers other three rules, where the rule associated to `modif` is not presented here. Rule 3 (line `08`) deals with *name* producing `class` *`class_id`* as output. Rules 4 (line `10`) and 5 (line `13`) deals with `body`.

Rule 4 deals with a non-inherited (not shown) and Rule 5 with an inherited class producing the text fragment "`extends ID {`"as output. Both rules delegate the processing of the abstract declarations (abstract variables and invariants) to Rule 6 (line `16`).

```
01: to-emfatic:
02:   ClassDecl(modif,name,body) ->  $[ [modif'] [name'] [body'] } ]
03:   with
04:     body'  := <to-emfatic> body;
05:     name'  := <to-emfatic> name;
06:     modif' := <to-emfatic> modif
07:
08:  to-emfatic: ClDeclName(x) -> $[class [x]]
09:
10: to-emfatic: ClBodyA(decls) -> $[{ [decls'] ]
11:       with decls' := <to-emfatic> decls
12:
13:  to-emfatic: ClBodyB(c, decls) -> $[ extends [c] { [decls'] ]
14:       with decls' := <to-emfatic> decls
15:
```

Abstract declarations. Rule 6 (line `16`) iterates over the abstract declarations of *PL* code, delegating to other rules (e.g. Rule 9) the remainder translation.

Rule 9 (line `19`) maps a variable declaration in *PL* to a class feature in EMFatic that follows the pattern: *modifiers featureKind emfType id*. For us, we can ignore *modifiers*. The *featureKind* is the kind of the class feature, that can be *attr* (an EAttribute), *val* (an EReference with containment = true) and *ref* (an EReference, containment = false). The *emfType* is the type of the feature and *id* is the identifier.

Rule 14 (line `42`) is responsible to start the generation of the EVL code. It defines the constraints over instances of the resulting metamodel. The `class_id` is the name of the class (in *PL*) taken as source of the translation rules.

```
16:  to-emfatic: AbsMbrsDecls(d*) -> $[ [d'*] ]
17:       with d'* := <to-emfatic> d*
18:
19:  to-emfatic: AbsVarDecl(x, t) ->
20:    $[
21:       [f] [t'] [x];
22:     ]
23:    with
24:       f  := <feature-kind> t;t' := <to-type> t
25:
```

```
26:   feature-kind: f -> x
27: where
28:     switch !f
29:       case !f => PredefType(t)  : x := "attr"
30:       ...
31:     end
32:
33:   to-type: PredefType(t) -> x
34: where
35:     switch !t
36:      case "bool"     : x := "boolean"
37: ...
38: end
39:
40:   to-type: ClAsType(t) -> t
41:
```

Invariants. Rule 14 generates the EVL code that defines the context in which the list of invariants `LstPredicates(p*)` applies to.

For each invariant `inv`, Rule 15 (line `48`) generates a constraint section named *CONSTRAINT_NAME* to encapsulate `inv` and trigger Rules 16.x to translate `inv` in terms of EVL.

Rule 16.1 (line `57`) handles universal quantified invariant over a collection of items (in *PL*). In this rule, `x` is an item of the collection `coll`, `expr` is an boolean `Expression` involving `x`.

```
42:   to-evl: ClInv(LstPredicates(p*)) ->
43:      $[context class_id {
44:        [p'*]
45:      }]
46:      with p'* := <to-evl> p*
47:
48: to-evl: Predicate(Expression(inv)) ->
49:      $[
50:        constraint CONSTRAINT_NAME {
51:           check: [inv']
52:           message: "Put an error message here."
53:        }
54:      ]
55:      with inv' := <to-expr> inv
56:
57:   to-expr: ForAllExpr(BoundVarDecl(x, coll),
58:        Expression(expr)) -> $[ [coll'].forAll([x] | [expr']) ]
59: with
60:        coll' := <to-expr> coll; expr' := <to-expr> expr
```

4.4 Tool

The artifacts presented in Sects. 4.2 and 4.3 is used as input for the Spoofax Workbench to generate our translation tool. After Spoofax generates the *PL2EMF* tool, we now can use to it translate a *PL* formal specification given as input to generate the corresponding metamodel and constraints. For example, in Fig. 7(i), we show the *PL2EMF* tool using as input the Fault Tree formal specification presented in Sect. 6.

As mentioned in Sect. 3, only the syntax and static semantics (SS_L) of a DSL L are considered in the translation. The removal of the *PL* elements that are not considered in the extraction is responsibility of a preprocessing phase.

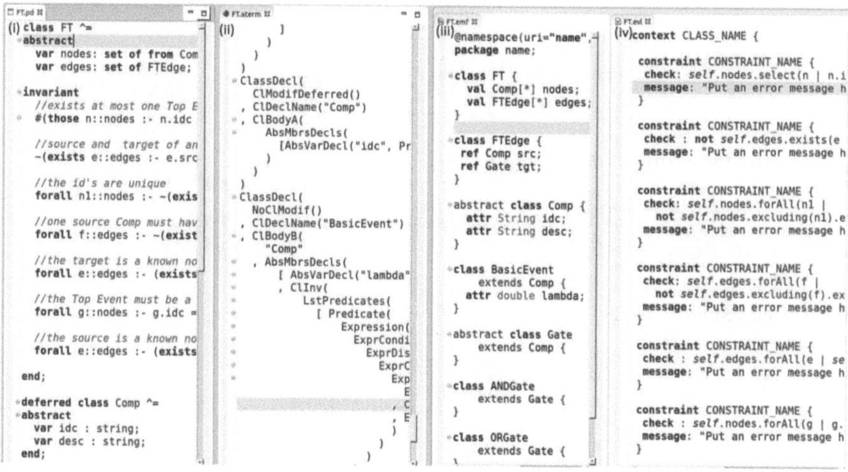

Fig. 7 (i) Fault tree formal specification; (ii) its abstract syntax using ATerms; (iii) Generated FT metamodel and (iv) Constraints

In Fig. 7(ii), we can see the abstract syntax tree of our *PL* expressed using the Annotated Term Format (ATerms). The notation ATerms is the format adopted by Spoofax for representation of abstract syntactic trees. Recall from Sect. 4.3 that rules defined using the Stratego language also act over ATerms of a *PL* specification.

Fig. 7 shows the (iii) metamodel and (iv) constraints resulting from the application of the translation rules over the FT formal specification. It is important to note that the developer (specifier) can add EuGENia annotations manually in the generated metamodel to improve its graphical concrete syntax. Annotations are responsible to give, for example, the shapes of a circle and an arrow to *BasicEvent* and *FTEdge* class, respectively. See our case study in Sect. 6. A significant name for the constraints, `CONSTRAINT_NAME`, is also a manual task.

5 Link

In this section we show how to link front (GMF) and back-ends (PD). This is accomplished via a plugin that makes the communication of these artifacts transparent to the end user.

This plugin uses the Epsilon Wizard Language [15], an extension of EOL. From within GMF editors we can execute EWL wizards to access underlying elements of models (instances of a metamodel), shown graphically in the GUI, and identify the user's graphical requests.

We see each wizard as an operation that the user can request via the GUI. This request triggers EOL operations that read the part of the model related to the request, manipulating it to deliver the correct data for the back-end corresponding operation. When the back-end returns the results, these communications occur in the reverse order.

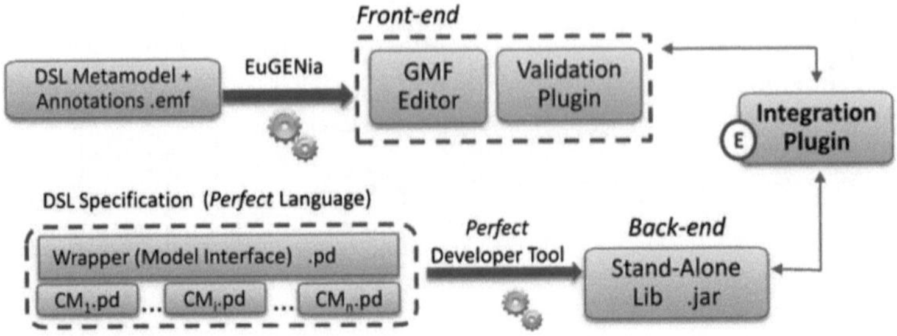

Fig. 8 Detailed link strategy

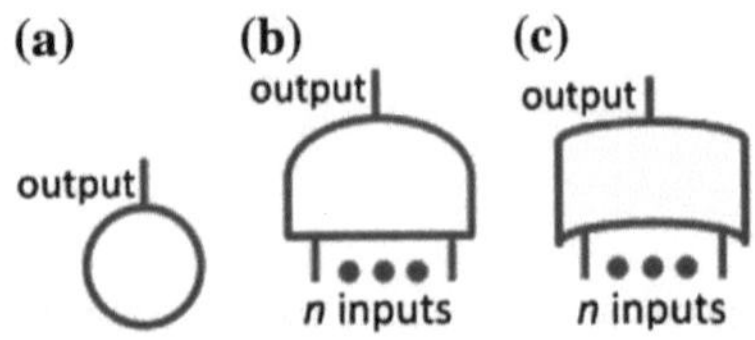

Fig. 9 Components of an FT

Figure 8 shows our integration strategy with more detail. From a *Perfect* Specification, we generate, using PD, the back-end as a stand-alone library (a jar file). Our experience has shown that the back-end as a library is more appropriate for our purposes. The integration plugin can view this library as one of its own. This facilitates the interaction, since from PD we can generate the back-end using the same target programming language that the integration plugin has been built. To provide a common access point for all operations provided by a DSL specification, we define a wrapper. It is able to handle any problem that can occur resulting from an incorrect external call. That is, it acts as a firewall between the back-end and external calls.

6 Case Study

This section presents a Fault Tree Analysis Tool.

FT Overview and Formal Specification. Recall from Sect. 2 that an FT is a directed acyclic graph (a tree), where each vertex is an FT component. Components can be basic events or gates. Gates can be *AND*- or *OR*- logical ports. We declare the class *Comp* as ***deferred*** to allow dynamic binding, see Figs. 9 and 10. Variables of type *Comp* can receive any instance from any of its descendant classes. In *Comp*, the functions *idc* and *desc* have the same name of their class attributes to allow public access.The class *Gate* specializes the class *Comp*, without any additional attributes or operations (Fig. 10). It contains two constructors (not shown) that calls the component's constructors. *AND-Gates*, and similarly *OR-Gates*, are just a specialization of *Gate* and *Comp*.

```
deferred class Comp ^=
abstract
  var idc : string;
  var desc : string;
interface
  operator = (arg);
  function idc;
  function desc;
  build{!idc: string}
    post desc! = "";
end;
class ANDGate ^= inherits Gate
interface
  build{idEv: string}
   inherits Gate{idEv};
end;
class ORGate ^= inherits Gate
interface
  build{idEv: string}
   inherits Gate{idEv};
end;
```

```
deferred class Gate ^=
    inherits Comp
  ...
end;

class BasicEvent ^=
    inherits Comp
abstract
  const E : real ^= 2.718281828;
  var lambda : real;
  invariant lambda > 0.0 ;
interface
  function lambda;
  function probf(t : real):real
  pre t >= 0.0
    ^=(1 - E^(-(lambda*t)));
  build{idEv: string, lb: real}
  pre lb > 0.0
    inherits Comp{idEv};
  post lambda! = lb;
end;
```

Fig. 10 FT components specification

We assume that basic events have an exponentially distributed life time. Then, its failure probability at time t is computed by using the function $probf(t) = 1 - e^{-\lambda t}$, $t \geq 0$ where λ is its failure rate and $e \approx 2.718281828$. As we can see, the class *BasicEvent* extends *Comp* by adding the attribute *lambda* (λ) that holds its failure rate. Additionally, a constant (E) is declared to represent the e constant and the function $probf$ is translated according to PL notation. As the main goal of this paper is to present our methodology (Sect. 3), we assume simplifications concerning the failure probability calculation of the top event: (i) in *AND-Gates*, input events are considered to be independents and (ii) in *OR-Gates*, their input events are mutually exclusive.

Then, traversing the FT (Fig. 3), and making multiplications (AND-gates) and sums (OR-gates) we can calculate the top event failure probability (indeed an approximation) of a fault tree. The function definition that makes this calculation can be found in [22]. For more details about FT Quantitative Analysis, please see [19].

Formal Verifications. For the Fault Tree specification, PD generated 71 verification conditions, from which 70 were confirmed automatically and the last one with user aid. For more details about these verification conditions, please see [22].

```
class FTEdge {                    | context FT {
  ref Comp src;                   |  constraint inv1 {check : self.nodes.select
  ref Comp tgt;                   |   (n | n.idc = "TOP_EVENT").size()<= 1}
}                                 |  constraint inv2 {check : not self.edges.
abstract class Comp {             |   exists(e | e.src.idc = e.tgt.idc) }
  attr String idc;                |  constraint inv3 {check: self.ftnodes.forAll(
  attr String desc;               |     n1 | not self.ftnodes.excluding(n1).
}                                 |         exists(n2 | n1.idc = n2.idc))}
class FT {                        |  constraint inv4 {
  val Comp[*] nodes;              |   check : self.edges.forAll(f |
  val FTEdge[*] edges;            |    not self.edges.excluding(f).exists(e |
}                                 |     f.src.idc = e.src.idc))}
                                  | }
```

Fig. 11 FT EMF model and constraints

FT GMF Editor. By applying the translation rules (Sect. 4) with the aid of our tool $PL2EMF$ [22], on our FT Formal Specification (Sects. 2 and 6) we obtain automatically the metamodel and constraints showed in Fig. 11. From this metamodel, we apply EuGENia that automatically generates the graphical editor. Figure 12a presents the FT Analysis Tool obtained following our rigorous proposed methodology. In the left-hand side we have a palette, where we can select components (gates and basic events) and connection edges.

Calculating Top Event Failure Probability. Consider a fault-tolerant multiprocessor computer system with multiple shared memory modules connected by a bus [19]. The system is operational if at least one processor (in a total of three), one memory module (in a total of two) and the bus are operational. This system is modeled by the FT presented in the right-hand side of Fig. 12b. We assume exponentially distributed failure time. Let $\lambda_P = 0.00125$, $\lambda_M = 0.00643$ and $\lambda_B = 0.00235$ be the failure rate of each processor, memory and bus, respectively. From this, the failure time distribution function of a processor is given by $probf(t)_P = 1 - e^{-0.00125t}$, $t \geq 0$.

To calculate the Top Event Failure Probability using our FT Analysis Tool, we right-click in the drawing blank area with the mouse and select the option *Wizards → Calculate Top Event Failure Probability*. As a result, an input dialog (not shown) is presented where the user can provide the time (t). Suppose that we provide 10.0 as input for t. After clicking on the OK button, the failure probability of the FT Top Event is calculated and presented as can be seen in Fig. 12b. This calculation happens in operations located in the back-end part of the tool. These operations are called by means of the integration plugin that links the back and front-end of the FT Analysis Tool (Fig. 12).

Validating the Fault Tree Tool. The work reported in [16] presents a system in which software application that can read, write and modify the content of the storage device *Disk1*. The system periodically replicates the produced data of one storage device (*Disk1*) in two storage replicas (*Disk2* and *Disk3*) to allow recovering in case of

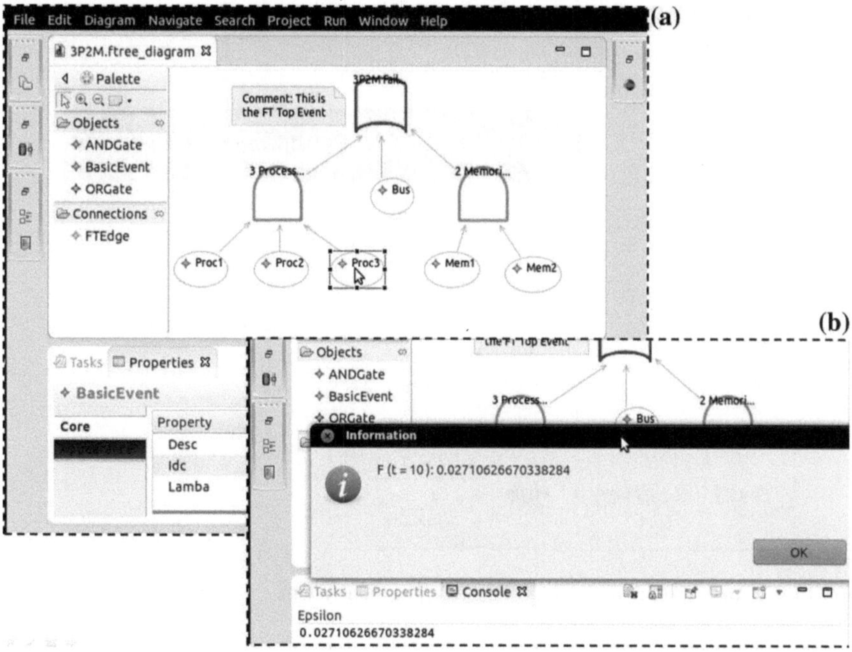

Fig. 12 FT Tool used to model a fault-tolerant multiprocessor computer system

data loss or data corruption. The system is also composed of one *Server* and *Hub* that connects the *Disk2* and *Disk3* to the server (left corner of Fig. 13). The system is considered to have failed if it is not possible to read, write or modify data on *Disk1* and if no data replica is available. Hence, if *Disk1* or the *Server* or the *Hub*, or either replica storages are faulty, the system fails. The respective FT is presented on right-hand side of Fig. 13.

We use this model to show how constraints validation work through the front-end. Note that we added two extra edges: one extra edge from the gate that represents the top event *SysFail* to itself (Fig. 13) and other from the basic event that represents the *Hub* to the gate that represents the failure of replicas. Selecting the option *Validate* in the *Edit* menu, we get an error message saying that the invariants *inv2* (source and target of an edge must be different) and *inv4* (one source component must have only one target components) (Fig. 13). This small example shows that the invariants that hold in the FT formal specification, also hold in the GUI created from its meta-model. We checked other invariant violation possibilities and all of them matched the expected results as characterized in the *P*L formal specification.

FT Tool Link. In Fig. 14, we show how the link of the FT tool was implemented. In the left-hand side we see the library, *backendftree.jar*, generated automatically, by PD, from the FT formal specification. In the top right-hand side we see the EWL wizard. It is responsible to interact with the GUI, when the user requests an FT failure probability.

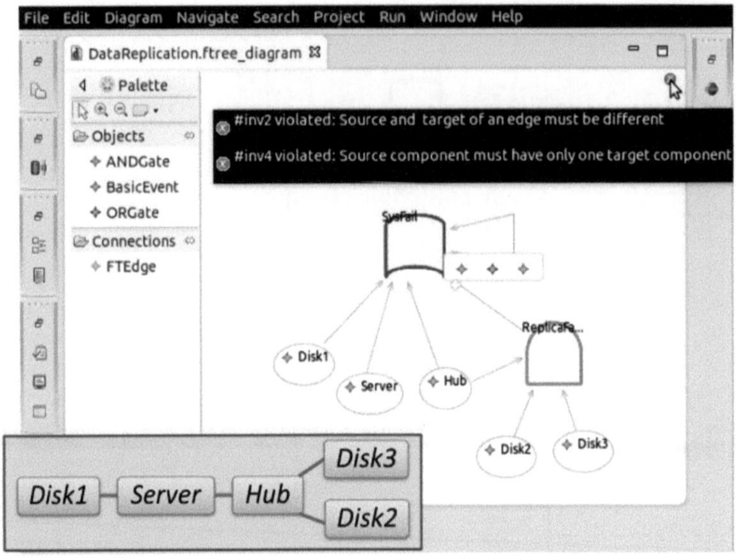

Fig. 13 FT Analysis tool used to model a data replication system

7 Related Works

The work [6] presents the benefits of integrating formal methods with MDE. It discusses the advantages of both and how they can be used to cover or weaken the disadvantages of the other. It proposes a strategy to integrate both worlds using the Abstract State Machine (ASM) with EMF. ASMs are always used to provide semantics to languages defined in the MDE. In relation to our work, the two main differences are: (i) our semantic model is not strict to an specific formalism; (ii) our metamodel is obtained automatically.

The goal of [27] is to take advantage of the strengths of heterogeneous formal methods in particular situations bridging them using MDE. The key point is to represent heterogeneous formal models into MDE as domain specific languages by metamodeling. Transformation rules developed based on the built metamodels of the formal models are used to define the semantic mapping between them. Both graphical and textual formal models are covered and the metamodels of formal models can be reused in different bridge constructions.

The work [18] presents how CSP-OZ can be integrated with UML and Java in the design of distributed reactive system. The advantages of such an integration lies in the rigor of the formal method and in checking adherence of implementations to models. The integration starts by generating a significant part of the CSP-OZ from the UML model. From this specification, properties can be verified. This CSP-OZ specification is also the basis for generating JML contracts (complemented by CSP_{jassda}) for the final implementation. Tools for runtime checking are used to supervise the adherence of the final Java implementation to the generated contracts. Large parts of

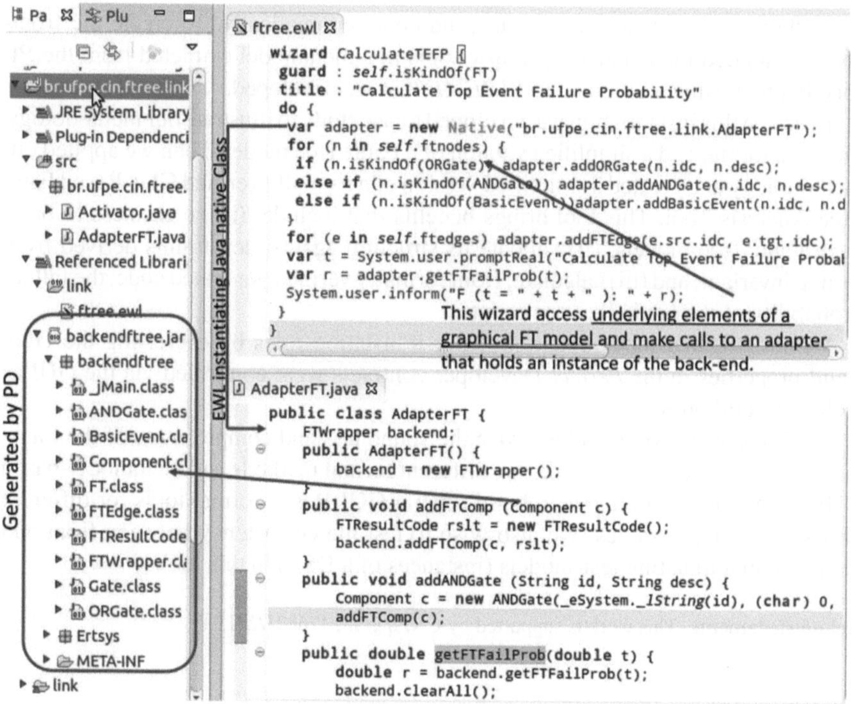

Fig. 14 FT tool link plugin

the integration approach are automated. However, as the development of tools that support the approach was not the focus, they are only prototypes. The work reported in [2] presents a practical and generic solution to define the precise dynamic semantics of DSLs by means of an experiment where Abstract State Machines (ASMs) are used to give the dynamic semantics of Session Programming Language (SPL). SPL is a DSL defined for the development of telephony services over the Session Initiation Protocol (SIP). This experiment is performed in the context of an MDE framework called AMMA (Atlas Model Management Architecture). However, unlike our proposed methodology, in [2] there is no formal verification of properties over the dynamic semantics of a DSL.

8 Conclusion and Future Work

This work proposed a methodology for creating GUI-based formally verified tools through the combination of metamodel-based GUI generators (an approach that follows MDE principles) with executable back-ends automatically generated from formal specifications.

We presented an instantiation of our methodology using the *Perfect* Language (and tool). With *PL* we describe a Mathematical model used in Engineering and with PD tool we synthesize the back-end code automatically. To create the front-end, we applied metamodeling eclipse tools to a metamodel extracted from the *PL* specification using a tool named *PL2EMF* that we developed.

Based on this instantiation we developed a case study to illustrate our methodology. First, we formalized a simplified version of a Fault Tree model. Then we applied our methodology and created a simplified version of a formally verified GUI-Based Fault Tree Analysis Tool. This tool brings benefits that include: (i) the easiness to build huge Fault Tree models; (ii) validate its structure against constraints derived from formal invariant; and (iii) calculate, from formally verified generated code, the failure probability of the Fault Tree top event.

We investigated the soundness of our translation rules by exercising the GUI. Valid properties in the *Perfect* Developer contracts were confirmed via the GUI as well as invalid ones.

As future work we intend to prove the soundness and completeness of the translation rules defined in this work, use different formal methods and metamodel-based GUI generators technologies to build several GUI-Based formal tools for different DSLs of real case studies. We also wish to test the consistency between front and back-end manipulating test-models (instances of a DSL) in this tool.

Acknowledgments This work is supported by CNPq grant 476821/2011-8.

References

1. Chalin, P., Kiniry, J.R., Leavens, G.T., Poll, E.: Beyond assertions: advanced specification and verification with JML and ESC/Java2. In: Boer, F.S., Bonsangue, M.M., Graf, S., Roever, W.-P. (eds.) Formal Methods for Components and Objects. Lecture Notes in Computer Science, vol. 4111, pp. 342–363. Springer, Heidelberg (2006). http://dx.doi.org/10.1007/11804192_16
2. Di Ruscio et al., D.: A practical experiment to give dynamic semantics to a DSL for telephony services development. Technical report, Laboratoire d'Informatique de Nantes-Atlantique (LINA) (2006).
3. Escher: Escher verification studio v5.0 (Academic license). http://www.eschertech.com. Accessed Oct 2012
4. Esterel Technologies: SCADE suite product. http://www.esterel-technologies.com/products/scade-suite/. Accessed Jan 2012
5. Eysholdt, M., Behrens, H.: Xtext: implement your language faster than the quick and dirty way. In: Proceedings of the ACM International Conference Companion on Object Oriented Programming Systems Languages and Applications Companion, SPLASH '10, pp. 307–309. ACM, New York, USA (2010)
6. Gargantini, A., Riccobene, E., Scandurra, P.: Integrating formal methods with model-driven engineering. In: Software Engineering Advances, 2009. ICSEA '09. 4th International Conference on, pp. 86–92 (2009)
7. Gargantini, A., Riccobene, E., Scandurra, P.: A semantic framework for metamodel-based languages. Autom. Softw. Eng. **16**(3–4), 415–454 (2009). http://dx.doi.org/10.1007/s10515-009-0053-0
8. GMF: Graphical modeling framework project (GMF). http://www.eclipse.org/modeling/gmp/. Accessed Jan 2013
9. Heering, J., Hendriks, P.R.H., Klint, P., Rekers, J.: The syntax definition formalism sdfreference manual. SIGPLAN Not. **24**(11), 43–75 (1989)

10. Kats, L.C.L., Visser, E.: The spoofax language workbench: rules for declarative specification of languages and IDEs. In: Cook, W.R., Clarke, S., Rinard, M.C. (eds.) In: Proceedings of the 25th Annual ACM SIGPLAN Conference on Object-Oriented Programming, Systems, Languages, and Applications, OOPSLA 2010, pp. 444–463. ACM, Reno/Tahoe, Nevada (2010)
11. Kats, L.C., Visser, E., Wachsmuth, G.: Pure and declarative syntax definition: paradise lost and regained. SIGPLAN Not. **45**(10), 918–932 (2010)
12. Kolovos, D.S., Paige, R.F., Polack, F.A.C.: On the Evolution of OCL for Capturing Structural Constraints in Modelling Languages. In: Abrial, J.-R., Glässer, U. (eds.) Rigorous Methods for Software Construction and Analysis. Lecture Notes in Computer Science, Vol. 5115, pp. 204–218. Springer, Heidelberg (2009). http://dx.doi.org/10.1007/978-3-642-11447-2_13
13. Kolovos, D.S., Rose, L.M., Abid, S.B., Paige, R.F., Polack, F.A.C., Botterweck, G.: Taming EMF and GMF using model transformation. In: Proceedings of the 13th International Conference on Model Driven Engineering Languages and Systems: Part I, MODELS'10, pp. 211–225. Springer, Heidelberg (2010)
14. Kolovos, D., Paige, R., Polack, F.: The epsilon object language (EOL). In: Rensink, A., Warmer, J. (eds.) Model Driven Architecture Foundations and Applications. Lecture Notes in Computer Science, vol. 4066, pp. 128–142. Springer, Heidelberg (2006)
15. Kolovos, D., Rose, L., García-Domínguez, A., Paige, R.: The Epsilon Book. http://www.eclipse.org/epsilon/doc/book/ (2012). Accessed 4 July 2012
16. Maciel, P.R.M., Trivedi, K.S., Jr., R.M., Kim, D.S.: Dependability modeling. In: Cardellini, V., Casalicchio, E., Kalinka, R.L., Júlio C.E., FranciscoJ.M. (eds.) On performance and dependability in service computing: concepts, techniques and research directions. pp. 53–97. IGI Global, Hershey (2012). doi:10.4018/978-1-60960-794-4.ch003
17. Mernik, M., Heering, J., Sloane, A.M.: When and how to develop domain-specific languages. ACM Comput. Surv. **37**(4), 316–344 (2005)
18. Moller, M., Olderog, E.R., Rasch, H., Wehrheim, H.: Integrating a formal method into a software engineering process with uml and java. Form. Asp. Comput. **20**(2), 161–204 (2008)
19. NASA: Fault Tree Handbook with Aerospace Applications. NASA Office of Safety and Mission Assurance, Washington (2002)
20. OMG: Object constraint language (OCL). http://www.omg.org/spec/OCL/2.3.1/PDF. Accessed Jan 2013
21. SDF: Stratego/XT manual. Chapter 6. Syntax Definition in SDF. http://hydra.nixos.org/build/5114850/download/1/manual/chunk-chapter/tutorial-sdf.html. Accessed May 2013
22. Silva, R.: GUI-based DSL Formal Tools Project. http://www.cin.ufpe.br/~rss7/mscproj/ (2013)
23. Spoofax: The spoofax language workbench v1.1. http://spoofax.org/ (2013). Accessed Feb 2013
24. Steinberg, D., Budinsky, F., Paternostro, M., Merks, E.: EMF: Eclipse Modeling Framework, 2nd edn. Addison-Wesley, Boston (2008)
25. Stratego: Stratego/XT manual. Part III. The Stratego language. http://hydra.nixos.org/build/5114850/download/1/manual/chunk-chapter/stratego-language.html. Accessed May 2013
26. Voelter, M., Solomatov, K.: Language modularization and composition with projectional language workbenches illustrated with MPS. In: van den Brand, M., Malloy, B., Staab, S. (eds.) Software Language Engineering, 3rd International Conference, SLE 2010. Lecture Notes in Computer Science. Springer (2010)
27. Zhang, T., Jouault, F., Bézivin, J., Zhao, J.: A MDE based approach for bridging formal models. In: Proceedings of the 2008 2nd IFIP/IEEE TASE, pp. 113–116. TASE '08. IEEE Computer Society, Washington, USA (2008)

Formal Modeling and Analysis of Learning-Based Routing in Mobile Wireless Sensor Networks

Fatemeh Kazemeyni, Olaf Owe, Einar Broch Johnsen and Ilangko Balasingham

Abstract Limited energy supply is a major concern when dealing with wireless sensor networks (WSNs). Therefore, routing protocols for WSNs should be designed to be energy efficient. This chapter considers a learning-based routing protocol for WSNs with mobile nodes, which is capable of handling both centralized and decentralized routing. A priori knowledge of the movement patterns of the nodes is exploited to select the best routing path, using a Bayesian learning algorithm. While simulation tools cannot generally prove that a protocol is correct, formal methods can explore all possible behaviors of network nodes to search for failures. We develop a formal model of the learning-based protocol and use the rewriting logic tool Maude to analyze both the correctness and efficiency of the model. Our experimental results show that the decentralized approach is twice as energy-efficient as the centralized scheme. It also outperforms the power-sensitive AODV (PS-AODV), an efficient but non-learning routing protocol. Our formal model of Bayesian learning integrates a real data-set which forces the model to conform to the real data. This technique seems useful beyond the case study of this chapter.

F. Kazemeyni (✉) · O. Owe · E. B. Johnsen
Department of Informatics, University of Oslo, Oslo, Norway
e-mail: kazemeyni@gmail.com

O. Owe
e-mail: olaf@ifi.uio.n

E. B. Johnsen
e-mail: einarj@ifi.uio.no

F. Kazemeyni · I. Balasingham
The Intervention Center, Oslo University Hospitals, University of Oslo, Oslo, Norway

I. Balasingham
Department of Electronics and Telecommunication, Norwegian University of Science and Technology, Trondheim, Norway
e-mail: ilangkob@medisin.uio.no

T. Bouabana-Tebibel and S. H. Rubin (eds.), *Integration of Reusable Systems*, Advances in Intelligent Systems and Computing 263, DOI: 10.1007/978-3-319-04717-1_6,

Keywords Wireless sensor networks · Mobility · Learning · Routing · Formal modeling and analysis · Probabilistic modeling · Rewriting logic · Maude

1 Introduction

Wireless sensor networks (WSNs) consist of sensor nodes which collaborate with each other to gather data and transmit the data to sink nodes. Sensor nodes usually have limited energy resources. They use routing protocols to find the path to transmit data to a designated sink node. Transmission is one of the most energy expensive activities in WSNs. So, routing protocols should be designed to be efficient. The nodes store the total paths in a routing table. Having such routing tables is not always feasible, especially in large-scale ad-hoc networks. If nodes move frequently, the cost of finding paths increases dramatically. In addition, the rate of packet loss increases because of the frequent disconnection between the nodes. In these cases, learning methods could increase the system's performance. These methods predict the best action, by learning from previous actions.

Protocol evaluation is usually done through simulators, e.g. NS-2. Instead, we use formal techniques, which provide us with an abstract way to model protocols. Formal techniques can prove the protocol's correctness, by inspecting all reachable states of a system. Simulators cannot prove the correctness of protocol's properties, due to non-deterministic behaviors of WSNs (e.g. mobility and timing), but provide quantitative information about the protocol.

This chapter introduces an approach to formally model the learning methods in WSN routing protocols. For this purpose, we aim to have a general model for learning-based routing protocols. This model is used to analyze and validate these kinds of protocols both quantitatively and formally. The motivation is to have a simplified model which includes the important features of the learning and adapting protocols and additionally the movement patterns of the nodes. We define routing protocols which estimate the most probable routing path for mobile WSNs, considering the Nodes' frequent movement. Each node's movement may have a pattern which is calculated, for every pair of nodes, as the probability of being connected. These probabilities are used to find the most probable existing path. We define a centralized and a decentralized approach for the routing protocol. In the centralized protocol, the routing paths are computed centrally by so-called *processing* nodes, using Dijkstra's Shortest Path Algorithm [1]. They have information of the whole network. Nodes request the best and most available routing path from the processing node. The processing node uses a modification of Dijkstra's Shortest Path Algorithm [1]. The notion of best is determined as a function of the cost of the links, i.e., communication links between the nodes of the network. In our model, these costs are determined based on the likelihood of availability and reliability of these edges. The best route is the one with the highest probability of successful message delivery. The centralized protocol is not very efficient because of the interactions between nodes and the processing node(s). In the decentralized protocol, nodes are responsible for finding the path. We model a protocol in which each node learns which neighbor has the best transmission history by using the Bayesian learning method [2].

We design and develop a formal, executable model of our protocol as a transition system in a SOS style. This model is analyzed using Maude [3], i.e., a formal modeling tool based on rewriting logic [4]. To have a more realistic analysis of our model, we introduce the concept of facts in our model. Facts are deterministic actions that must occur according to prescheduled times, complementing the non-deterministic actions in the model. Probability distributions are applied to abstractly model the probabilistic behaviors of the network. We feed our model with the real data-set as a fact-base. By feeding a model with real mobile traces, we obtain a more reliable analysis, in addition to taking advantage of the formal analysis. Note that using facts does not restrict the actions that can be performed in the model, it only adds extra actions that are expected to happen. Our model is used to analyze both performance and correctness.

This chapter is an extended version of a previous workshop publication [5]. The extensions include improving the presented formal model by modifying some of the rewrite rules, and adding more details about the implementation of the model in Maude. Also more experiments and more in depth analysis on the results of the model's performance are included in this chapter.

Related Work. The efficiency of routing protocols are discussed in several recent research studies. The authors of [6] provide an overview of features and mechanisms of energy-aware routing protocols. Studies, such as [7] and [8], aim at reducing the power consumption of the routing process, using cluster head election and multi-hop transmission. Another energy aware localized routing protocol is introduced in [9], and is based on distance-based power metrics. In [10], an energy-aware routing protocol for disruption-tolerant networks is introduced. In the case of a network disaster, this protocol uses the rescue vehicles (nodes) to reduce the required number of message transmissions. The routes are chosen based on the transmission delays. There are a few works which consider frequent movements of the nodes and its role in the successful message transmission and the energy efficiency of the network. Q-probabilistic routing [11] is a geographical routing algorithm which uses Bayesian algorithm to find the routing path. This protocol achieves a trade-off between the network's energy usage and the number of message retransmission. Parametric Probabilistic Sensor Network Routing Protocol [12] is another protocol which uses probabilities, based on the nodes' distances, for message retransmission. The PROPHET protocol [13] introduces the delivery predictability factor to choose the messages which are sent to the other nodes, in networks which do not guarantee fully connected networks. Recently, reinforcement learning methods have been used in the network protocols to improve their efficiency and adaptability. Papers [14] and [15] propose WSN protocols that adaptively learn to predict optimal strategies, in order to optimize their efficiency. There are centralized and also decentralized routing protocols. We have defined an adaptive and learning-based centralized and decentralized protocols, to be able to propose a general framework to model the learning observables in WSNs and performing the required analyses and comparisons. These protocols are based on the main features of the energy-efficient and learning-based routing protocols. In the routing protocol used in this chapter, we consider a WSN with mobile nodes. The links between nodes may exist with different probabilities. This protocol uses the Bayesian reinforcement learning method. The most similar protocol to the protocol used in this chapter is the Q-probabilistic routing protocol [11] which uses the Bayesian algorithm. This protocol achieves a trade-off between the energy usage and the number of retransmissions. The main difference between our protocol and the Q-probabilistic protocol is that we assume that nodes are mobile and do not have any information about their location, whereas the Q-probabilistic protocol does not talk about mobility of nodes and assumes that each node knows its location and the distance to the destination.

Formal methods are much less explored with respect to analysis of WSNs, but recently start to appear. The TinyOS operating system has been modeled as a hybrid automaton [16], and the UPPAAL tool has been used to model the LMAC protocol [17] and also to verify the temporal configuration parameters of radio communication [18]. A CREOL extension for heterogeneous environments includes radio communication [19]. Ölveczky and Thorvaldsen show that Maude is a well-suited modeling tool for WSNs [20]. Maude is used to model and analyze the efficiency of the OGCD protocol [20] and the LMST protocol [21]. A process algebra has been introduced specifically for active sensor processes such as sensing [22]. We follow this line of research and use Maude as a tool to develop a learning-based routing protocol for WSNs. There are very few works which study learning techniques in formal methods. A reinforcement method (Q-learning) which is used in a Q-routing protocol is modeled in [23], using the SPIN model checker. In contrast to our work, this chapter does not introduce a generalization for learning rules in the context of formal models. The results of the formal model in [23] are only qualitative, while we provide a realistic analysis both qualitatively and quantitatively, using a real dataset.

Paper Overview. Section 2 introduces reinforcement learning. Section 3 presents the generalized routing protocol which is modeled later in this chapter. Section 4 presents the main contribution of the chapter, which is integrating learning of observables in a probabilistic model. Section 5 discusses the modeling of the routing protocols, using the learning rules. Section 6 describes the implementation, the case study and the analysis results. Section 7 concludes the chapter.

2 Reinforcement Learning

Reinforcement learning is concerned with predicting the actions which maximize the accumulated rewards for agents [24]. Agents learn which actions to choose by observing the rewards obtained from previous actions. Figure 1 represents the interactions of an agent and the environment. The actions are chosen by agents and change the states. As a result, agents receive a *reinforcement learning signal* in the form of a reward. Reinforcement learning is formulated as a *Markov Decision Process*, which

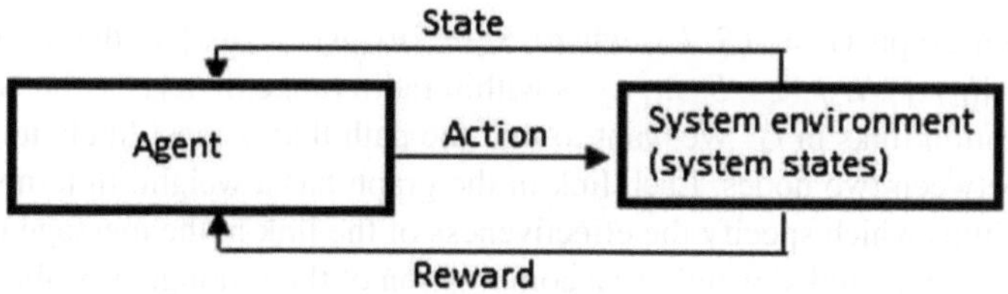

Fig. 1 A learning flowchart

consists of a tuple (ST, A, P, R) where ST is a set of environment states, A is a set of actions, $P(st'|st, a)$ is the *probability* of being in state st' after choosing action a in state st, and $R(st, a, st')$ is the *reward* associated with this action.

Bayesian inference is an extension of reinforcement learning which updates the probability estimation of actions (called hypotheses), based on *Bayes* rules [25]. In Bayesian inference, a set of actions is defined and agents learn to estimate the probability of the actions and choose the action with the highest probability. Assume that there are k actions $A = \{a_1, a_2, ...a_k\}$, and D is the set of training actions (including the information which is obtained from the accumulated history of observations). $P(a)$ denotes the prior probability of action a (before any observation from the set of training actions), $P(D|a)$ is the probability of D given action a (indicating the perceived likelihood of observing the action which is actually observed), and $P(D)$ is the prior probability of the set training actions which is obtained from the previous actions ($P(D) = \sum_{a \in A} P(D|a)P(a)$). Then, $P(a|D)$ is the probability of action a given the set of actions D. The Bayesian method finds the maximum posteriori action a_{MAP}, which is the most probable action $a \in A$:
$a_{MAP} = argmax_{a \in A} P(a|D)$, which is equal to $argmax_{a \in A}(\frac{P(D|a) \times P(a)}{P(D)})$, and consequently

$$a_{MAP} = argmax_{a \in A}(P(D|a) \times P(a)), \tag{1}$$

where, $argmax_{a \in A} P(a|D)$ returns the action $a \in A$ for which $P(a|D)$ attains its maximum value. Note that the action a which maximizes the value of $P(D|a) \times P(a)$, also maximizes the value of $\frac{P(D|a) \times P(a)}{P(D)}$. We use the Bayesian method for predicting the best action for routing.

3 The Selected Routing Protocol

This section explains the general learning-based routing protocols for WSNs, based on the node movement patterns and the probability associated to the communication links. Assume that we have a WSN with nodes which move frequently. We want to find the best routing path between the sensor and the sink. We consider the best path to be the one which is the most reliable and has the highest probability of availability during the message transmission. We assume that a WSN is a graph $G = (S, L)$, where $S = \{n_1, n_2, ..., n_N\}$ is the set of nodes and $L = \{l_{i,j} \mid \forall i, j \in S \text{ s.t. } n_i \text{ is within radio range of } n_j\}$ is the set of the communication links in G. We want to find the path that is most likely to deliver a message between two nodes. Each link in the graph has a weight, in terms of a set of probabilities which specify the effectiveness of the link in the message transmission. We define the links' weight as a combination of the in-range probability of the link and the reliability of the link. The link's reliability is important, as well as its availability. If a link is available but it is not reliable, messages may get lost. Each link $l_{i,j} \in L$ has a probability which we call the *in-range probability* or $EP_{l_{i,j}}$. This probability is calculated using the history of the availability of the network links. We note that if two nodes are not within radio range from each other, a link between them is not available, because nodes cannot receive each others' messages. Nodes meet when they enter the radio communication range of each other. Assume that an initial probability is defined for each link $l_{i,j}$, between two nodes i and j. The chance that two moving nodes meet each other (be in each others' range) during a period of time could increase or decrease the initial probability using the following definition similar to [26]:

$EP_{l_{i,j}} = EP_{l_{i,j}} \times (1 - f) + f$ if i and j meet within a predefined time interval T,
$EP_{l_{i,j}} = EP_{l_{i,j}} \times (1 - f)$ otherwise,
where f, the *adapting factor* $0 < f < 1$, is a predefined value which specifies the rate of probability update. The factor f, the initial probability of the links, and the time interval T, depend on the specific properties of a network. The specification can range from conservative, e.g. $EP_{l_{i,j}} = 0 \;\; for\ all\ i, j \in S$, to optimistic, e.g. $EP_{l_{i,j}} = N \times (\pi\omega^2/c) \;\; for\ all\ i, j \in S$, where ω and c are the radius of the radio range and the total area under study, respectively. Here, we assume all nodes are uniformly distributed within the area of study. The reliability of a link depends on different factors. For example, the physical nature of the environment between two nodes changes the reliability of their link in time. For simplicity, we consider the average historical reliability of the link $l_{i,j}$ as a probability $RP_{l_{i,j}}$, which is obtained from the network's history based on the observed reliability so far in the communication between the nodes i and j. The probability which is assigned as weight to each network link $l_{i,j}$, denoted as $P_{l_{i,j}}$, is the combination of these two independent probabilities: $RP_{l_{i,j}}$ and $EP_{l_{i,j}}$, namely $P_{l_{i,j}} = RP_{l_{i,j}} \times EP_{l_{i,j}}$.

When the total information of the routing paths is not available, specially in the decentralized approaches where the knowledge of each node for transmitting a message is limited to its neighbors, we use acknowledgment (ACK) messages to infer the probability of availability of the routing paths to each destination node. The source node s can calculate the acknowledgment probability $AP_{s,e,d}$ for the messages to the destination d through the neighbor node e, as $AP_{s,e,d} = \frac{ackNum^e_{s,d}}{msgNum^e_{s,d}}$, where $msgNum^e_{s,d}$ and $ackNum^e_{s,d}$ are the total number of the sent data messages from node s to d and the total number of the ACK messages from node d to s, respectively, in both cases through the neighbor node e. In the case of not having the total information of the routing path, we use the Bayesian learning method to predict which path has the highest chance to deliver a message successfully, based on the acknowledgment probability. Assume $M = \{e_1, e_2, ..., e_m\}$ is the set of neighbors of node s. According to the Bayesian learning method, there are m hypotheses $H = \{h_{e_1}, h_{e_2}, ..., h_{e_m}\}$, each refers to selecting neighbor node $e_k \in M$. We need to find h_{MAP} based on Formula 1. In our protocol, for each source node s and destination node d, $P(D|h_{e_k})$ is the rate of successful message transmission, if node s chooses neighbor node e_k to transmit the message toward d, which means $P(D|h_{e_k}) = AP_{s,e_k,d}$. and $P(h)$ is initially assigned to $P_{l_{s,e_k}} = RP_{l_{s,e_k}} \times EP_{l_{s,e_k}}$.

Each time that node s wants to send a message, it calculates $P(D|h_{e_k})$ for all neighbors $e_k \in M$ and chooses the hypothesis (neighbor node) with the highest probability. If the message is transmitted successfully, an acknowledgment is sent from d to the source node s. By receiving this message, s is rewarded by updating the reward table ($ackNum^e_{s,d}$). This reward affects the next predictions by improving $AP_{s,e,d}$. This ratio can give us an estimate about the probability of the availability of a path. The accuracy of this estimation improves during the time, by observing more transmissions. Consequently, this ratio shows the chance of a successful transmission through that path. We assume that the learning time is less than the movement time to avoid possible learning failures (e.g. missing a movement).

To compare centralized and decentralized learning, we define two versions of this protocol, one centralized and another, more robust and flexible, which is decentralized. There is a trade-off between using these approaches. In the centralized approach, central nodes needs to have information about the total system. Gathering and storing the information are costly, and sometimes impossible. The decentralized approaches are often less accurate, but more efficient. When processing nodes are not reachable, the decentralized protocol is the only feasible option.

3.1 The Centralized Approach

In the centralized approach, we use a modified Dijkstra's algorithm to find the most probable path. This modification returns all the nodes in the best path, while Dijkstra's algorithm only returns the best path's weight. To calculate a link's weight, we use the links' probability, instead of the nodes' distance which is used by the original Dijkstra algorithm. This is a centralized process which is performed in the processing node. When sensor nodes move, they inform the processing node about their new position. Each time a node needs a path, it sends a message to the processing node and asks for the path. The processing node sends a reply message to the node which includes the result path.

3.2 The Decentralized Approach

In the decentralized protocol, each node decides locally which node should be chosen as the next node in the routing path. Nodes only have the information of themselves and their neighbors, not the total information of the network. There are some methods which can be used in these situations, such as the "ant colony" [27]. But these methods frequently rebroadcast messages asking for routing paths, which makes them less energy-efficient. Instead, we let each node collect information from previous transmissions to learn how to predict the best route. We note that nodes do not have all the information ideally needed to decide about the total path, but use what is available to them, to pass a message to the node about which path to the destination is the most probable. In the decentralized approach, each node uses the Bayesian learning method to predict the neighbor with the highest chance of successful message delivery.

4 Integrating Learning of Observables in a Probabilistic Model of WSNs

This section integrates the reinforcement learning aspects discussed in the Sect. 3.2 with a probabilistic WSN model, presented in SOS style [28]. An SOS rule has the general form

$$\text{(RULE NAME)}\quad \frac{\mathit{Conditions}}{\mathit{conf} \longrightarrow \mathit{conf}'}$$

This rule locally transforms configurations or sub-configurations which match *conf* into *conf′* if the premises *Conditions* hold. In a probabilistic rule the *Conditions* or *conf′* may depend on a sample drawn from a probability distribution. Non-determinism may be controlled by such rules. As an example, such a rule may look like

$$\text{(RULE NAME)}\quad \frac{\begin{array}{c}\mathit{Conditions}\\ \mathit{smp} = \mathit{sample}(\pi)\end{array}}{\mathit{conf} \longrightarrow \textbf{if } \mathit{smp} \textbf{ then } \mathit{conf}' \textbf{else } \mathit{conf}}$$

where π is a Boolean valued probability distribution. This rule locally transforms configurations or sub-configurations which match *conf* into *conf′* if the *Conditions* hold and if the sample drawn from π is true.

By controlling all sources of non-determinism by probabilistic rules, one may obtain a model which gives rise to representative runs. One may then perform statistical model checking on the model, simply by statistically analysing the results of a sufficient number of representative runs. This is the approach taken in this chapter. The presence of non-deterministic rules may result in non-representative rules (as typically is the case with Maude simulations) and we therefore avoid such rules in our model. As a consequence we need to control the independent speed of distributed nodes probabilistically. We do this by including the scheduling of nodes in the model.

For our purposes, a configuration is a multiset of nodes, messages and time information. Following rewriting logic conventions, we let whitespace denote the associative and commutative constructor for configurations, and we assume pattern matching modulo associativity and commutativity. Distributed computing is modeled by rules involving only one node, together with messages and time information. Nodes and messages have respectively the form

- $obj(o, n, rt)$, where o is the node's identifier, n is the list of neighbors, and rt is the reward table.
- $msg(m, s, d)$, where m is the content of the message (its name and parameters), and s and d are the source and the destination nodes.

In order to handle scheduling of the nodes, a discrete *time* is added to the global configuration of the system, and *execution marks* and *scheduled execution marks* sorts are added as subsorts of *Configuration*. The global time has the form $time(t)$, where t has a floating point value. Execution marks have the form $exec(o)$ and scheduled execution marks have the form $[t, exec(o)]$ where o is the node's identifier and t is the scheduled time.

The auxiliary function *sampleBe*(δ) samples the Bernoulli distribution with rate δ; and *sampleExp*(δ) samples an exponential distribution with rate δ. These distributions are assumed to be predefined. The *sampleBe*(δ) function returns a Boolean value and the *sampleExp*(δ) function returns a float value, by considering the exponential probability distribution with rate δ. Their definition can be found in standard textbooks, e.g. [29].

Generic Execution Rules for Enabled Transitions We use a ticketing system to obtain probabilistic scheduling of enabled transitions, adapting an approach used with PMaude [30]. The idea is that a node needs a ticket to execute, after which its ticket is delayed for some time determined by an exponential probabilistic distribution. The probabilistic scheduling works in combination with time advance: time cannot advance when a node has an enabled ticket, and time can only advance until the next scheduled time/ticket in the configuration. Object execution is captured by a set of rules with the general form of

$$\text{(RULE NAME)}\quad \frac{Conditions}{exec(o)\ obj(o, n, rt) \longrightarrow done(o)\ obj(o, n', rt')}$$

where the left-hand and/or the right-hand side may additionally involve messages and the global time. We let *done*(o) indicate that node o has been executing and can be rescheduled. If more than one transition is needed to describe the activity of o at a given time, one may use additional rules with a conclusion of the form $exec(o)\ obj(o, n, rt) \ldots \longrightarrow exec(o)\ obj(o, n', rt') \ldots$, taking care that any nondeterminism is resolved probabilistically, and that there are rules producing $done(o)$

when the activity of o at time t finishes. (In a Maude implementation, one may use equations to describe such activity, leaving $exec(o) \longrightarrow done(o)$ as a rule, which then will be used after the equations.)

The scheduling is done in the PROBABILISTIC TICKETING rule, where δ is a predefined value used in the exponential probabilistic distribution. The tickets are enabled one by one, using the ENABLING rule. The TICK rule advances the time, to that of the next execution ticket, as defined by the function *maxTimeAdv*. We let $maxTimeAdv(conf)$ return the maximum possible time advance (i.e., the minimum scheduled time), but such that it returns 0 if there are any $exec$ messages in $conf$. Note that this rule must consider the whole system in order to not miss any scheduled execution tickets. Thus we need a way to express that *conf* is the whole system, not just a subconfiguration. For this purpose we define a system to have the form $\{conf\}$, i.e., a configuration *conf* enclosed in curly brackets, where (as above) *conf* is a multiset of nodes, messages, and time information, and with exactly one global time, $time(t)$. The TICK rule is the only rule about the whole system in our model.

(PROBABILISTIC TICKETING)

$$\frac{t' = t + sampleExp(\delta)}{done(o)\ time(t) \longrightarrow [t', exec(o)]\ time(t)}$$

(TICK)

$$\frac{t' = maxTimeAdv(conf),\ t' > t}{\{conf\ time(t)\} \longrightarrow \{conf\ time(t')\}}$$

(ENABLING)

$$\frac{}{[t, exec(o)]\ time(t) \longrightarrow exec(o)\ time(t)}$$

The Rule for Lossy Links Links may be lossy. It means that in the model, each link has an associated probability which reflects the link's reliability. Therefore, the chance of losing messages sent between nodes o and o' depends on the link between o and o'. Message loss is captured by rule LOSSY LINK. Messages are lost if $sampleBe(\delta)$ is true, otherwise they are not lost.In our model, $send(M, o, o')$ represents messages not yet delivered, whereas $msg(M, o, o')$ represents messages being successfully delivered.

(LOSSY LINK)

$$\frac{\begin{array}{c} haslinkProbability(conf, o, o',) \\ noLoss = sampleBe(\delta) \end{array}}{send(M, o, o')\ conf \longrightarrow \textbf{if}\ noLoss\ \textbf{then}\ msg(M, o, o')\ conf\ \textbf{else}\ conf}$$

In the rule, $haslinkProbability(conf, o, o',)$ expresses that information about reliability of the link from o to o' can be found in *conf* and that the link probability is δ. We will not model a general theory for link probability, and how such information can be embedded in a configuration, but we give details on this for our case studies (letting each link have a fixed probability).

In order to ensure that applications of the LOSSY LINK rule are not delayed, which may give rise to non-determinism, we redefine the *maxTimeAdv* function such that *maxTimeAdv*(*conf*) gives 0 if there are any *send* messages in *conf*, thereby ensuring that a *send* message produced by o at time t is lost or delivered at time t. As no other object o' can execute at time t, the message cannot be received at time t, which means that message delivery to o' may not happen while o' is active (which could cause non-determinism).

4.1 The Rule for Reinforcement Learning

Regarding the learning theory (Fig. 1), we define three categories of rules to model a learning process.

The Rule for Forcing Observed Facts In contrast to non-deterministic configuration changes, there are some deterministic actions in the system (we call them *facts*), that must be forced to perform according to prescheduled points in time. In this model, facts are specified in a fact-base FB. These facts represent actions that are observed from the environment (e.g. the movement of nodes in a real network). So we schedule them to happen in the model as they happen in reality. These actions should be performed at the prescheduled times. Actions in FB have the general form of $action(o, data, t)$, where $data$ is defined based on the system which is under study. In this case study, actions in FB are defined as node movements which cause a node to meet new nodes. Here, $data$ represents the neighbors which node o meets at time t. In the FACTS rule, node o modifies its neighbor list n when receiving an action message, by appending the new neighbors mentioned in $data$ to n. By including FB in the configuration, the *maxTimeAdv*(*conf*) function (in the TICK rule) considers the smallest time of unexecuted actions in FB as well as the nodes' execution tickets to choose the maximum possible time advance. Thereby we avoid missing any action in FB.

$$\text{(FACTS)}\quad \frac{\begin{array}{c} FB' = FB\backslash\{action(o, data, t)\} \\ action(o, data, t) \in FB \\ n' = append(n, data) \end{array}}{obj(o, n, rt)\ time(t)\ FB \longrightarrow obj(o, n', rt)\ time(t)\ FB'}$$

The Rule for Predicting Actions When node s wants to send data to node d, it generates a message. This message is transmitted through a path of nodes leading to d. Each time a node o in the path transmits a message (with source node s and destination d), it needs to predict which neighbor $o' \in n$ has the best chance to pass the message successfully. This prediction is based on the Formula 1 and by having the previous reward information (in the reward table rt) which is updated by the rewards that were obtained from the previous successful transmissions (through acknowledgment messages). The reward table rt has a row corresponding to the delivery rate to destination d if choosing neighbor o'. The value stored, i.e., $rt[(o, d)]$ is a pair (u, v) where u is the number of successful transmissions and v is the total number of transmissions.

We use two auxiliary functions, *reward* and *incSum*, to update a reward table rt of an object o for a given neighbor o' and destination d, increasing the number of successful transmissions and the total number of transmissions, respectively:

$$\begin{aligned} reward(rt, o', d) &= rt[(o', d) \mapsto (u + 1, v)] \\ incSum(rt, o', d) &= rt[(o', d) \mapsto (u, v + 1)] \end{aligned}$$

where u and v are given by $rt[(o', d)] = (u, v)$. The rule for prediction is then

$$\text{(PREDICT)} \quad \frac{\begin{array}{c} o'' = prediction(n, rt, d) \\ rt' = incSum(rt, o'', d) \end{array}}{\begin{array}{l} msg(data(s, d, path), o', o)\ exec(o)\ obj(o, n, rt) \\ \longrightarrow\ send(data(s, d, (path; o)), o, o'')\ done(o)\ obj(o, n, rt') \end{array}}$$

More details on the *prediction* function can be found in the Sect. 5.

The Rule for Rewards Each node is rewarded after a successful action when receiving an acknowledgment message. The reward table of a node stores the information of the message transmissions, captured by the REWARD rule.

$$\text{(REWARD)} \quad \frac{\begin{array}{c} o' = next(o, path) \\ rt' = reward(rt, o', d) \end{array}}{msg(ack(path), d, o)\ exec(o)\ obj(o, n, rt) \longrightarrow done(o)\ obj(o, n, rt')}$$

In the case that the path contains loops, one should not reward a node several times, and the *next* function should be relative to the last occurrence of o in *path*, returning the next node in the path. This will be detailed in the Sect. 5.

The application of these general rules are shown in the model of the centralized and the decentralized protocols.

5 Application of the Learning Rules

We have applied the learning rules in a case study, which includes the generalized learning-based routing protocol of Sect. 3. The case study is implemented in rewriting logic, using the Maude tool. In this section, we present SOS style rules for the decentralized protocol reflecting the learning process in a more abstract way. In this model nodes have the form

$$\textbf{node}(o, e, n, rt)$$

where o is the node's identifier, e is a pair of $(power, energy)$, where $power$ is the power capability and $energy$ is the total remaining energy in the node, n is the node's neighbor information, and rt is the reward table. The execution rules for enabled transitions, the rules for lossy links, and the FACTS rule, can be used in this case study as they are presented in Sect. 4. The PREDICT and REWARD rules depend on the protocol and need some modifications to reflect the details and requirements of different protocols. According to the requirements of this case study, the prediction rule is split into three rules. The PREDICT1 rule performs when the destination node is in the neighborhood. The second rule is related to the situation that the destination is not in the neighborhood and the node uses the learning method to choose a node to pass the message. In the PREDICT3 rule, there is no neighbor to transmit the message, and the data being sent is lost. Each unicast message transmission is sent by consuming the minimum message transmission power of the nodes. When the destination node receives a data message, it broadcasts an *ack* message to all the nodes in the routing path, using its maximum transmission power (the ACKNOWLEDGMENT rule). We assume that destination nodes (e.g. sink nodes in a WSN) have unlimited power available to be able to send *ack* messages using their maximum transmission power. The *ackAll* function is used in the ACKNOWLEDGMENT rule to send *ack* messages to useful transfers, letting a node o be rewarded with respect to a neighbor o' if there is a tail part of the path going from o to o' and reaching the destination without revisiting o. The REWARD rule updates the reward table of nodes after receiving an *ack* message.

(PREDICT1)

$$\frac{\begin{array}{c} d \in n \\ rt' = incSum(rt, d, d) \\ e' = (power, energy - power^{min}) \end{array}}{\begin{array}{l} msg(data(id, s, d, path), o', o)\ exec(o)\ node(o, e, n, rt) \\ \longrightarrow send(data(id, s, d, (path; o)), o, d)\ done(o)\ node(o, e', n, rt') \end{array}}$$

(PREDICT2)

$$\frac{\begin{array}{c} n \backslash \{o'\} \neq nil \\ d \notin n \\ o'' = prediction(n, rt, d, o, o', t, 0, 0) \\ rt' = incSum(rt, o'', d) \\ e' = (power, energy - power^{min}) \end{array}}{\begin{array}{l} msg(data(id, s, d, path), o', o)\ exec(o)\ node(o, e, n, rt) \\ \longrightarrow send(data(id, s, d, (path; o)), o, o'')\ done(o)\ node(o, e', n, rt') \end{array}}$$

$$\frac{n \backslash \{o'\} = nil}{msg(data(id, s, d, path), o', o)\ exec(o)\ node(o, e, n, rt) \longrightarrow done(o)\ node(o, e, n, rt)} \quad \text{(PREDICT3)}$$

$$\frac{\begin{array}{c} e' = (power, energy - power^{min}) \\ rt' = reward(rt, o', d) \end{array}}{msg(ack(id, o'), d, o)\ exec(o)\ node(o, e, n, rt) \longrightarrow done(o)\ node(o, e', n, rt')} \quad \text{(REWARD)}$$

$$\frac{e' = (power, energy - power^{max})}{msg(data(id, s, d, path), o, d)\ exec(d)\ node(d, e, n, rt) \longrightarrow ackAll(id, d, path)\ done(d)\ node(d, e', n, rt)} \quad \text{(ACKNOWLEDGEMENT)}$$

$$\begin{array}{ll} ackAll(id, d, [o]) & = \emptyset \\ ackAll(id, d, (o; o'; path)) & = \textbf{if } o \in path \textbf{ then } ackAll(id, d, (o'; path)) \\ & \quad \textbf{else } msg(ack(id, o'), d, o)\ ackAll(id, d, (o'; path)) \end{array}$$

The PREDICT2 rule uses the *prediction* function. It is a recursive function which chooses one of the node's neighbors with the best estimation regarding the related link reliability(lr) and in-range probability(ep) and the accumulated rewards in the reward table regarding each neighbor. It also calls the *calcRew* function which calculates the stored rewards in the reward table for each neighbor node. The ; symbol connects an element or a sublist to a list, and $[o]$ denotes the singleton list of o alone. Functions *firstNode*(rt), *secondNode*(rt), *numOfRecMsg*(rt), and *numOfSentMsg*(rt), return the node's neighbor, the destination, the number of messages sent through this neighbor, and the number of these messages received by the destination, respectively.

$$\begin{array}{l} \textbf{function } prediction((n_1; n_2), rt, o, o', o'', t, i, f) = \\ \quad \textbf{if } f' > f \textbf{ and } t < time(n_1) \\ \quad \textbf{then } prediction(n_2, rt, o, o', o'', t, nodeID(n_1), f') \\ \quad \textbf{else } prediction(n_2, rt, o, o', o'', t, i, f) \\ where\ f' = (calcRew(rt, i, o) \times (\mathrm{lr}(n_1) \times \mathrm{ep}(n_1))). \end{array}$$

function $calcRew((rt_1; rt_2), o_1, o_2) =$
if $firstNode(rt_1) = o_2$ **and** $secondNode(rt_1) = o_2$
then $numOfRecMsg(rt_1)/numOfSentMsg(rt_1)$
else $calcRew(rt_2, o_1, o_2)$.

6 Implementation and Analysis

In order to implement and analyze the proposed model, we transform the SOS style model to rewriting logic (RL) [4]. It can be executed on the rewriting logic tool Maude, which provides a range of model verification and analysis facilities. As before, a *system configuration* is a multiset of objects and messages inside curly brackets. Following rewriting logic (RL) conventions, whitespace denotes the associative and commutative constructor for configurations. In our Maude model, the term

⟨`O: Node | Attributes`⟩

denotes a `Node` object, where `O` is the object identifier, and `Attributes` a set of attributes of the form `Attr : X` where `Attr` is the attribute name and `X` the associated value. RL extends algebraic specification techniques with transition rules: The dynamic behavior of a system is captured by rewrite rules supplementing the equations which define the term language. From a computational viewpoint, a rewrite rule $t \longrightarrow t'$ may be interpreted as a *local transition rule* allowing an instance of the pattern t to evolve into the corresponding instance of the pattern t'. In our model, successful messages have the general form

`(M from O to O')`

where `M` is the message body (possibly with parameters), `O` the source, and `O'` the destination. In broadcast messages, the destination is replaced by `all` which is a constructor indicating that the message is sent to *all nodes within range*. In our model, each link has a probability which reflects the link's reliability. Therefore, the chance of losing messages depends on that link, captured by the rule

```
rl (send M from O to O')
⟶ if sampleBernoli(LinkP(O,O')) then (M from O to O') else None fi .
```

letting `send M from S to O'` represent messages not yet delivered, and where `LinkP(O,O')` is the probability reflecting the reliability of the link between Nodes `O` and `O'`, and `None` is the empty configuration.

A complete set of Maude rules of both the centralized and the decentralized protocol is presented in [31].

We used the dataset of iMotes devices in the Intel Research Cambridge Corporate Laboratory [32] to evaluate the performance of the centralized and the decentralized models. An iMote is a small and self-contained device, which is able to communicate with the other iMotes through radio links. This dataset represents Bluetooth sightings of the iMotes nodes which are carried by some 16 admin staff and researchers of Intel's lab for six days, in January 2005. One iMote is placed at a kitchen as the stationary node. Only the information of 9 iMotes are available, because the others had been reset frequently. Each row of the dataset includes a pair of these iMote nodes which meet (are within each others' radio communication range), in addition to the starting time and the length of this meeting. The information about the reliability of links is not available in the dataset. So, we assigned the same value of 3/4 to all the links between all the nodes. That is, we assume that the links are fairly reliable, which is a reasonable assumption, otherwise the very attempt to use such a WSN should be questioned. Note that as mentioned in Sect. 3, links may be available but not reliable (e.g. because of environmental conditions such as foggy weather). In other words, a message that is transmitted between two nodes which are in each others' range, may still be lost because of link unreliability. In the decentralized protocol, a sender node chooses a neighbor node from its current set of neighbors. In the centralized protocol, the path is already chosen by the processing node. If the next node in the path is not in the neighborhood, the message will be lost.

We assume that nodes use a given transmission power to reach their neighbors (the node's minimum power), and also that this transmission power is two times larger than the cost of receiving a message; broadcasting is done with the maximum power. Using such power ratio frees our experiment from any particular radio model. In our experiments, we assume all nodes can reach a processing node and nodes (except the stationary one) move according to the dataset's specification. Every node has an initial energy budget of 1000 units. Finally, we run experiments to investigate the protocol's efficiency (energy-cost), effectiveness (delivery rates) and also to investigate some of its formal properties.

6.1 Investigating the Protocol's Efficiency

In the first experiment, each node sends one single data message to the sink. The purpose of this experiment is to study the overhead cost of the learning phase in these protocols, as well as the cost incurred by the nodes in the centralized approach in updating the processing node every time they move. Figure 2 compares the average remaining energy of all the nodes which send the data message by using centralized and decentralized protocols after running the simulation for 3000 time ticks. The figure shows that the decentralized protocol consumes nearly 1/3 less energy than the centralized protocol at the end of simulation run. Since there is one message being delivered, we claim that the process of keeping the processing node updated is to be blamed as a non-trivial consumer of energy. One could argue that we should have compared our protocols with another protocol, for instance the power-sensitive AODV [33] (a modified version of AODV which finds the cheapest routing path instead of the shortest one). We argue that we would not learn anything because the

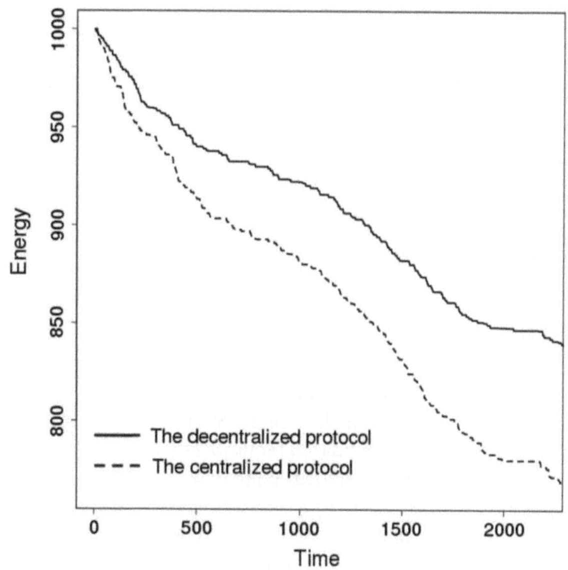

Fig. 2 The comparison of the average energy consumption of the centralized and the decentralized routing protocols for one data message

power-sensitive AODV does not have a learning phase and since there is only one message being sent, therefore the comparison of the power-sensitive AODV would not be realistic.

We did another experiment to study the power consumption of the network when all the nodes regularly send data messages. This experiment which studies several data messages, includes a comparison to the power-sensitive AODV protocol. In this experiment, each node sends 50 data messages. The data messages are sent from different nodes to the sink node, at random times during the simulation. Figure 3 represents the average of the remaining energy of all nodes during running the decentralized, the centralized and the power-sensitive AODV routing protocol models. Again, each simulation runs for 3000 time ticks, and we assume there is no message interference in the network. This experiment, which is more faithful to the normal operation of a WSN, allows one to better appreciate the cost of the route requests in the centralized protocol (in addition to the cost of maintaining the processing node up-to-date).

For comparison with our protocols, we now add the power-sensitive AODV protocol to our experiment. In the power-sensitive AODV, we assume that the links are not probabilistic and they always exist and the messages always reach the sink. Note that those are rather *optimistic* assumptions. In the power-sensitive AODV, each time that a node needs to send a message after it moves, it needs to request a new routing path to the sink. This process is done by broadcasting several route request messages through intermediary nodes. These messages are followed by route reply messages.

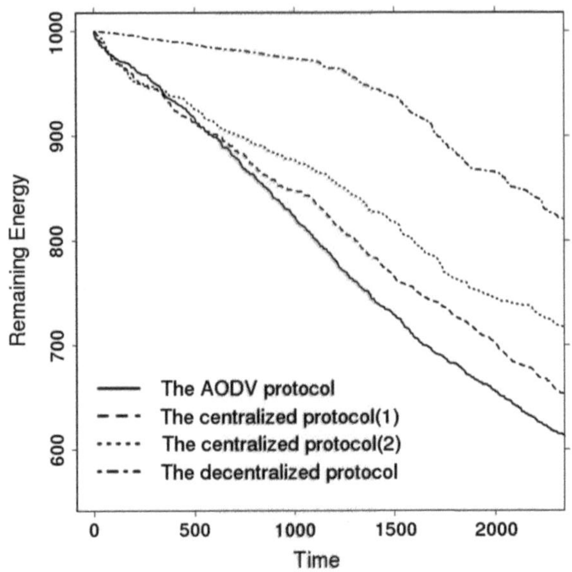

Fig. 3 The comparison of the average remaining energy of nodes in the decentralized protocol, the centralized protocol (1) and the centralized protocol (2)

This routing path discovery process is costly for the network. We assume that nodes also use their minimum power to broadcast the messages in the power-sensitive AODV. As before, both the centralized and the decentralized protocols spend some time in the beginning for learning. We also investigated two scenarios for the centralized protocol. In the first scenario (denoted as "centralized protocol (1)"), we assume, as in the previous experiment, that communicating with the processing node is as costly as communicating with a neighbor node. The second scenario (denoted as "centralized protocol (2)") has similar situation as the first scenario except that nodes use twice as much power to communicate with the processing node. While this improves the chances that a processing node is reached, it also consumes more energy, i.e., these scenarios investigate the power consumption of the network when the cost of reaching the processing node increases. The results show that in the long term the decentralized protocol consumes 45.9, 53.5 and 59.7 % less power than the first and second scenarios of the centralized protocol and the power-sensitive AODV protocol, respectively. In addition, we note that (*i*) the centralized and the decentralized protocols consume less energy than the power-sensitive AODV protocol, and (*ii*) the decrease in the average amount of remaining energy is relatively less pronounced for the decentralized protocol, which is a desirable feature as well. For a better representation of the protocols' energy consumption characteristics, we generated graphs which show the maximum, average and minimum amount of the energy consumed by all the nodes in a simulation. The scenario is the same as considered in Fig. 3. These graphs are generated for the decentralized protocol (Fig. 4), the centralized protocol

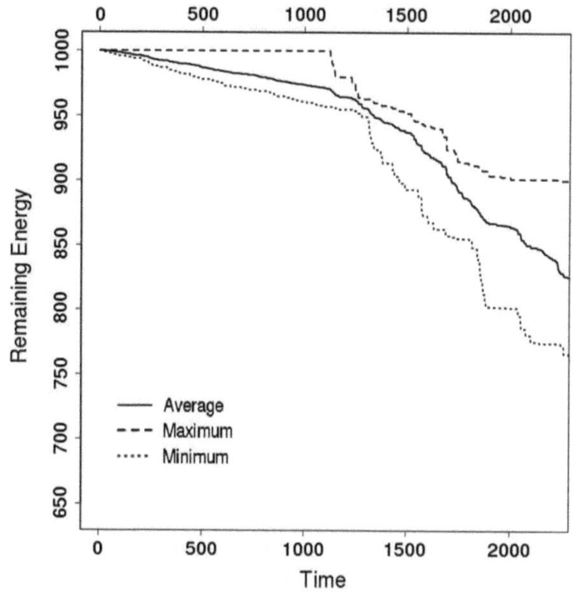

Fig. 4 The comparison of maximum, average and minimum amount of the remaining energy of the nodes in the decentralized protocol

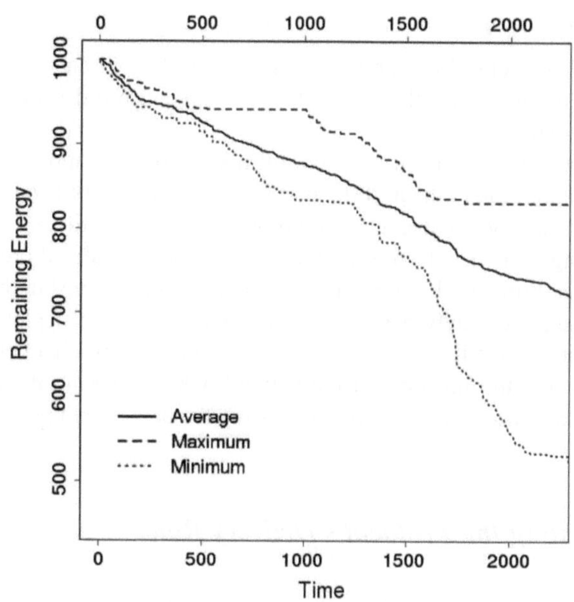

Fig. 5 The comparison of maximum, average and minimum amount of the remaining energy of the nodes in the centralized protocol (1)

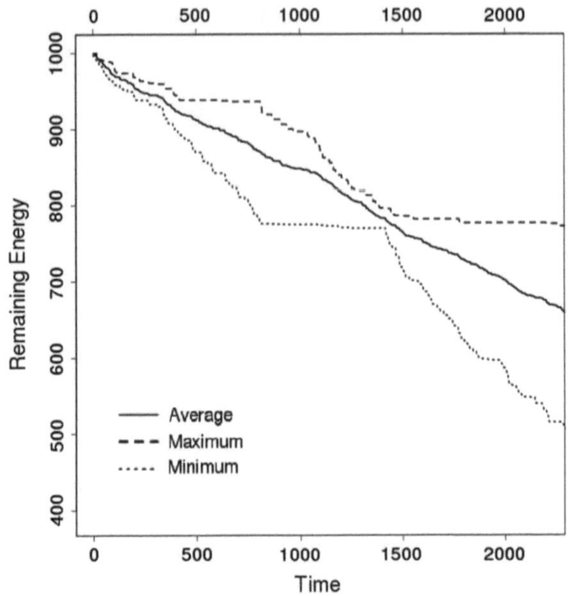

Fig. 6 The comparison of maximum, average and minimum amount of the remaining energy of the nodes in the centralized protocol (2)

(1) (Fig. 5) and the centralized protocol (2) (Fig. 6). As expected the gap between maximum and minimum is relatively tighter in the centralized version, suggesting that it is more robust.

In our last experiment in this section, all nodes continue moving and sending messages at random intervals until the first node dies, i.e., it runs out of energy. In both the centralized and the decentralized protocols, we show the minimum amount of energy among all nodes. Here we compare only the centralized protocol (1) and the decentralized protocol. Figure 7 shows that the first node dies in the centralized protocol after approximately 5000 time ticks, while the most deplete node in the decentralized protocol still has almost half of its initial energy budget. This experiment suggests that the life time of the network when applying the decentralized protocol is at least twice as long than when using the centralized protocol.

6.2 *Investigating the Protocol's Delivery Rate*

While the above experiments explain the energy efficiency of the discussed routing protocols we also need to consider their effectiveness, i.e., what are their actual delivery rates. After all a highly efficient protocol that does not lead to a good delivery rate is of very little practical use.

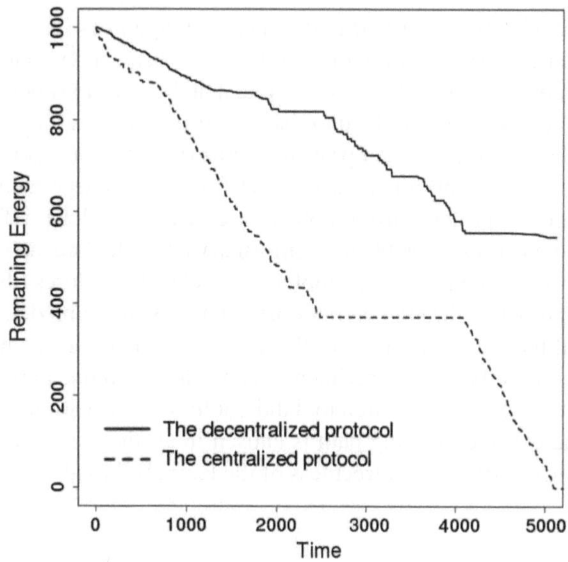

Fig. 7 The comparison of minimum remaining energy of nodes in the decentralized and the centralized protocol

Considering the successful message transmission rate as the metric of interest, the centralized protocol, as expected, is more effective than the decentralized. On average, the successful message transmission rate of the centralized protocol is 87.1 %, while it is 83.6 % for the decentralized protocol. According to the average power consumption in Fig. 3, the decentralized protocol uses nearly 50 % less power than the centralized protocol at the cost of being merely about 4 % less effective. We believe this is quite acceptable trade-off. Furthermore, we note that even though the decentralized protocol made wrong decisions in a few cases, in other cases the centralized approach was not able to deliver its message due to unreachability of the processing nodes. That is, the decentralized protocol revealed itself as a very good alternative to the centralized one, as well to the power-sensitive AODV protocol.

6.3 Investigating the Formal Properties of the Protocol

In addition to the quantitative analysis of the performance of the discussed routing protocols, we took the advantage of the formal modeling methods to prove the validity of the model, according to a correctness property. The correctness of the routing protocols is formalized as a *Linear Temporal Logic (LTL)* formula

$$\Box(MSG(m, s, d) \Rightarrow MPP(m, s, d))$$

The predicate *MSG*(m, s, d) is true if a data message m is sent from node s to node d. The predicate *MPP*(m, s, d) is true if the chosen routing path for m is the most probable path between nodes s and d. Note that this formula does not guarantee the delivery of messages (due to the unreliability of links). This formula means that in all states of the system, the most probable paths (based on the information up to the current time) are chosen for messages. *MPP* is defined separately for the two versions of protocol. The centralized protocol uses available global information and the decentralized one uses available local information to calculate the value of *MPP*. The Maude environment has a *search* tool that searches for failures (the negation of the correctness property) through all possible behaviors of the model. We used this tool to search all the reachable states of the system to find out if the most probable path is always chosen. For this experiment, nodes only send one message to have a limited search state space. The search tool did not find any violation of the property, meaning that the expected routing path is chosen in all the reachable traces of the model's run, thus asserting the correctness of the learning-based routing protocol.

7 Conclusion

In this chapter, we have proposed a model for learning-based protocols. We have integrated the rules required for learning the observables with a probabilistic model of WSNs. The model includes the concept of a fact-base, which consists of the scheduled actions that happen in certain orders. We have defined a routing protocol which includes the Bayesian learning method to predict the best routing path. This protocol has centralized and decentralized versions. We have used Maude to model the routing protocol and to formalize the Bayesian learning method. Besides applying Maude facilities to analyze the model, we provided more realistic analysis by feeding the model by a real dataset and managing quantitative data analysis. The results show a trade-off between two versions of the protocol. The centralized one is slightly more effective (higher delivery rate), while the decentralized one is much more efficient. Both versions of the protocol outperform the PS-AODV protocol in term of efficiency and are also more robust, especially the decentralized one. We have qualitatively analyzed the protocol and proved that the protocol satisfies its correctness property.

In future work, we are going to refine the protocol by capturing more detailed patterns from the node movements, e.g. temporal patterns. We also plan to perform probabilistic reasoning of the model via statistical model checking.

References

1. Dijkstra, E.W.: A note on two problems in connection with graphs. Numer. Math. **1**, 269–271 (1959)
2. Mitchell, T.M.: Machine Learning (ISE Editions). McGraw-Hill, Boston (1997)

3. Clavel, M., Durán, F., Eker, S., Lincoln, P., Martí-Oliet, N., Meseguer, J., Quesada, J.F.: Maude: specification and programming in rewriting logic. Theoret. Comput. Sci. **285**, 187–243 (2002)
4. Meseguer, J.: Conditional rewriting logic as a unified model of concurrency. Theoret. Comput. Sci. **96**, 73–155 (1992)
5. Kazemeyni, F., Owe, O., Johnsen, E.B., Balasingham, I.: Learning-based routing in mobilewireless sensor networks: applying formal modeling and analysis. In: Proceedings of IEEE 14th International Conference on Information Reuse and Integration—Workshop on Formal Methods Integration (FMi'13), pp. 1–8. IEEE (2013)
6. Olagbegi, B.S., Meghanathan, N.: A review of the energy efficient and secure multicast routing protocols for mobile ad hoc networks. CoRR, abs/1006.3366 (2010)
7. Liu, M., Cao, J., Chen, G., Wang, X.: An energy-aware routing protocol in wireless sensor networks. IEEE Sens. **9**(1), 445–462 (2009)
8. Wang, J., Cho, J., Lee, S., Chen, K-C., Lee, Y-K.: Hop-based energy aware routing algorithm for wireless sensor networks. IEICE Trans. **93-B**(2), 305–316 (2010)
9. Stojmenovic, I., Lin, X.: Power-aware localized routing in wireless networks. IEEE Trans. Parallel Distrib. Syst. **12**(11), 1122–1133 (2001)
10. Uddin, M.Y.S., Ahmadi, H., Abdelzaher, T., Kravets, R.: A low-energy, multi-copy inter-contact routing protocol for disaster response networks. In: Proceedings of 6th Annual IEEE communications society conference on Sensor, Mesh and Ad Hoc Communications and Networks (SECON'09), pp. 637–645. IEEE Press (2009)
11. Arroyo-Valles, R., Alaiz-Rodriguez, R., Guerrero-Curieses, A., Cid-Sueiro, J.: Q-probabilistic routing in wireless sensor networks. In: Proceedings of 3rd International Conference on Intelligent Sensors, Sensor Networks and Information (ISSNIP'07), pp. 1–6 (2007)
12. Barrett, C.L., Eidenbenz, S.J., Kroc, L., Marathe, M., Smith, J.P.: Parametric probabilistic routing in sensor networks. Mob. Netw. Appl. J. **10**, 529–544 (2005)
13. Lindgren, A., Doria, A., Schelén, O.: Probabilistic routing in intermittently connected networks. ACM SIGMOBILE Mob. Comput. Commun. Rev. **7**, 19–20 (2003)
14. Wang, P., Wang, T.: Adaptive routing for sensor networks using reinforcement learning. In: Proceedings of the Sixth International Conference on Computer and Information Technology (CIT '06), pp. 219–219. IEEE Computer Society (2006)
15. Pandana, C., Liu, K.J.R.: Near-optimal reinforcement learning framework for energy-aware sensor communications. IEEE J. Sel. Areas Commun. **23**, 209–232 (2002)
16. Coleri, S., Ergen, M., Koo, T.J.: Lifetime analysis of a sensor network with hybrid automata modelling. In: Proceedings of first ACM International Workshop on Wireless Sensor Networks and Applications (WSNA'02), pp. 98–104. ACM (2002)
17. Fehnker, A., van Hoesel, L., Mader, A.: Modelling and verification of the LMAC protocol for wireless sensor networks. In: Proceedings of the 6th International Conference on Integrated Formal Methods (IFM'07). Lecture Notes in Computer Science, vol. 4591, pp. 253–272. Springer (2007)
18. Tschirner, S., Xuedong, L., Yi, W.: Model-based validation of QoS properties of biomedical sensor networks. In: Proceedings of the 8th ACM & IEEE International conference on Embedded software (EMSOFT'08), pp. 69–78. ACM (2008)
19. Johnsen, E.B., Owe, O., Bjørk, J., Kyas, M.: An object-oriented component model for heterogeneous nets. In: Proceedings of the 6th International Symposium on Formal Methods for Components and Objects (FMCO 2007). Lecture Notes in Computer Science, vol. 5382, pp. 257–279. Springer (2008)
20. Ölveczky, P.C., Thorvaldsen, S.: Formal modeling, performance estimation, and model checking of wireless sensor network algorithms in real-time maude. Theoret. Comput. Sci. **410**(2–3), 254–280 (2009)
21. Katelman, M., Meseguer, J., Hou, J.C.: Redesign of the LMST wireless sensor protocol through formal modeling and statistical model checking. In: Proceedings of the 10th International Conference on Formal Methods for Open Object-Based Distributed Systems (FMOODS'08). Lecture Notes in Computer Science, vol. 5051, pp. 150–169. Springer (2008)

22. Dong, J.S., Sun, J., Sun, J., Taguchi, K., Zhang, X.: Specifying and verifying sensor networks: an experiment of formal methods. In: 10th International Conference on Formal Engineering Methods (ICFEM'08), Lecture Notes in Computer Science, vol. 5256, pp. 318–337. Springer (2008)
23. Kulkarni, S.A., Rao, G.R.: Formal modeling of reinforcement learning algorithms applied for mobile ad hoc network. Int. J. Recent Trends Eng. (IJRTE) **2**, 43–47 (2009)
24. Kaelbling, L.P., Littman, M.L., Moore, A.W.: Reinforcement learning: a survey. J. Artificial Intell. Res. **4**, 237–285 (1996)
25. Box, G.E.P., Tiao, G.C.: Bayesian Inference in Statistical Analysis (Wiley Classics Library). Wiley-Interscience, New York (1992)
26. Shakya, S., McCall, J., Brown, D.: Using a Markov network model in a univariate EDA: an empirical cost-benefit analysis. In: Proceedings of the 2005 Conference on Genetic and Evolutionary Computation, GECCO'05, pp. 727–734. ACM (2005)
27. Dorigo, M., Stutzle, T.: Ant Colony Optimization. MIT Press, Cambridge (2004)
28. Plotkin, G.D.: A structural approach to operational semantics. J. Logic and Algebraic Program. (JLAP) **60–61**, 17–139 (2004)
29. Kalbfleisch, J.G.: Probability and Statistical Inference, Vol. 1: Probability (Springer Texts in Statistics). Springer, Secaucus (1985)
30. Agha, G., Meseguer, J., Sen, K.: PMaude: rewrite-based specification language for probabilistic object systems. Electron. Notes Theoret. Comput. Sci. **153**(2), 213–239 (2006)
31. Kazemeyni, F., Owe, O., Johnsen, E.B., Balasingham, I.: Learning-based routing in mobile wireless sensor networks: formal modeling and analysis for WSNs. Technical Report ISBN 82-7368-390-7, Department of Informatics, University of Oslo (2013)
32. Scott, J., Gass, R., Crowcroft, J., Hui, P., Diot, C., Chaintreau, A.: CRAWDAD trace: cambridge/haggle/imote/intel (v. 2006–01-31). http://crawdad.cs.dartmouth.edu/cambridge/haggle/imote/intel (2006)
33. Kazemeyni, F., Johnsen, E.B., Owe, O., Balasingham, I.: Formal modeling and validation of a power-efficient grouping protocol for WSNs. J. Logic Algebraic Program. (JLAP) **81**(3), 284–297 (2012)

On the Use of Anaphora Resolution for Workflow Extraction

Pol Schumacher, Mirjam Minor and Erik Schulte-Zurhausen

Abstract In this chapter we present three anaphora resolution approaches for workflow extraction. We introduce a lexical approach and two further approaches based on a set of association rules which are created during a statistical analysis of a corpus of workflows. We implement these approaches in our generic workflow extraction framework. The workflow extraction framework allows to derive a formal representation based on workflows from textual descriptions of instructions, for instance, of aircraft repair procedures from a maintenance manual. The framework applies a pipes-and-filters architecture and uses Natural Language Processing (NLP) tools to perform information extraction steps automatically. We evaluate the anaphora resolution approaches in the cooking domain. Two different evaluation functions are used for the evaluation which compare the extraction result with a golden standard. The syntactic function is strictly limited to syntactical comparison. The semantic evaluation function can use an ontology to infer a semantic distance for the evaluation. The evaluation shows that the most advanced anaphora resolution approach performs best. In addition a comparison of the semantic and syntactic evaluation functions shows that the semantic evaluation function is better suited for the evaluation of the anaphora resolution approaches.

Keywords Workflow extraction · Process oriented case-based reasoning · Information extraction · Anaphora resolution

---

P. Schumacher (✉) · M. Minor · E. Schulte-Zurhausen
Goethe Universität Frankfurt - Institut für Informatik - Lehrstuhl für Wirtschaftsinformatik, 60325 Frankfurt am Main, Germany
e-mail: schumacher@cs.uni-frankfurt.de

M. Minor
e-mail: minor@cs.uni-frankfurt.de

E. Schulte-Zurhausen
e-mail: eschulte@cs.uni-frankfurt.de

T. Bouabana-Tebibel and S. H. Rubin (eds.), *Integration of Reusable Systems*, Advances in Intelligent Systems and Computing 263, DOI: 10.1007/978-3-319-04717-1_7,

1 Introduction

Traditionally, *workflows* are "the automation of a business process, in whole or part, during which documents, information or tasks are passed from one participant to another for action, according to a set of procedural rules" [1]. Recently, a broader notion is emerging, where a workflow describes any flow of activities in the sense of an instruction whose execution is supported automatically. For instance, a repair instruction for an aircraft can be described by a workflow including some activities. The workflow participants are one or several technicians who conduct the repair activities. This is not any more a classical workflow supporting a business process.

However, the notion of a workflow for such an instruction provides many benefits: Graphical workflow modeling languages can be used to visualize the instructions making them easy to understand. Workflow management systems can be used for workflow execution providing a step-by-step guidance for the users. The workflow execution can be tracked easily to fulfill regulatory requirements with respect to quality management, traceability issues etc. The workflows can be reused and adapted to new situations. This can be supported by reasoning methods like those of case-based reasoning (CBR) [2]. Recently, process-oriented Case-Based Reasoning (POCBR) emerged as a new branch of CBR [3]. This field of research investigates new reasoning approaches to handle procedural knowledge. Several approaches for retrieval [4–6] and adaptation [7] of workflows have been presented.

Workflow modeling is a laborious task that consumes a lot of human effort for different artifact A workflow consists of a control-flow and a data-flow. A set of *activities* combined with *control-flow-structures* like sequences, parallel or alternative branches, and loops forms the control-flow. In addition, activities consume *resources* and create certain *products* which both can be physical matter (such as cooking ingredients or a screw) or information. The data-flow describes the interaction of activities with resources and products. Automated assistance can be achieved by transforming textual descriptions of instructions into formal workflow models. A large amount of such instruction texts from different domains is readily available, e.g., on cooking recipes, on computer-how-tos or on the processing of an order from a customer. Recently, the extraction of workflows from text has been investigated in research on process-oriented Case-Based Reasoning (POCBR) [8, 9].

Workflow extraction has to face several challenges. It has to identify the activities and to organize them in a control-flow. Further, the control-flow has to be enriched with the data-flow. Frequently, the data-flow has to be completed since the textual description of the data-flow is incomplete due to the human economy of language. For instance, the references from a component of an aircraft to its constituents are not expressed explicitly in the text even if the constituents have been mentioned before. Further, the extraction process has to fulfill non-functional requirements like robustness and scalability. The issue of domain-dependency requires special attention. On the one hand side, a framework for workflow extraction should cover several domains to create some impact. On the other hand, some components of the extraction process need to be domain-specific to achieve valuable results. An optimal balance between both is a non-trivial research goal.

The chapter is an extended and revised version of our earlier work [10]. We presented a novel, extensible pipes-and-filters framework for the extraction of workflows from textual process descriptions in an arbitrary domain. In this chapter we perform an additional semantic evaluation. We compare a syntactic evaluation which is based on an exact match (based on a lexical comparison) of the case with a golden standard and a semantic evaluation which allows a partial match (by means of an ontology based similarity measure) of the test case and the golden standard. We investigate the impact of the evaluation method on the result. The chapter is organized as follows. In the next section we describe the pipes-and-filters framework. The third section presents how the framework is applied to set-up a sample domain. The subsequent section introduces the data-flow creation with a focus on anaphora resolution approaches. In an evaluation, different anaphora resolution approaches are compared. We present two function to measure the performance of the data-flow creation. One function is based on syntactic information and the other one uses syntactic and semantic features. The chapter closes with a discussion of related work, a brief conclusion and an outlook on future work.

2 Workflow Extraction Framework

Systems which process natural language need to be robust and scalable. A large corpus of frameworks for generic NLP (Natural Language Processing) tasks exist (e.g., GATE,[1] Stanford Parser[2] or OpenNLP[3]). We have developed a generic framework on top of these that is dedicated especially to workflow extraction from text. It focuses on the workflow specific structures and operations. The capabilities of the framework include the identification of activities, organizing them in a control-flow, and enriching the control-flow by a data-flow. The framework is based on a pipes and filter architecture [11], i.e., it consists of subsequent extraction components called filters. The architecture of the framework is domain independent. However, some of the particular filters are domain specific. The language in process descriptions has a specific style, e.g., the steps are in the correct order and mainly active voice is used [12]. For instance, aircraft maintenance manuals are written in simplified technical English which is a controlled language whose rules are defined in a specification [13].

[1] http://www.gate.ac.uk

[2] http://www.nlp.stanford.edu/software/

[3] http://www.opennlp.apache.org

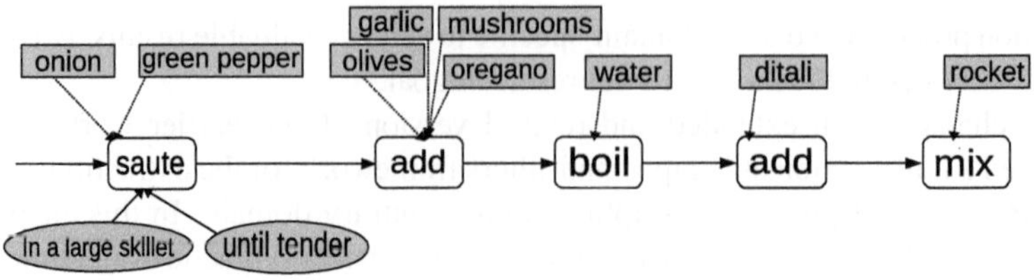

Fig. 1 Sample workflow for cooking rocket pasta with vegetable

2.1 Workflow Representation

The target of workflow extraction is a formal representation of the workflow in a workflow description language [1]. We have chosen an XML-based language [14] that is part of the CAKE[4] system. It is a block-oriented workflow description language: The control-flow can contain sequences, XOR-, AND-, LOOP-blocks. These building blocks cannot be interleaved but they can be nested. In addition an activity has a set of semantic-descriptors, resources and products. A semantic descriptor is for example the name of the task or additional information which describes "how" a task should be performed, e.g., "for 10 min". Please note that this information is stored as a text rather than a formal temporal notation. Resources and products contain a set of semantic information, these describe additional information about the resources, e.g., amounts or if a resource should be preprocessed like "chopped". Figure 1 shows a sample workflow which was extracted of the recipe of Listing 1.1. The rounded rectangles are the activities, the grey rectangles are the products, and the grey ovals are semantic-descriptors. The first activity is "saute" with the resources "onion" and "green pepper". In addition, it contains the semantic-descriptors "In a large skillet" and "until tender". The activities "saute", "add", "boil", "add" and "mix" are aligned in a sequence. The data-flow of the sample is not complete as it only contains resources.

2.2 Information Extraction Software

Our framework uses the NLP software SUNDANCE (Sentence UNDerstanding ANd Concept Extraction) developed by Riloff and Phillips [15]. SUNDANCE performs the usual NLP task like tokenization or part of speech tagging. Our previous work has shown that SUNDANCE has good balance between coverage and robustness [8].

SUNDANCE has been inspired by the conceptual dependency theory published by Schank [16]. The theory aims at illustrating how people think and process information. It is based on three kinds of concepts, the nominal, the action, and the modifier. A sentence or a thought can be represented by those three types of concepts. Nominals

[4] Collaborative Agile Knowledge Engine.

are things that can be thought of without the need for relating them to other concepts. It is usually used to represent things or people. The concept of an action is what a nominal is doing. These actions are usually represented by a verb. The last type of concept is the modifier, which specifies properties of a nominal or an action. A sentence is built of one or more concepts and a concept may depend on another one to get the whole meaning [16].

SUNDANCE allows to specify extraction patterns for the system called case frames [17]. These patterns are similar to the concepts described in the concept dependency theory. The SUNDANCE parser assigns syntactic roles (subject, direct object, and indirect object) to text snippets based on a heuristic. Then the SUNDANCE information extraction engine tries to fill a case frame as follows. Each case frame specifies a trigger phrase. If the trigger phrase is detected in a sentence, then the according case frame is activated. This means that the activation functions of the frame try to match the specified linguistic patterns with the syntactic roles. A slot specifies a syntactic role whose content is to be extracted from the text. A filled case-frame can be mapped to a task. The trigger phrase is the task-name and the filled slot contains a list of products- or resources-names.

2.3 Extraction Pipeline

The framework is based on a pipes and filters [11] architecture. Such an application is a sequence of filters which are connected by pipes. A filter is a self-contained element which performs a data-transformation-step on the data-stream. The pipes channel the data-stream from the output of a filter to the input of the subsequent filter. A data-stream is sent through this pipeline and each filter is applied to the stream. Filters can be added or deleted without affecting the other filters in the pipeline.

While a classical data-stream is pulsed using bits or bytes, our case-stream uses cases as a pulse. A case is the smallest unit which can be processed by a filter. At the beginning of the pipeline, the case initially consists of the textual process description. While the case passes through the pipeline it is enriched with additional structure. At the end of the pipeline we have a complete case consisting of the textual process description and the formal workflow representation.

Our framework extends the original pipes and filters architecture. We allow two different types of filters. The first one, the so called *local filters* operate with a focus on one case. The second one, the *window filters* collect a part of the case-stream (e.g., 5000 cases) and operate on that. The model of a window filter is necessary, because the framework processes a stream of cases which is potentially infinite. The intention is to use statistical methods for a larger number of textual process descriptions. The statistical approaches benefit from the pipes and filter principle because we use them on processed data. This intermediate data is the result of the preceding steps of the extraction pipeline. It contains less noise and has more structure than the raw input-data. The filters are hand-crafted by analyzing the textual process descriptions.

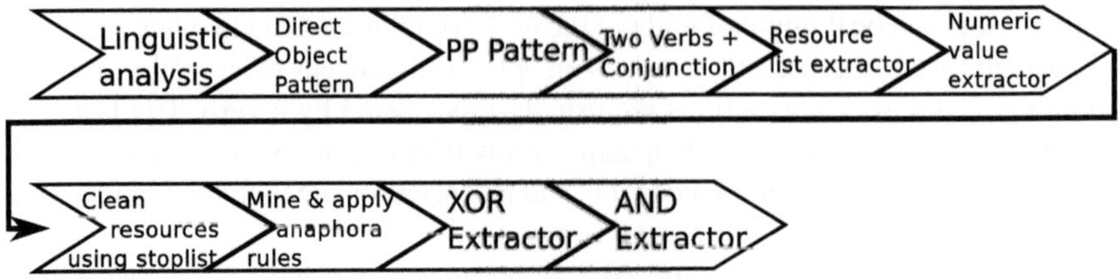

Fig. 2 Overview of the extraction pipeline for the cooking domain

We refrain from using machine learning techniques to create the filters because the effort to create a training-set is higher than the effort to build the filters by hand.

3 Workflow Extraction

The following section describes the different sub-tasks of the workflow extraction. We describe the sub-tasks by using the cooking domain to illustrate the general framework in a sample domain. Figure 2 illustrates the respective pipeline with a sequence of filters. The filters can not be classified by means of sub-tasks, e.g., the main task of the "AND Extractor"-filter is to detect a parallel control-flow but it also creates new activities.

3.1 Linguistic Analysis

In the first filter, the linguistic analysis we use the SUNDANCE natural-language-processing-system [15] to perform the standard language processing. During this processing the recipe text is split into sentences (end of sentence detection). In the next step the sentences are divided into tokens (tokenization). At the end the tokens are tagged with their part-of-speech and syntactic role (subject, direct object, etc.). SUNDANCE uses a set of heuristics to perform these tasks.

3.2 Recognition of Activities

They use pattern-matching to extract the activities. The first pattern ("Direct Object Pattern") searches for filled case-frames with a verb phrase in active voice and direct object as slot. The second pattern ("PP Pattern") is a filled case-frame with an active verb phrase and a prepositional phrase. If no case-frame can be filled, it is checked whether a verb or two verbs combined by a conjunction is at the beginning of the sentence, for instance "Cover and cook.". Sentences like this are quite difficult because they do not fit in a normal grammatical context. Therefore a special pattern ("Two Verbs + Conjunction") is needed for this frequently occurring structure.

3.3 Recognition of Resources

The "Resource list extractor" extracts the resources and the "Numeric value extractor" extracts the related amounts. The resources are cleaned using an expert created stop-list ("Clean resources using stop-list"). The step "Mine & apply anaphora rules is explained in Sect. 4. We assume that the noun phrases which are related to activities are resources. This leads to a quite high recall for the resources. The precision is too low because a lot of items which are not a resource or product are extracted by the filters, for instance cookware or tools. After analyzing the data we discovered that most tools are used with an indefinite article ("a skillet") while resources usually have no article or definite article. For resources with an indefinite article we check if we find a corresponding item in the ingredients list, if so it is considered as resource. In the case we don't find any corresponding item we eliminate that resource.

3.4 Building the Control-Flow

Currently our system is able to extract a sequential, parallel ("AND Extractor") or disjunctive control-flow ("XOR-Extractor") for process descriptions of the cooking domain. Extraction of a sequential control-flow is a straight forward procedure. Usually textual descriptions describe a process sequentially [12]. For the extraction of disjunctive of disjunctive or parallel control-flow we analyzed the process descriptions and looked for patterns which suggest a control-flow. We found that a small number of patterns occurs very frequently which facilitates the extraction of a non-sequential control-flow. There are different types of patterns. A very simple pattern is for example the keyword "*occasionally*". In a sentence like "*Cook, stirring occasionally for about* 10 min." we can extract the activity"*cook*" and "*stir*" which are performed in parallel (see Fig. 3). Instructions which force to choose one of two ingredients induce a disjunctive control-flow. The sentence "*Add butter or margarine*", e.g., suggest two activities "*add*" with either "*butter*" or "*margarine*" as ingredient (see Fig. 4). These two activities are on alternative paths. We found also some patterns which use the additional structure which was provided by the description authors. The authors usually divide the descriptions into steps and we preserve this information during the extraction process. If we find the keyword "*Meanwhile*" as first word of such a step, we can deduce that the activities of the previous step and the activity of the step in which "*Meanwhile*" was found are executed in a parallel control-flow.

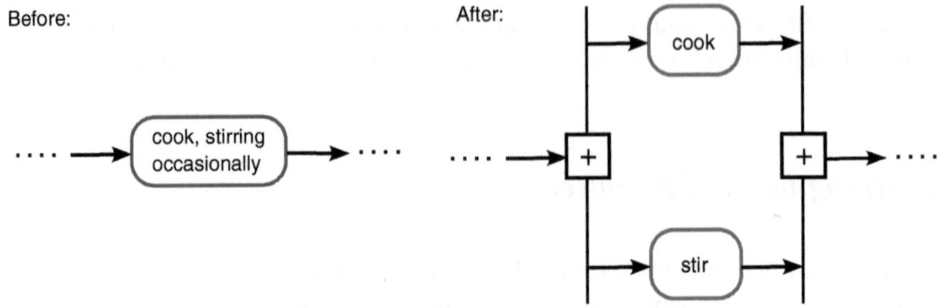

Fig. 3 Illustration of the application of the filter *AND Extractor*

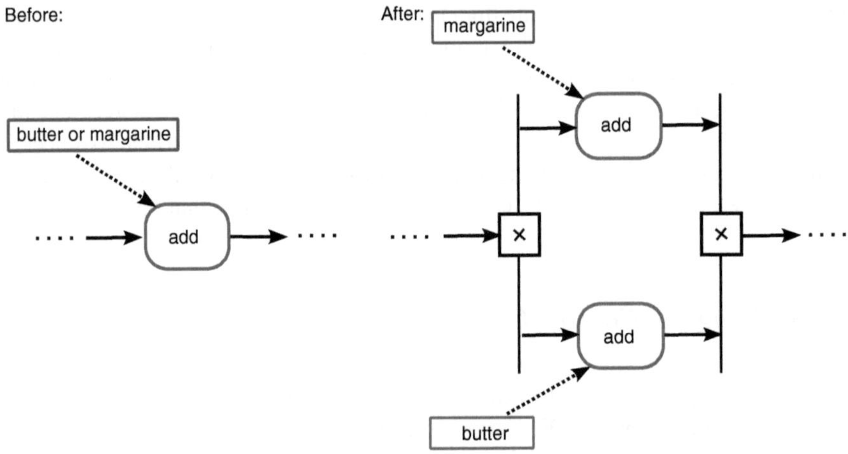

Fig. 4 Illustration of the application of the filter *XOR Extractor*

4 Data-Flow Creation and Evolutive Anaphora Resolution

The data-flow describes the flow of resources and products through the workflow. The creation of a data-flow is a complex problem which is tied to the linguistic problem of anaphora resolution. An anaphora is a linguistic entity which indicates a referential tie to some other entity in the same text [18]. An evolutive anaphora is an anaphoric reference to an object that has been destroyed or created [19]. In workflow terminology, this is a reference to resources that have been consumed by an activity. For example, the object "dough" is created by the activity "mix flour, water and yeast". In the context of workflow extraction anaphora resolution is the determination of the activity in which an object is created. The resources consumed by that activity are called constituents. In the above sample, "flour", "water" and "yeast" are the constituents of the anaphoric reference "dough", produced by means of the activity "mix". Now, we introduce our method which is used to resolve evolutive anaphoras in workflows. We use data-mining techniques to mine anaphora-rules which enables us to determine the activity an anaphora is referring to. The left side of an anaphora-rule is a set of constituents and the right side contains the anaphora.

Listing 1.1 Sample recipe in XML format

```
<recipe>
<title>Rocket Pasta with vegetable</title>
<resources><res>ditali</res>
<res>rocket</res><res>water</res>
<res>herbs</res><res>onion</res>
<res>olives</res><res>mushrooms</res>
<res>garlic</res><res>oregano</res>
<res>green pepper</res>
</resources>
<steps><step>In a large skillet, saute onion
and green pepper until tender. Add garlic
mushrooms, olives and oregano.</step>
<step>Boil water. Add ditali.</step>
<step>Mix the cooked vegetable  with the ditali and the rocket.</step>
</steps></recipe>
```

Listing 1.2 Sample transactions for workflow in Fig. 1.

```
WorkflowId,TransactionTime,{Items}
0,0,{Ditali,rocket,water,herbs,onion,
    mushrooms,garlic,oregano,green pepper}
0,1,{onion,green pepper}
0,2,{garlic,mushrooms,olives,oregano}
0,3,{water}
0,4,{ditali}
0,5,{rocket}
```

4.1 Mining Anaphora-Rules

We use a method for sequential pattern mining that was presented by Agrawal [20] to mine anaphora-rules, because it is a well established method and straight forward transfer to the workflow domain is possible. We are going to introduce the problem. We derive workflow transactions with the fields *workflow-id*, *task-position* and the items used in the transaction. The mining of anaphora-rules precedes the extraction of control-flow, therefore the workflow is sequential at this stage. Quantities are omitted. An *item-set* is a non-empty set of items. A *sequence* is an ordered list of item-sets. A set of items can be mapped to a set of contiguous integers. Let $o = (o_1, o_2, ..., o_m)$ be an item-set where o_i is an item. Let $s = (s_1, s_2, ..., s_n)$ be a sequence where s_i is an item-set A sequence $\langle a_1, a_2, ..., a_n \rangle$ is contained in another sequence $\langle b_1, b_2, ..., b_m \rangle$ if there exist integers $i_1 < i_2 ... < i_n$ such that $a_1 \subset b_{i_1}, a_2 \subset b_{i_2}, ..., a_n \subset b_{i_n}$. A workflow supports a sequence s if s is contained as a sequence of activities in this workflow. The *support* for a sequence is defined as the fraction of workflows in total which support this sequence. The problem of mining sequential patterns is to find the maximum sequences which have a certain user-specified minimum support. By restricting the length of the sequential pattern to two item-sets, we get anaphora-rules. In addition we create a transaction with an item-set corresponding to the ingredient list and the transaction-time zero. Listing 1.2 displays the transactions which were created from the workflow in Fig. 1. It assumes that the workflow has the id 0. The items at the transaction at time 0 are extracted from the corresponding ingredient list of the original recipe. For the details of the sequential pattern mining algorithm we refer to the original paper [20]. In addition, the sequences of the sequential patterns do not need to be maximal. The algorithm delivers a set of anaphora rules with a corresponding support value. The minimum support value which is used is 0.005. This value is domain dependent and must be tuned during the adaptation of the filter for a new domain.

Listing 1.3 Top 10 patterns by support value.

```
0.025: <item=butter><item=dough>
0.024: <item=butter><item=mixture>
0.024: <item=flour><item=batter>
0.021: <item=eggs><item=batter>
0.021: <item=flour><item=dough>
0.018: <item=yeast><item=dough>
0.016: <item=baking powder><item=batter>
0.016: <item=butter><item=batter>
0.015: <item=garlic><item=mixture>
0.015: <item=vanilla><item=batter>
```

4.2 Creation of Data-Flow

The rule-set is improved by using two observations about evolutive anaphoras which:

1. Anaphoras are not enumerated in the resource list.
2. Constituents of an anaphora are used before the anaphora or are part of the resource list.

The first observation enables us to delete a lot of wrong rules. The algorithm creates rules whose right side contains items which are initial resources. Initial resources are all the resources which are not produced by an activity in the workflows. Usually textual process descriptions contain a list with the initial resources. In the domain of cooking this is the ingredients list, in the aircraft maintenance domain it is the list of parts needed to perform a reparation. All rules whose right side contains such an initial resource are omitted. The second observation is essential for the application of anaphora-rules. During the application, first a candidate list is generated by selecting all the rules whose right side is matching the anaphora then we check if the second observation holds. All rules for which the second observation does not hold are dropped from the candidate list.

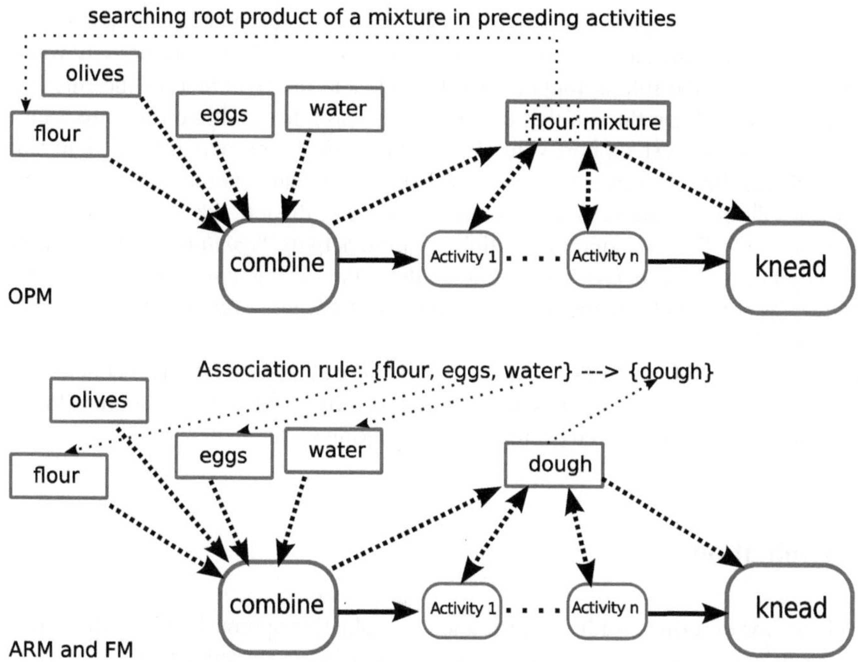

Fig. 5 Illustration of OPM, ARM and FMS

The creation of the data-flow begins with extraction of the first use of a resource. In a second step, the products of the activities have to be determined. It is necessary to perform anaphora resolution to detect the correct products. At the end, the resources of the activities are completed with the products of the respective preceding activity. We implemented three different approaches which we are going to introduce.

One pattern method (OPM) (see Fig. 5) is based on the observation, that a lot of anaphoras contain the token "*mixture*". This approach searches for resources which contain this token. If such a resource is found, we perform a backward search for the name-giving resource of a mixture e.g., "flour" is the name-giving resource of "flour mixture". For those activities it is checked, if they contain multiple resources. A sole product *name-giving resource* + "*mixture*" is created for that activity. At the end, the data-flow is completed by copying the products as resources to the next activity. For Fig. 5 the resource "*flour mixture*" is found at the activity "*knead*". After deleting the token "*mixture*" we get the resource "*flour*". We search for the resource "flour" in the preceding activities and find it at the activity "*combine*". This activity uses four resources. We assume now, that the activity combine produces the product "*flour mixture*".

Anaphora-rule method (ARM) (see Fig. 5) includes OPM in addition it iterates over all activities, starting at the first. For each resource of the activity, it is checked, if there is an association rule with a matching right side. If such a rule is found, it is looked if one of the previous activities has a matching resource for the left side of the rule. We assume then that the anaphora enters the workflow at the activity where the left side of the rule is found. Therefore we add the right side of the rule (the anaphora) as sole product of that activity. In the case that multiple rules are found, the one with the best support value is chosen. In Fig. 5 the search for a right side of a rule resulted with the resource "*dough*". When a match is found, the algorithm searches the preceding activities for the resources of the left side of the rule "*flour, eggs, water*". These resources are linked to the activity "*combine*". Although the activity has more resources than the left side of the rules enumerates (the resource "*olives*" is not included in the rule) the sole product "*dough*" is created for the activity "*combine*".

Filter method (FM) includes ARM and a domain specific list of items is added. This list is used to filter out items which are incorrectly extracted as resource. In the cooking domain this list is used to filter out cooking ware.

5 Evaluation

This section describes our hypothesis and our evaluation approach. The performance of the different methods was measured using the evaluation functions which are based on Precision, Recall and F-measure [21]. We developed a syntactic and semantic version of Precision, Recall and F_1-Measure dedicated to workflow extraction. The syntactic evaluation function uses a simple lexical comparison to decide whether an element of the extracted workflow is matching a golden standard workflow or not. The semantic evaluation function uses a similarity measure which derives a similarity value of an ontology. Each hypothesis is tested with both types of evaluation functions. We tested the following three hypotheses:

1. The data-flow created by OPM has the best precision in comparison to ARM and FM.
2. The data-flow created by ARM has a better recall than OPM.
3. The data-flow created by FM has the highest F-measure in comparison to OPM and ARM.

5.1 Experimental Set-Up

The experiment was performed on a set of 37 recipes. These recipes were selected randomly from a set of 36 898 recipes which were crawled from a cooking community website.[5] A human expert modeled the data-flow for the recipes in the test set. This means that the human expert added the resources and products to every task in a

[5] www.allrecipes.com

given workflow. These test cases serve as the golden standard for the evaluation. This evaluation aims at the data-flow therefore the expert got the control-flow which was automatically extracted as framework for the golden standard workflow. This approach eliminates the paraphrasing and granularity problem of the control-flow. The paraphrasing problem is the problem that the same process can be described by different workflows. The granularity problem is the problem of handling the different levels of abstraction which can be used to formalize a process using a workflow. The expert was allowed to use all resources and products that she thought should be in the data-flow, even if they were not mentioned in the text. So we got a semantically correct data-flow. The only constraint was that the expert was not allowed to use synonyms for products which were mentioned in the original recipe texts. If a product was mentioned in the text, this term must be used in the data-flow. This restriction should reduce the paraphrasing problem for the data-flow.

5.2 Syntactic-Based Evaluation

We adapted recall, precision, and F-measure to our scenario. Every activity in the golden standard workflow had per definition a corresponding activity in the evaluated workflow. We are handling the elements in the products and resources sets separately, if a product is missing once in the input and once in the output set, it counts twice even if it's the same item. We are going to define the evaluation function formally.

Let T and T' be sets of activities of the workflows W and W'. W' is the golden standard workflow. Each activity $t \in T$ has a corresponding activity $t' \in T'$ which are equal except for the resource and product sets. Let I and I' be resource sets and O and O' product sets for the activities t and t'. The precision for an activity $t \in T$ is defined as:

$$precision(t) = \frac{\mid I \cap I' \mid + \mid O \cap O' \mid}{\mid I \mid + \mid O \mid}$$

The recall for an activity is defined as:

$$recall(t) = \frac{\mid I \cap I' \mid + \mid O \cap O' \mid}{\mid I' \mid + \mid O' \mid}$$

This leads to the evaluation functions for a workflow:

$$precision(W) = \frac{1}{\mid T \mid} \sum_{i=1}^{|T|} precision(t_i)$$

$$recall(W) = \frac{1}{\mid T \mid} \sum_{i=1}^{|T|} recall(t_i)$$

The F_1 measure is defined as:

$$F_1(W) = 2\frac{precision(W) * recall(W)}{precision(W) + recall(W)}$$

5.3 Semantic-Based Evaluation

Syntactic evaluations approaches can not handle paraphrases. As simple lexical distances are used to decide whether products or resources are equal. These distances lead to the problem that two items are classified as not matching even if they are semantically very close. For example *broth* and *soup* would be classified as unequal. We extended the syntactic evaluation approach to differentiate between semantically close mismatches and those which are not.

Precision and recall are commonly used evaluation functions in information extraction and there are multiple other function which are derived from them (e.g., F_1 measure). This leads to the requirement that a semantic aware evaluation function should be based on precision and recall. The evaluation of ontology alignments and the evaluation of data-flows is very similar, in both cases it is necessary to determine the semantic overlap of sets of items. We adapt a generalized version of precision and recall which was presented for the evaluation of ontology alignments [22]. The generalized precision on an activity is defined as:

$$precision_\omega(t) = \frac{\omega(I, I') + \omega(O, O')}{\mid I \mid + \mid O \mid}$$

The generalized recall is defined as:

$$precision_\omega(t) = \frac{\omega(I, I') + \omega(O, O')}{\mid I' \mid + \mid O' \mid}$$

$\omega(I, I')$ is the overlap function which is defined as follows:

$$\omega(I, I') = \sum_{\forall i \in I} \begin{cases} 1, & i \in I' \\ 0, & \max_{i' \in I'} sim(i, i') < threshold \\ \max_{i' \in I'} sim(i, i'), & i \notin I' \end{cases}$$

The sim-function is a function which measures the semantic similarity between two concepts. The sim function can return small values which indicate that the two concepts are semantically different. The sum of all these small values would influence the result, therefore we use a threshold to filter them out. For our experiment in the cooking domain we used a threshold of 0.25. The position of a concept in an ontology is used to derive a similarity value. An obvious approach is to count the edges between two concept but the degree of semantic similarity implied by a single edge is not consistent [23]. Therefore we use a more sophisticated approach. We use an information-theoretic definition of the similarity which was inferred from an axiomatic system by Lin [24]. The similarity between two concepts is defined as:

$$sim(A, B) = \frac{log P(common(A, B))}{log P(description(A, B))}$$

$common(A, B)$ is the information content needed to state the commonality of A and B and $description(A, B)$ is the information content to fully describe A and B. The information content of a concept can be quantified as negative of the log likelihood of a concept [25]. Let c be a concept of an ontology and r be a randomly chosen concept of the same ontology. Then the likelihood of the concept c is the probability that $c = r \vee c \in subsumers(r)$. This quantification is intuitive because it means that as probability increases, informativeness decreases, so the more abstract a concept is, the lower the information content [25]. By specifying the general definition of $common$ and $description$ we can infer a similarity for two concepts A and B in a taxonomy:

$$sim(A, B) = \frac{2IC(lcs(A, B))}{IC(A) + IC(B)}$$

$lcs(A, B)$ is the concept of the least common subsumer. To measure the information content (IC) we use an approach which was presented for ontologies [26]. The information content (IC) of a concept is defined as:

$$IC(A) = -\log \frac{\frac{|leaves(A)|}{|subsumers(A)|} + 1}{| leaves(root) | +1}$$

In a taxonomy $leaves(A)$ is the set of all leaves below the concept A. The set $subsumers(A)$ contains all concepts which are a generalization of the concept A. Sanchez et al. presented a benchmark of different combinations of methods for calculating the similarity and the information content [26]. The combination of the similarity measure and the information content function which we use is according to [26] the best performing combination benchmark (excluding approaches which use a corpora). For our evaluation we used the implementation which is included in the semantic measures library.[6] In addition we used the ontology of the Taaable project. The ontology is specialized on food and cooking. It can be extended by the community as it is available in a Semantic Media Wiki We used an owl export from 30th August 2013, which contains 2166 different ingredients. These ingredients are linked with different information e.g., nutritional value, compatible diets etc.[7]

[6] http://www.semantic-measures-library.org

[7] http://www.wikitaaable.loria.fr

Table 1 Results of the evaluation for selected cases

Recipe	P_{OPM}	P_{ARM}	P_{FM}	R_{OPM}	R_{ARM}	R_{FM}	F_{OPM}	F_{ARM}	F_{FM}
Mexican Egg	0.55	0.55	0.55	0.67	0.67	0.67	0.61	0.61	0.61
Classic Thumbprint	0.58	0.37	0.37	0.32	0.29	0.29	0.41	0.33	0.33
Cranberry Glazed Roast	0.48	0.46	0.52	0.31	0.31	0.33	0.38	0.37	0.41

5.4 Results

Figures 6a, c and e show that the results of the syntactic evaluation for the three methods are very close. The best average precision is achieved by Method OPM. For 7 cases the application of additional filter (Method ARM) reduces the precision compared to Method OPM. The best average recall is performed by Method FM. We examine the case of the "*Mexican Egg Bake*" recipe (Table 1) for which the results are equal for the three methods. This can happen, when no matching association rule is found and when no cookware item is filtered out by the stop-list. For the sample of the "*Classic Thumbprint Cookies*" recipe (Table 1) a wrong association rules is chosen, therefore the precision and the recall is lower for Method OPM & Method ARM as it is for Method OPM. A very interesting case is the one of "*Cranberry Glazed Roast Pork*" (Table 1). There we see a drop in precision from Method OPM to Method ARM. First a wrong rules is used in Method ARM which is corrected by Method FM, which ended in a higher precision for Method FM. The values of the recall might raise the question, how it is possible that a filter step produces a higher recall. Due to the fact that the filter step is preceding the association rule mining step in the pipeline , we can get a different set of association rules. If the rule mining step would precede the filter step, then indeed the results would show a higher precision and a recall which cannot be higher than before the filtering. Fig. 6b, d and f show the summary of the results for the semantic evaluation. Comparing the results of the syntactic and semantic evaluation a slight rise of the performance is observed. Table 2 shows selected cases of the semantic evaluation. The selected cases are the top three cases based on the difference between the F_1-Measure of the syntactic and the F_1-Measure of semantic evaluation for Method FM. In the case of "Cheesy Italian Tortellini", "ground meat" and "Italian sausage" is processed in a task "crumble". After this task, the recipe and the expert reference only "ground meat" but the automatically extracted workflow references both items. The semantic evaluation took into account that "Italian sausage" is a kind of meat. This had a large impact on the result because those items were used in a lot of tasks. In the case of "Slow Cooker Chicken Cordon Bleu" the semantic evaluation recognized that "chicken breast" is related to "chicken" and that "maple syrup" is a kind of "syrup". For the case "Chocolate Peanut Bars" the semantic evaluation recognized the similarity of "butter" and "peanut butter".

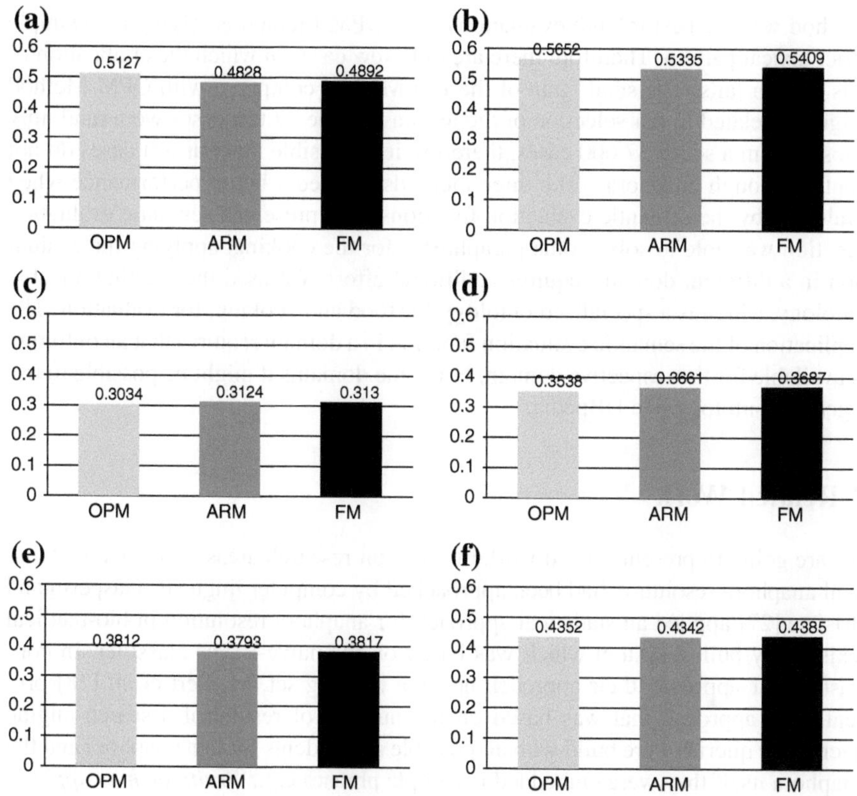

Fig. 6 Summary of the average results of the evaluation. **a** Syntactic Precision. **b** Semantic Precision. **c** Syntactic Recall. **d** Semantic Recall. **e** Syntactic F_1-Measure. **f** Semantic F_1-Measure

Table 2 Three cases with the highest difference ($F_{sem} - F_{syn}$)

	F_{sem}	F_{syn}	Difference
Chocolate Peanut Bars	0.6281	0.3654	0.2627
Slow Cooker Chicken Cordon Bleu	0.5710	0.2542	0.3169
Cheesy Italian Tortellini	0.6914	0.3482	0.3432

F_{sem} is the F_1-Measure resulting from the semantic evaluation, F_{syn} is the F_1-Measure resulting from the syntactic evaluation of Method FM

6 Discussion

The results of the data-flow-evaluation are promising given the complexity of the data-flow-creation. Both, the syntactic and semantic evaluation indicate that the benefit of the statistical anaphora resolution is low compared to the result which was achieved with the simple approach of OPM. Especially the precision of the OPM

Method was the best in both evaluations. The OPM Method is relying on a simple strict lexical pattern. Therefore there are only few cases in which the application of this pattern fails. The small gain of the FM Method compared with OPM Method might be related to our selection of the test cases. The 37 test cases were randomly chosen from a set of 37 000 cases, therefore it is possible that our test cases do not contain enough anaphoras. The three methods showed a better performance when evaluated by the semantic evaluation functions. The presented semantic evaluation function was able to solve some paraphrases for the cooking, applying the evaluation in a different domain requires additional effort. We used the existing Taaable ontology, which is a specialized ontology for food and cooking, for evaluation. The application of the semantic evaluation function in a domain requires that an ontology is available for the respective domain. For some domains it might be possible to use a general ontology like DBpedia.[8]

7 Related Work

We are going to present related work of different research areas. The area of statistical anaphora resolution had been approached by computer linguists. Gasperin and Briscoe [27] applied an statistical approach for anaphora resolution in bio-medical texts. They built a system which was based on the naive-Bayes classifier. In contrast to our approach their approach needs a training set. Markert et al. [28] presented an approach that was based on the number of results of a search engine query. The queries were build with all possible antecedents for that anaphora and the anaphora itself they were embedded in sample phrases e.g., "*fruits such as apples*". Our approach makes use of the workflow format, which provides an order of task inducing an order of their resources and products. The user-specified list of products which is part of the textual process descriptions (e.g., ingredient list in the cooking domain) is used by our approach. This list is more precise than textual queries based on word occurrence. TellMe [29] system allows the user to define procedures trough utterances in natural language that are interpreted by the system and transformed to formal workflow steps. In comparison to our system, the process of the TellMe system is interactive; the user might get feedback from the system and can specify her input. Zhang et al. [30] describe a method to extract processes from instructional text. Unlike our approach, their process models do not follow the workflow paradigm, as their control-flow is strictly sequential and they are not supporting a data-flow. Friedrich et al. [31] developed a system to extract a formal workflow representation of a textual process descriptions. To avoid noise, a manual preprocessing step is applied. Our approach is capable to process textual content as it is. The work of Dufour-Lussier et al. [9] is very similar to ours. They extract a tree representation of recipes from text. Contrary to our method, the focus is on food components that are transformed by cooking activities to other food components. The TAAABLE ontology is applied to resolve references in the text by matching sets of food components,

8 www.dbpedia.org

e.g., "blueberry" and "raspberry" with "fruits". They presented in [9] an evaluation for the data-flow. Their system delivered very good result. However, the test set was not representative and they were only counting the first occurrence of a product within a recipe.

8 Conclusion and Future Work

This chapter addresses the problem of workflow extraction from textual process descriptions and presents a framework to support the development of extraction applications by a pipes-and-filters architecture. The framework is illustrated by an application to the cooking domain. Multiple data-flow-creation approaches are evaluated with a focus on the anaphora resolution problem. Three different anaphora resolution approaches are presented. Two approaches are based on association rules which were created during an analysis of the corpus of workflows. The experimental evaluation uses workflows as a golden standard which have a semantically correct data-flow created by a human expert. A syntactic and a semantic evaluation function was presented. These functions which are based on precision, recall, and F-measure were used to asses the quality of the data-flow. The results show that the data-flow has been improved slightly by anaphora resolution. However, the anaphora resolution methods should be improved by the use of further semantics. It is promising that the method with the most sophisticated semantics (the FM method) achieved the highest value of the F-measure. The successful application in the cooking domain highlights that a pipes-and-filters framework is a good means to combine domain-specific and domain-independent filters. To prove that our approach is not limited to cooking recipes, we are going to implement a new domain.

Acknowledgments This work was funded by the German Research Foundation, project number BE 1373/3-1.

References

1. Workflow Management Coalition: Workflow management coalition glossary and terminology. http://www.wfmc.org/standars/docs/TC-1011_term_glossary_v3.pdf (1999). Accessed 23 May 2007
2. Aamodt, A., Plaza, E.: Case-based reasoning: Foundational issues, methodological variations, and system approaches. AI commun. **7**(1), 39–59 (1994)
3. Minor, M., Montani, S., Recio-Garca, J.A.: Process-oriented case-based reasoning. Inf. Syst. **40**, 103–105 (2014)
4. Bergmann, R., Gil, Y.: Similarity assessment and efficient retrieval of semantic workflows. Inf. Syst. **40**, 115–127 (2014)
5. Minor, M., Tartakovski, A., Schmalen, D., Bergmann, R.: Agile workflow technology and case-based change reuse for long-term processes. Int. J. Intell. Inf. Technol. **4**(1), 80–98 (2008)
6. Kendall-Morwick, J., Leake, D.: Facilitating representation and retrieval of structured cases: Principles and toolkit. Inf. Syst. **40**, 106–114 (2014)
7. Minor, M., Bergmann, R., Görg, S.: Case-based adaptation of workflows. Inf. Syst. **40**, 142–152 (2014)

8. Schumacher, P., Minor, M., Walter, K., Bergmann, R.: Extraction of procedural knowledge from the web. In: Proceedings of the Workshop WWW'12, Lyon, France (2012)
9. Dufour-Lussier, V., Le Ber, F., Lieber, J., Nauer, E.: Automatic case acquisition from texts for process-oriented case-based reasoning. Inf. Syst. **40**, 153–167 (2014)
10. Schumacher, P., Minor, M., Schulte-Zurhausen, E.: Extracting and enriching workflows from text. In: Proceedings of the 2013 IEEE 14th International Conference on Information Reuse and Integration, pp. 285–292 (2013)
11. Zhu, H.: Software Design Methodology: From Principles to Architectural Styles. Butterworth-Heinemann (2005)
12. Langer, G.: Textkohärenz und Textspezifität. Europäische Hochschulschriften, vol. 152. Peter Lang, Ireland (1995)
13. AeroSpace and Defence Industries Association of Europe: ASD simplified technical English. http://www.asd-ste100.org/ (2013). Accessed 19 Sept 2013
14. Minor, M., Schmalen, D., Bergmann, R.: XML-based representation of agile workflows. In: Bichler, M., Hess, T., Krcmar, H., Lechner, U., Matthes, F., Picot, A., Speitkamp, B., Wolf, P. (eds.) Multikonferenz Wirtschaftsinformatik 2008, pp. 439–440. GITO-Verlag, Berlin (2008)
15. Riloff, E., Phillips, W.: An introduction to the sundance and autoslog systems. Technical report, Technical report UUCS-04-015, School of Computing, University of Utah, (2004)
16. Schank, R.: Conceptual dependence: A theory of natural language understanding. Cogn. Psychol. (**3**)4, 532–631 (1972)
17. Fillmore, C.: The case for case reopened. Syntax Semant **8**(1977), 59–82 (1977)
18. Tognini-Bonelli, E.: Corpus Linguistics at Work. John Benjamins Publishing, Amsterdam (2001)
19. Higginbotham, J., Pianesi, F., Varzi, A.C.: Speaking of events. Oxford University Press, New York (2000)
20. Agrawal, R., Srikant, R.: Mining sequential patterns. In: Proceedings of the Eleventh International Conference on Data Engineering, pp. 3–14. IEEE Press, New York (1995)
21. Kowalski, G.: Information system evaluation. In: Information Retrieval Architecture and Algorithms, pp. 253–281. Springer, Berlin (2011)
22. Euzenat, J.: Semantic precision and recall for ontology alignment evaluation. In: Proceedings of the IJCAI, pp. 348–353 (2007)
23. Pedersen, T., Pakhomov, S.V., Patwardhan, S., Chute, C.G.: Measures of semantic similarity and relatedness in the biomedical domain. J. Biomed. Inform. **40**(3), 288–299 (2007)
24. Lin, D.: An information-theoretic definition of similarity. ICML **98**, 296–304 (1998)
25. Resnik, P.: Using information content to evaluate semantic similarity in a taxonomy. In: Proceedings of the IJCAI, vol. 1, pp. 448–453. Morgan Kaufmann, San Francisco (1995)
26. Sánchez, D., Batet, M., Isern, D.: Ontology-based information content computation. Knowl. Based Syst. **24**(2), 297–303 (2011)
27. Gasperin, C., Briscoe, T.: Statistical anaphora resolution in biomedical texts. In: Proceedings of the 22nd International Conference on Computational Linguistics, vol. 1. COLING '08, Stroudsburg, USA, Association for Computational Linguistics, pp. 257–264 (2008)
28. Markert, K., Modjeska, N., Nissim, M.: Using the web for nominal anaphora resolution. In: Proceedings of the EACL Workshop on the Computational Treatment of Anaphora, Budapest, pp. 39–46 (2003)
29. Gil, Y., Ratnakar, V., Fritz, C.: Tellme: learning procedures from tutorial instruction. In: Proceedings of the 15th International Conference on Intelligent User Interfaces, pp. 227–236 (2011)
30. Zhang, Z., Webster, P., Uren, V.S., Varga, A., Ciravegna, F.: Automatically extracting procedural knowledge from instructional texts using natural language processing. In: Calzolari, N., Choukri, K., Declerck, T., Dogan, M.U., Maegaard, B., Mariani, J., Odijk, J., Piperidis, S. (eds.) LREC. European Language Resources Association (ELRA), Istanbul, pp. 520–527 (2012)
31. Friedrich, F., Mendling, J., Puhlmann, F.: Process model generation from natural language text. In: Rolland, C. (ed.) Advanced Information Systems Engineering, vol. 6741, pp. 482–496. Springer, Heidelberg (2011)

A Minimum Description Length Technique for Semi-Supervised Time Series Classification

Nurjahan Begum, Bing Hu, Thanawin Rakthanmanon and Eamonn Keogh

Abstract In recent years the plunging costs of sensors/storage have made it possible to obtain vast amounts of medical telemetry, both in clinical settings and more recently, even in patient's own homes. However for this data to be useful, it must be annotated. This annotation, requiring the attention of medical experts is very expensive and time consuming, and remains the critical bottleneck in medical analysis. The technique of Semi-supervised learning is the obvious way to reduce the need for human labor, however, most such algorithms are designed for intrinsically *discrete* objects such as graphs or strings, and do not work well in this domain, which requires the ability to deal with *real-valued* objects arriving in a streaming fashion. In this work we make two contributions. First, we demonstrate that in many cases a surprisingly small set of human annotated examples are sufficient to perform accurate classification. Second, we devise a novel parameter-free stopping criterion for semi-supervised learning. We evaluate our work with a comprehensive set of experiments on diverse medical data sources including electrocardiograms. Our experimental results suggest that our approach can typically construct accurate classifiers even if given only a single annotated instance.

Keywords MDL · Semi-supervised learning · Stopping criterion · Time series

N. Begum (✉) · B. Hu (✉) · T. Rakthanmanon · E. Keogh
Department of Computer Science and Engineering, Kasetsart University, University of California, Riverside, CA, USA
e-mail: nbegu001@ucr.edu

B. Hu
e-mail: bhu002@ucr.edu

E. Keogh
e-mail: eamonn@cs.ucr.edu

T. Bouabana-Tebibel and S. H. Rubin (eds.), *Integration of Reusable Systems*,
Advances in Intelligent Systems and Computing 263, DOI: 10.1007/978-3-319-04717-1_8,

1 Introduction

Over the past decade, the data mining community has turned a significant fraction of its attention to time series data. Such data is available in almost all aspects of human endeavor, including motion capture [3], medical science [38], finance [27], aerospace, meteorology [18], entertainment, etc. Given such ubiquity of time series data, it is not surprising that a plethora of algorithms for time series classification have been proposed in recent years [10, 29, 32]. Virtually all these algorithms assume the availability of plentiful labeled instances [31, 37]. However, this assumption is often unrealistic, particularly in the medical community, as it is expensive and time consuming to manually annotate large medical datasets [5, 9]. Nevertheless, huge amounts of *unlabeled* data are readily available. For example, the PhysioBank archive contains over seven hundred GB of data [11], only a tiny fraction of which are labeled. Similarly, Google's Total Library [15] contains millions of digitized books available, which could be mined after conversion to time series space, given that plentiful labeled data is available.

Given the enormous intra-patient variability of cardiological signals such as electrocardiograms (ECGs), it is difficult to consider classification problems independent of patients. For example [35] notes that "*...age, sex, relative body weight, chest configuration, and various other factors make ECG variant among persons with same cardiological conditions*". Thus we cannot easily generalize labeled ECG data from one patient to another. Therefore, constructing a classifier in order to detect some medical phenomena, (e.g. ECG arrhythmia) is a non-trivial problem.

In spite of all these challenges, we would like to be able to build accurate classifiers in the face of *very* scarce knowledge, as little as a *single* labeled instance.

Given such circumstances, Semi-supervised Learning (SSL) seems like an ideal paradigm, because it can leverage the knowledge of both labeled and unlabeled instances. Although many SSL models have been studied in the literature, the critical subroutine of finding a robust stopping point for these models has not been widely explored.

The vast majority of the literature [24, 25, 33, 41] addressing this problem possess some flaws that have limited their widespread adoption. For example, the approach proposed in [41] suffers from the problem of *early stopping*. That is, it is overly conservative and may refuse to label examples that are (at least *subjectively*) obvious enough to be labeled. The approach introduced in [33] fails to identify class boundaries accurately, because it tends to produce too many false negatives.

Our work is motivated by the observation of the critical and underappreciated importance of the stopping criteria. If an otherwise useful algorithm is too conservative (or too liberal) in its stopping criteria, it is doomed to produce many false negatives (or false positives), and thus it is almost certainly never going to be deployed in the real world. We illustrate this observation in Fig. 1.

From Fig. 1 *left*, we can see that in a dataset with a training set size as small as a single instance, we cannot avoid an unwanted production of lots of false positives with

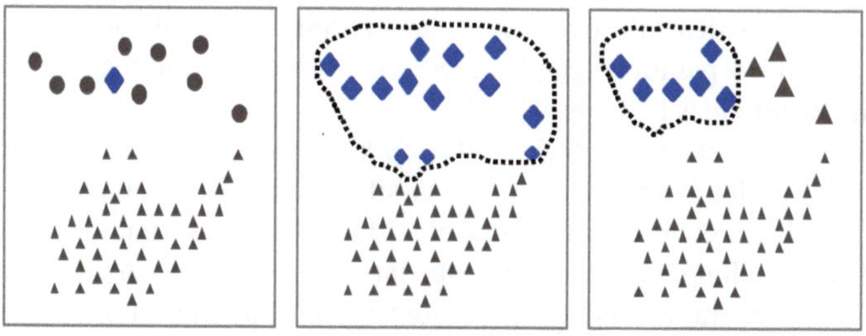

Fig. 1 *left*) The initial positive instance (*blue/square*) along with unlabeled positive (*grey/circle*) and negative (*grey/small triangle*) instances. *center*) A too liberal stopping criterion produces lots of false positives (*blue/small square*). *right*) A too conservative stopping criterion produces lots of false negatives (*grey/large triangles*)

a too liberal stopping criterion (Fig. 1 *center*). Similarly, if we have a too conservative stopping criterion, we must invariably encounter lots of false negatives (Fig. 1 *right*).

In response to the noted limitations of the state-of-the-art stopping criteria for SSL, we propose a semi-supervised learning framework with a novel stopping criterion using Minimum Description Length (MDL). MDL has been widely used in bioinformatics, text mining, etc. yet it is still under explored in time series domain. This is mainly because MDL is defined in discrete space. Some recent publications [17, 30] provide an in-depth analysis of the potential of applying MDL technique to real-valued time series data. In this work, we propose a parameter-free algorithm for finding a stopping criterion for SSL using the MDL technique. Its simplicity will allow adoption by experts in the medical community.

The rest of the chapter is organized as follows. In Sect. 2, we discuss related work. We give definitions and the notations in Sect. 3. We demonstrate how our semi-supervised learning algorithm works and then propose our stopping criterion using MDL in Sect. 4. In Sect. 5, we provide a detailed empirical evaluation of our ideas. Finally, we discuss conclusions and directions for future work in Sect. 6.

2 Related Work

Semi-supervised Learning (SSL) can be seen as intermediate between the techniques of Supervised Learning and Unsupervised Learning. SSL is generally provided with *some* supervision information in addition to unlabeled instances [4]. Labeled data is often hard to obtain, expensive, and may require human intervention. For example, there is often no replacement of a dedicated domain expert observing and labeling medical phenomena [26], human activities [28], insect behaviors [39], etc. for constructing annotated datasets. In contrast, unlabeled data is abundant and easy to collect. SSL exploits the use of unlabeled data in addition to labeled data for building more accurate classifiers [44]. Because of this, SSL is of great interest in both theory and practice. There have been many SSL techniques studied in the literature.

These can be classified into five major classes: SSL with generative models, SSL with low density separation, graph based techniques, co-training techniques and self-training techniques [2, 4, 8, 41, 44].

Among all these classes of SSL, *Self-training* is the most common method used [1, 22, 34]. The idea is to first train the classifier with a small amount of data, and then use the classifier to classify unlabeled data. The most confidently classified unlabeled points, along with their predicted labels, are added to the training set. The classifier is then re-trained and the procedure repeats, i.e. the classifier uses its own predictions to teach itself [44]. Self-training approaches have been used for natural language processing tasks [21, 43] and object detection systems [34]. Because of their generality and simplicity, we use self-training as the foundation of our research efforts.

In our work, we focus on the question of finding the *correct* stopping point for our self-training classifier. The question of finding a *good* stopping criterion has tentative solutions based on MDL, Bayesian Information Criterion, bootstrapping [36, 44], etc. For time series domain, this question has been surprisingly under explored; to the best of our knowledge, only [24, 25, 33, 41] have considered it. In [41], a stopping criterion based on the *minimal nearest neighbor distance* was introduced. However, this conservative approach suffers from the shortcoming that it can result in an expansion of only a small number of positives.

In order to improve the algorithm of [41], the work in [33] proposed a stopping criterion by using the previously observed distances between candidate examples from the unlabeled set to the initial positive examples. Although this refinement can add more positive examples to the labeled set, it is unable to identify accurate positive and negative instances from the unlabeled set, because it produces too many false negatives, and therefore not accurate enough in most real life applications [6].

A recently proposed method [24] is called LCLC (Learning from Common Local Cluster), which is a cluster-based approach rather than an instance-based approach. Although this method more accurately identifies the decision boundaries, it has two drawbacks. First, it assumed that *"All the instances belonging to a local cluster have the same class label"*. The same examples within the same cluster received the same class label, and therefore, LCLC can misclassify some examples if the clustering subroutine does not produce pure clusters. Second, the LCLC algorithm uses K-means algorithm, and therefore inherits K-means lack of determinism and its assumption that the data can be represented by Gaussian "balls".

To overcome the drawbacks of the LCLC approach, the same research group proposed an ensemble-based approach called En-LCLC (Ensemble based Learning from Common Local Clusters) [25]. This algorithm performs the clustering process multiple times with different settings to obtain diverse classifiers. Each instance is assigned a "soft" probabilistic confidence score and based on these scores, potential noisy instances, which can confuse the classifier, are eliminated. An Adaptive Fuzzy Nearest Neighbor (AFNN) classifier is then constructed based on the "softly labeled" set of positive and negative instances. However, this approach is *extremely* complicated and requires many unintuitive parameters to be set. It is not clear how sensitive the settings of these parameters are.

A recent work on time series SSL [6] has somewhat an ad-hoc stopping criterion, in which the classification continues until the unlabeled dataset is exhausted of true positives.

In this chapter, we propose a general Minimum Description Length (MDL) based stopping criterion, which works at the instance level, is parameter free and leverages the intrinsic nature of the data to find the stopping point. In time series domain, MDL has been the cornerstone of several recent efforts to solve classification and clustering problems [17, 30]. However, to the best of our knowledge, this is the first work to address time series SSL using MDL.

3 Background and Notation

We begin by defining the data type of interest, *Time Series*:

Definition 1 *Time Series*: A time series $T = t_1, \ldots, t_m$ is an ordered set of m real-valued variables.

We are only interested in the local properties of a time series, thus we confine our interest to *subsequences*:

Definition 2 *Subsequence*: Given a time series T of length m, a subsequence C_p of T is a sampling of length $w \leq m$ of contiguous positions from T with starting position at p, $C_p = t_p, \ldots, t_p + w - 1$ for $1 \leq p \leq m - w + 1$.

The extraction of subsequences from a time series is achieved by the use of a *Sliding Window*.

Definition 3 *Sliding Window*: Given a time series T of length m, and a user-defined subsequence length of w, all possible subsequences can be extracted by sliding a window of size w across T and extracting each subsequence C_p.

The most common distance measure for time series is the *Euclidean distance*.

Definition 4 *Euclidean Distance*: Given two time series (or time series subsequences) Q and C both of length m, the Euclidean distance between them is the square root of the sum of the squared differences between each pair of the corresponding data points:

$$D(Q, C) \equiv \sqrt{\sum_{i=1}^{m} (q_i - c_i)^2}$$

We illustrate these definitions in Fig. 2.

Note that our use of Euclidean distance in this work does not preclude other distance measures; however, it has recently been forcefully shown that Euclidean distance is a very competitive distance measure [40].

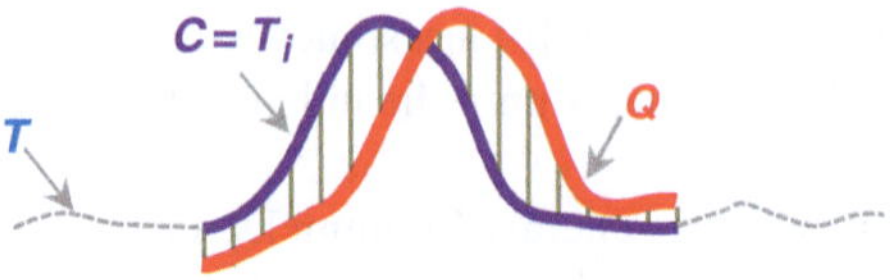

Fig. 2 A long time series T (*grey/dashed*) with a subsequence T_i (*purple/thin*) extracted and compared to a query Q (*red/bold*) under the Euclidean distance

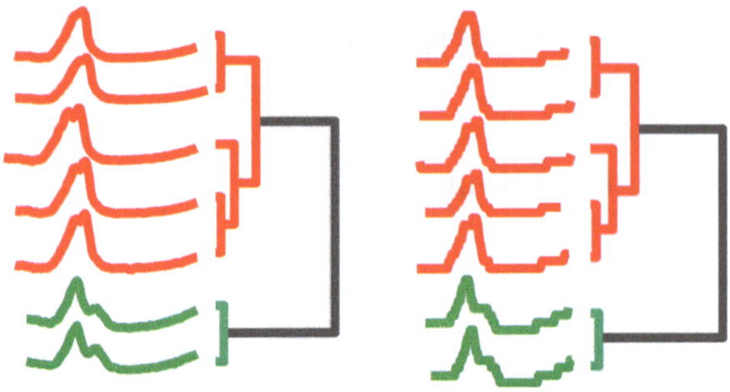

Fig. 3 A complete linkage clustering of time series incartdb I70 [11]. *left*) Original 32-bit data. *right*) The same data with cardinality reduced to eight values (3-bits)

Minimum Description Length (MDL) can be regarded as a formalization of Occam's Razor, which states that the best hypothesis for a given set of data is the one that leads to the best compression of the data [14]. Note that MDL is defined in the *discrete* space, but most time series are *real*-valued. To use MDL in time series domain, we need to cast the original real-valued data to discrete values. The reader may imagine that a drastic reduction in the precision of the data could sacrifice accuracy. However, recent empirical work suggests that this is not the case. For example, [17, 30] have performed extensive experiments comparing the effect of using the original real-valued data versus its quantized version. The results showed that even drastically reduced cardinality does not result in reduced accuracy. To see this in our domain we performed a simple time series clustering experiment, as shown in Fig. 3.

Figure 3 suggests that even significantly reducing the cardinality of time series has minimal effect.

To further test the claim that reducing the cardinality of the data has no *statistically significant* effect on classification accuracy, we show the error rates and ranks of the 1-NN classifier on six ECG datasets both in raw and reduced cardinality format from the UCR archive [19] in Table 1. For each dataset, we assign the best algorithm the lowest rank.

From Table 1 we can see that for the six datasets shown, classification using the raw data yields slightly better average rank. However, we cannot claim that the difference in average ranks of the 1-NN classifier for the two different cardinality representations is *statistically significant*, given of the following analysis.

Table 1 Classification error rates and ranks of 1-NN classifier for raw data (32 bit) and cardinality reduced data (4 bits). The best result for each dataset has been shown in bold/green

Dataset	Error rate (Original data)	Error rate (Cardinality 16)
ECG200	**0.12 (1)**	0.13 (2)
ECGFiveDays	0.2 (2)	**0.15 (1)**
TwoLeadECG	**0.25 (1)**	0.27 (2)
CinCECG_Torso	0.1 (2)	**0.07 (1)**
NonInvasiveFatalECG _Thorax1	**0.17 (1)**	0.24 (2)
NonInvasiveFatalECG _Thorax2	**0.12 (1)**	0.21 (2)
Average rank	1.33	1.67

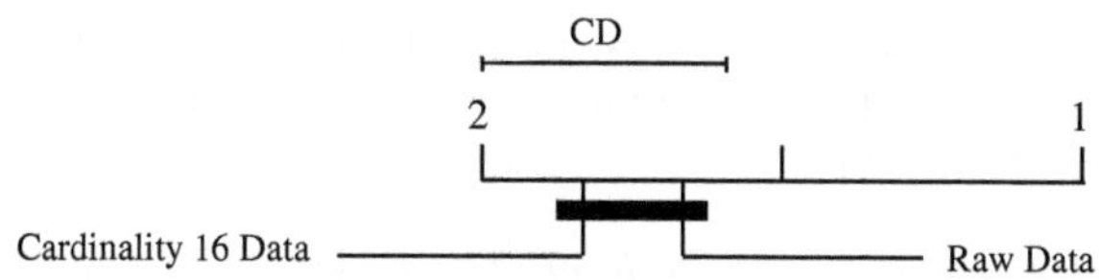

Fig. 4 Critical difference diagram representing the ranked error rates of the 1-NN classifier for the two cardinality representations of the data (Table 1)

The null hypothesis we test is, "*The performances of a classifier for different cardinality representations of the data are the same and the observed differences are merely random*". In order to compare the performances of the classifier as shown in Table 1, we denote the number of datasets and the number of different cardinality representations by N and k respectively. In our case, $N = 6$ and $k = 2$. We employ the two-tailed Nemenyi test [23] to test the null hypothesis. According to this test, the performances of a classifier for different cardinality representations are significantly different if the corresponding average ranks differ by at least the critical difference (CD):

$$CD = q_{\alpha}\sqrt{\frac{k\,(k+1)}{6N}}$$

where critical values q_{α} are based on the Studentized range statistic divided by $\sqrt{2}$ [7].

For a significance level α of 0.05, $q_{\alpha} = 1.96$, yielding a CD of 0.8. From Table 1, the difference in average ranks between the two different cardinality representations is 0.34, which is less than CD. Therefore, according to the Nemenyi test, we fail to reject the null hypothesis.

In Fig. 4, we show a critical difference diagram [7, 16] of the ranked error rates from Table 1.

From Fig. 4 we see that both variants of the 1-NN classifier for different cardinality representations are connected by a single clique, which means they are not significantly different from each other.

Based on the above and more extensive empirical work in [17, 30], we reduce time series data to its 16-value (4-bit) cardinality version, using a discrete normalization function:

Definition 5 *Discrete Normalization Function*: A *discrete* function *Dis_Norm* is the function to normalize a real-valued subsequence *T* into *b*-bit discrete value of range $[1, 2^b]$. It is defined as below:

$$Dis\_Norm\,(T) = round\left(\frac{T - min}{\max - min}\right) * \left(2^b - 1\right) + 1$$

where min and max are the minimum and maximum value in T respectively.

As we previously noted, MDL works in discrete space, therefore we are interested in determining how many bits are required to store a particular time series *T.* We call it the *Description Length* of *T.*

Definition 6 *Description Length*: A *description length DL* of a time series *T* is the total number of bits required to represent it.

$$DL\,(T) = w * \log_2 c$$

where *w* is the length of *T*, *c* is the cardinality.

One of the key steps of designing our stopping criterion using the MDL technique is to find a representation or model of our data, and use this model to compress the data in a lossless fashion. We call this representation, a *hypothesis:*

Definition 7 *Hypothesis*: A hypothesis *H* is a subsequence used to encode one or more subsequences of the same length.

Naively, the number of bits required to store a time series depends only on the data format and its length. However, we are interested in knowing the *minimum* number of bits to exactly represent the data. This depends on the *intrinsic structure* of the data. In general case, this number is not calculable, as it is equivalent to the Kolmogorov complexity of the time series [20], however, there are many ways to approximate this, e.g. Huffman encoding, Shanon-Fano coding, etc.

Ultimately, we interest in how many bits are required to encode *T given H*. We call this the *Reduced Description Length* of *T*.

Definition 8 *Reduced Description Length*: A *reduced description length* of a time series *T* given hypothesis *H* is the sum of the number of bits required in order to encode *T* exploiting the information in *H*, i.e. $DL(T|H)$, and the number of bits required for *H* itself, i.e. *DL(H)*. Thus, the reduced description length is defined as:

$$DL\,(T, H) = DL\,(H) + DL\,(T|H)$$

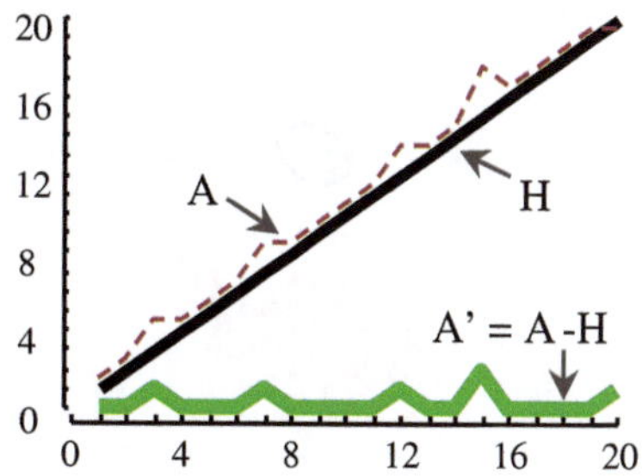

Fig. 5 Time series *A* (*maroon/dashed*) is represented by the sum of hypothesis *H* (*solid/black*) and the difference vector *A*' (*green/bold*)

We can encode *T* using *H* in many ways. One simple approach of this encoding is to store a *difference vector* between *T* and *H*, and we can easily generate the whole time series *T* from the hypothesis. Therefore, we use, $DL\,(T\,|\,H) = DL\,(T - H)$.

Given the definitions above, let us consider a toy example of our MDL based encoding. In Fig. 5, *A* is a sample time series of length 20:

$A =$ **1 2 4 4 5 6 8 8 9 10 11 13 13 14 17 16 17 18 19 19**

H is another time series, which is a straight line of length 20:

$H =$ **1 2 3 4 5 6 7 8 9 10 11 12 13 14 15 16 17 18 19 20**

Without encoding, the naive bit requirement to store *A* and *H* is, $2 * 20 * \log_2 20 =$ 173 bits. In contrast, using *H* as the model of the data, we can encode *A* as A', which is simply the difference vector obtained by subtracting *H* from *A*. We can see that, $A' = |\boldsymbol{A}\text{-}\boldsymbol{H}| =$ 0 0 **1** 0 0 0 **1** 0 0 0 0 **1** 0 0 **2** 0 0 0 0 **1**. From Fig. 5, the reader can see that *A* is a slightly corrupted version of *H* and that in the difference vector, there are five mismatches. The bit requirement is now just $20 * \log_2 20 + 5 * \left(\log_2 20 + \left[\log_2 20\right]\right) = 134$ bits, which is significantly less than the naive bit requirement.

We are now in a position to demonstrate how our semi-supervised learning approach works with the novel MDL based stopping criterion in the next section.

4 Algorithms

4.1 Semi-Supervised Time Series Classification

Semi-supervised learning technique can help build better classifiers in situations where we have a small set of labeled data, in addition to abundant unlabeled data [8]. We refer the interested reader to [4] and [44] for a more detailed treatment.

We use the self-training technique to build our classifier for the reasons noted in Sect. 2. In this work, we require only a *single* positive instance as the initial training set. During the training process, the training set will be augmented iteratively by labeling more unlabeled instances as positive.

Below are the two steps by which the classifier trains itself iteratively:

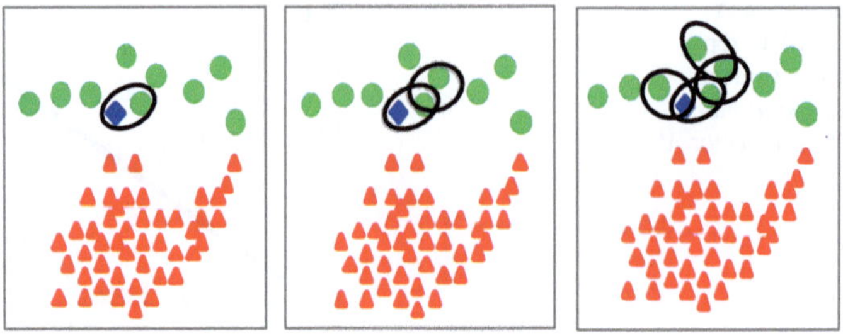

Fig. 6 *left*) The initial positive instance (*blue/square*) selects its nearest positive instance (*green/circle*). *center* and *right*) As we continue iterating, more unlabeled instances are labeled as positive and added to the training set

Step 1: We find the nearest neighbor of any instance of our training set from the unlabeled instances.

Step 2: This nearest neighbor instance, along with its newly acquired positive label, will be added into the training set.

The intuition of this idea is straightforward. The labeled positive example serves as a model which characterizes what a positive example should "look like". If an unlabeled instance is very similar to a positive example, its probability to have the positive class label is also assumed to be high. By adding such an example to the training set, the description of the positive class is generalized.

We give an illustrative example of the training procedure of our algorithm in Fig. 6. Each ellipse corresponds to labeling one unlabeled instance as positive. As we can see, there is a "chaining effect" of the semi-supervised learning: a set of positive examples helps labeling more positive examples.

For concreteness, we give the algorithm for self-training time series classification using just a single labeled instance in Table 2.

In order to demonstrate the benefit of our approach, we compare our method with the naive one-nearest-neighbor straw man, because the nearest neighbor algorithm is particularly suitable for time series data, and has the advantage of being parameter-free and allowing fair comparisons between methods [42]. The naive approach also—starts with one initial seed. Instead of finding the nearest neighbor of any instance of the updated training set, the algorithm only considers the nearest neighbors of this sole seed. Thus the algorithm can be imagined as expanding concentric circles around the initial seed.

However, this approach possesses a representational flaw as shown in Fig. 7. As the iteration continues, the probability of mistakenly labeling a negative instance (red/triangle) as positive also increases.

In contrast, Fig. 6 shows that the selection of an instance with the smallest distance from the current labeled positive set reduces this probability of misclassifying the instances too early. This example suggests that the expressiveness of our approach is greater than the naive nearest neighbor straw man. Note however that this greater

Table 2 Semi-supervised time series classification algorithm with only one labeled instance

	pos_Obj = **Self_Training_Classifier** (P,N)
Input	P, set of labeled/positive instances.
	//initially |P| = 1
	N, set of unlabeled/negative instances
Output	pos_Obj, set of labeled objects
	while (the stopping criterion)
	new_pos_obj = find the nearest neighbor for P
	P = P ∪ {new_pos_obj}
	N = N - {new_pos_obj}
	end while
	pos_Obj = P

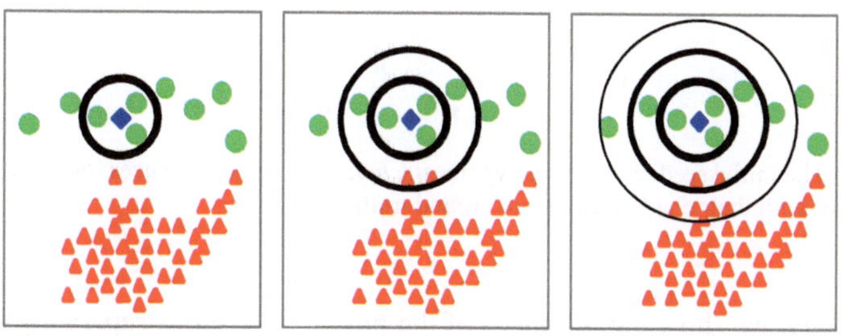

Fig. 7 Inner *circles* (*black/thicker*) correspond to the earlier iterations. *left*) The initial positive instance (*blue/square*) selects its nearest neighbor as a positive instance (*green/circle*). *middle* and *right*) False Positives (*red/triangle*) are added

representation power is all for naught unless it is coupled with the ability to stop adding instances at the correct time. We consider this issue next.

4.2 Stopping Criterion

In this section, we demonstrate how our algorithm finds the correct stopping point using an MDL based technique. As shown in Fig. 8, we consider a contrived, but real ECG dataset from [11]. Inspired by the compression-based approach of [30], our stopping criterion algorithm attempts to losslessly compress the data by finding repeated patterns, and it signals termination when further compression is not possible.

In the toy example in Sect. 3 we used a *straight line* hypothesis to explain the data. Here we use the seed heartbeat as the hypothesis instead. Our algorithm will attempt to compress the purple time series by expressing all its instances in terms of this hypothesis H and the (hopefully small) residual error A'. After compression, our original time series is supposed to have shorter length, which we call its *Reduced Description Length*.

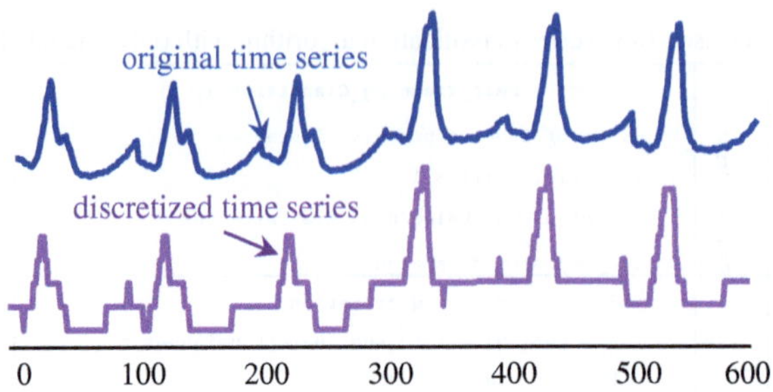

Fig. 8 An original time series is shown in *blue* and its discretized version is shown in *purple*

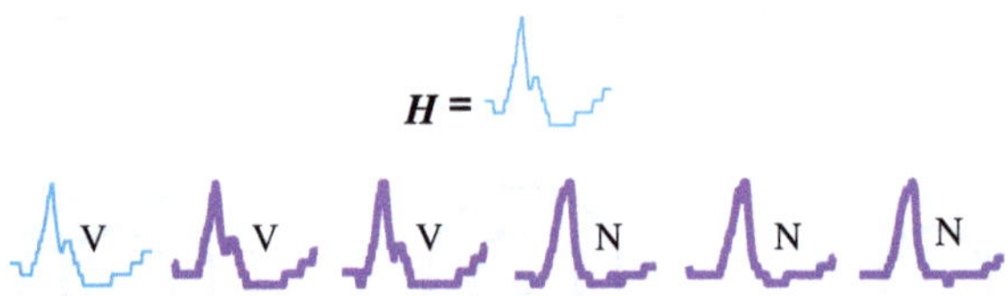

Fig. 9 *top*) The initial seed chosen as a hypothesis. *bottom*) The six instances (of length 100 each) of the dataset with their class labels (from ground truth)

For clarity, we show the instances of the purple time series separately with their class labels ('V' and 'N') in Fig. 9 *bottom.* Note that these instances are *not* labeled by the algorithm; these labels come from the ground truth and are given for the reader's introspection only.

As the first step, we randomly give an instance as the seed to our SSL module, which is the *only* member of the training set. According to the algorithm in Sect. 4.1, this module will augment the training set. Let us imagine we do not have a stopping criterion for this module. In that case, the algorithm will eventually ingest all the instances of the dataset in the training set.

Given this, let us see how we find a stopping criterion for our SSL module using MDL. As hypothesis, we take the same initial seed we gave to our SSL module (Fig. 9 *top*). We use this hypothesis to encode the remaining five instances of our purple time series.

Without any encoding, the description length for this time series is $DL\,(T) = 6 * 100 * \log_2 16 = 6 * 400 = 2400$ bits.

Our SSL module will select the nearest neighbor of any of the instance(s) in the training set from the dataset. We will encode this instance in terms of our heartbeat hypothesis and keep the rest of the dataset uncompressed. We find a total of 22 mismatches of the encoded instance with the hypothesis (Fig. 10 *top*). The reduced description length of the dataset becomes:

$$\begin{aligned}\mathrm{DL}(T, H) &= \mathrm{DL}(H) + \mathrm{DL}(T|H) + \mathrm{DL}(uncompressed)\\ &= 5^{*}400 + 22 \times (4 + \lceil(\log_2 100)\rceil) = \mathbf{2242}\text{ bits.}\end{aligned}$$

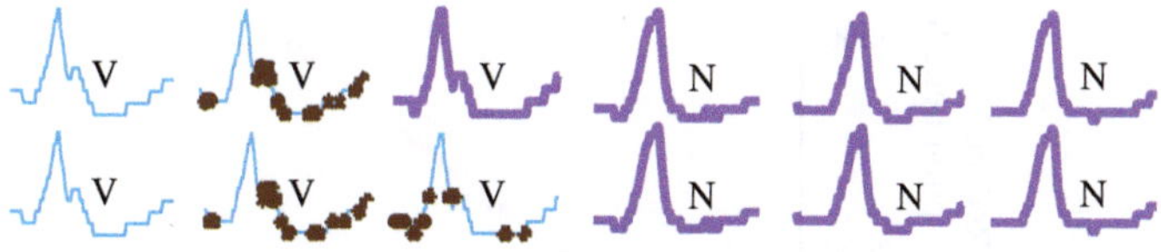

Fig. 10 The instances (*cyan/thin*) in the training set are encoded. The mismatches are colored as *brown**

Fig. 11 After correctly encoding the three 'V' instances, attempts to encode the 'N' instances require more bits than the raw data representation

This means we achieved data compression, which our algorithm interprets as an indication that the heartbeat in question really is a true positive.

We simply continue to test to see if unlabeled instances can be added to the positive pool by this "`reduces the number of bits required`" criteria. After the module adds the third instance in the training set, the cumulative mismatch count in terms of the hypothesis increases to 37, resulting in a reduced description length of **2007** bits, which means we are still achieving some compression (Fig. 10 *bottom*).

The compression indicates that the two instances encoded are very similar to the hypothesis. By visual inspection, the reader will notice that the last three instances in Fig. 9 are *dissimilar* to the initial seed. Without a stopping criterion, the SSL module will eventually include these instances in the labeled set. This means our labeled set now starts including true *negatives*. However, our MDL criteria can save us, at least in this toy example. If we encode the last three instances using our hypothesis, the overhead to store the mismatches will not allow further compression. This means adding the true positives, and *only* the true positives, is the *best* we can do in terms of achieving compression for this example.

The effect of encoding the last three 'N' instances is shown in Fig. 11. The cumulative mismatch count jumps to 115, 192 and 273, respectively for these instances. The resulting *increase* of the reduced description lengths to **2465**, **2912** and **3403** bits respectively reinforces the observation that attempting to compress data which is unlike the hypotheses is not fruitful, and this signals an appropriate stopping criterion.

We illustrate this observation in Fig. 12. As long as the SSL module selects instances similar to our hypothesis, we achieve data compression (shown in green). Once this module starts including instances dissimilar to the hypothesis, we can no longer achieve compression (shown in red). Therefore, this first occurrence of a subsequence that cannot be compressed with the hypotheses is the point where the SSL module should **stop.**

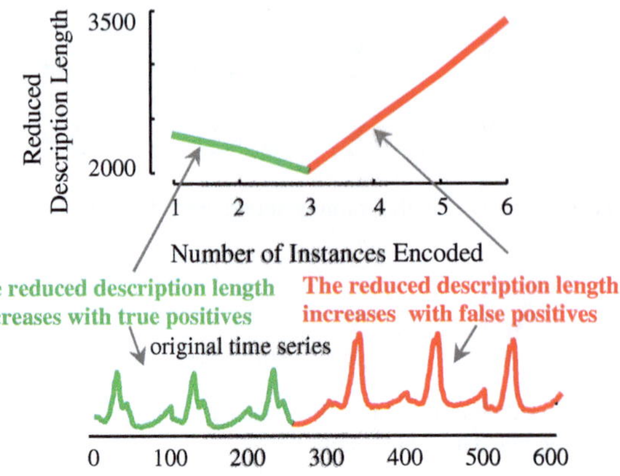

Fig. 12 *top*) Reduced description length curve after encoding the instances shown in *bottom*)

Table 3 Stopping criterion for self-training time series classification algorithm

```
stop_flag = find_stop_point (P, N)
Input:
    P, set of labeled/positive instances.
    // initially |P| = 1
    N, set of unlabeled/negative instances
Output:
    stop_flag, a Boolean flag indicating the stopping
point initialized to false

    hypothesis = P; // the initial seed is hypothesis
    DL_old = inf ;
    while (N is not empty)
      pos_Obj = Self_Training_Classifier (P, N);
      DL_cur = description length of pos_Obj;
      if DL_cur < DL_old
        DL_old = DL_current;
      else
        stop_flag = true ;
      break;
      end if
    end while
```

In Table 3, we modify the algorithm shown in Table 2 to allow this MDL based stopping criterion.

5 Experimental Results

We begin by stating our experimental philosophy. In order to ensure that our experiments are reproducible, we have built a website which contains all data and code [45]. In addition, this website contains additional experiments that are omitted here for brevity.

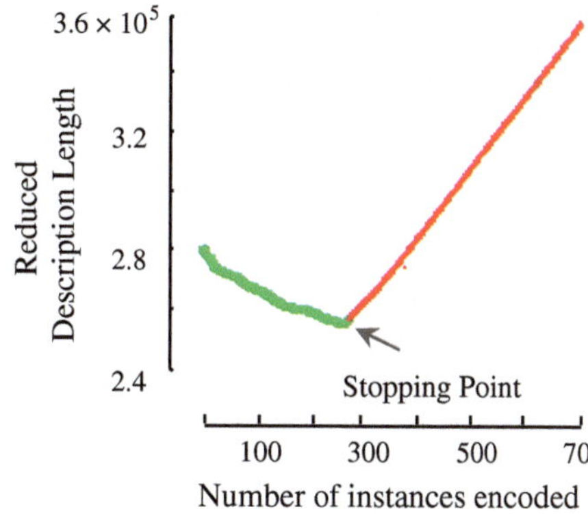

Fig. 13 Reduced description length versus number of encoded instances plot for the SVDB dataset. *green/thick* and *red/thin* points denote TP and TN instances respectively

5.1 MIT-BIH Supraventricular Arrhythmia Database

We consider an ECG dataset from the MIT-BIH Supraventricular Arrhythmia Database (SVDB) [12]. Each recording in this database includes one or more ECG signals and a set of beat annotations by cardiologists. Out of the 78 half-hour ECG recordings in this database beat annotations by cardiologists, we randomly chose record 801 and signal ECG1 for our experiment. Here we formulate the problem of classifying heartbeats as a 2-class problem and because identifying abnormal heartbeats are of more interest to cardiologists, our target class (i.e. the positive class) consists of only abnormal heartbeats. This dataset contains 268 abnormal heartbeats (Premature Ventricular Contraction or PVC, in this case) and our self-training classifier classifies 266 of them correctly.

Starting with a labeled set *S* with only one initial positive instance, our self-training classifier augments *S* by adding one instance at a time. As our algorithm iterates, we calculate the reduced description length DL_S of our labeled set. As expected, with the augmenting of the labeled set with more and more true positive (TP) instances, DL_S continues decreasing in our experiment. However, from the point we start adding true negatives (TN) in S, DL_S continues increasing. As a sanity check, we augmented the size of our labeled set up to 700, and found no change in the curve pattern of our DL_S versus number of time series encoded. The results are shown in Fig. 13. From this figure, we can see that we still allow a tiny number of TP instances beyond the stopping point suggested by our algorithm. Nevertheless, this does not significantly hurt the performance of our classifier in terms of accuracy.

5.2 St. Petersburg Arrhythmia Database

Next we consider classifying the heartbeats from the St. Petersburg Institute of Cardiological Technics 12-lead Arrhythmia Database (INCARTDB) [11] using our self-training classifier. This database has 75 annotated readings extracted from 32

Fig. 14 Reduced description length of the labeled set versus number of encoded time series plot for the INCARTDB dataset

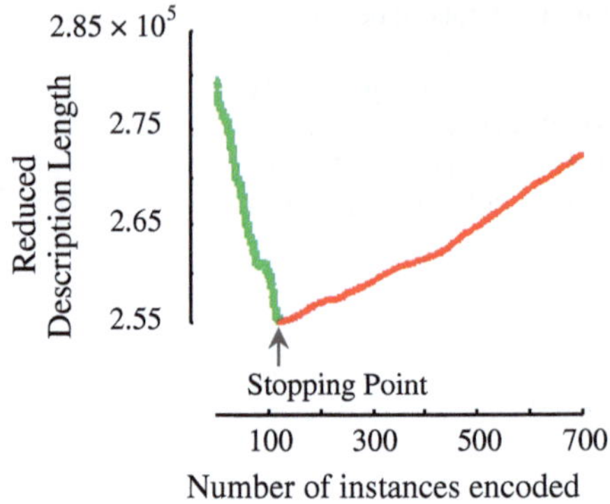

Holter records. Each record, sampled at 257 Hz, contains 12 standard leads. Over 175,000 beat annotations are stored in the reference annotation files. An automatic algorithm produced the annotations and then corrections were made manually. We randomly chose record I70 and signal II for our experiment. This dataset contains 126 Atrial Premature Beat, and our self-training classifier classifies 120 of them correctly.

Figure 14 shows that we obtained similar results to the last experiment.

As we augment our labeled set with TP instances, we obtain a decreasing trend of DL_S. The moment we exhaust the TP data and start ingesting TN instances, DL_S starts increasing. The result is obvious; the inclusion of similar patterns to the hypothesis pattern in the labeled set helps achieving further compression of the data, and we find decreasing DL_S. However, when we start including the TN instances in our labeled set, DL_S keep on increasing, because of the overhead associated with the encoding of the mismatched portions. From Fig. 14, we stop *exactly* at the instant where the TN instances start being included in the labeled set.

5.3 Sudden Cardiac Death Holter Database

This experiment classifies the heartbeats of the Sudden Cardiac Death Holter Database (SDDB) [13], which is archived by Physionet [11] to support research on a wide variety of substrates responsible for sudden cardiac death syndromes. Such sudden deaths happen to 400,000 Americans and millions more worldwide each year and therefore, is of great importance.

The Sudden Cardiac Death Holter Database [13] includes 18 patients with underlying sinus rhythm. All these patients had a sustained ventricular tachyarrhythmia, and most had an actual cardiac arrest. Out of the twenty three records, we randomly chose record number 52 and signal ECG (1st) for our experiment. This dataset contains 216 R-on-T Premature Ventricular Contraction, and our self-training classifier classifies 193 of them correctly.

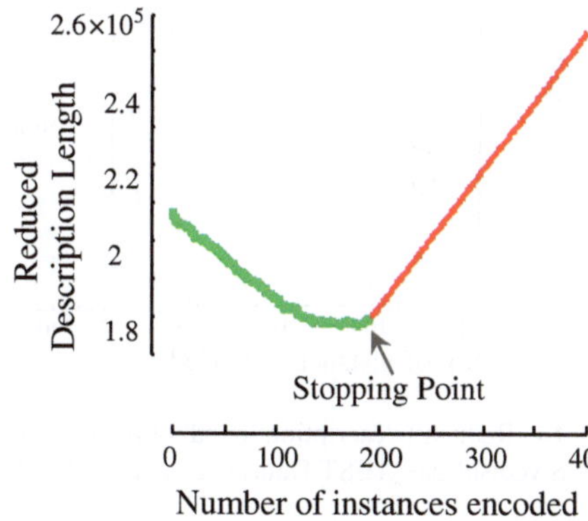

Fig. 15 Reduced description length of the labeled set versus number of encoded time series plot for the SDDB dataset

We show the experimental results for this dataset in Fig. 15. We obtain similar results to the two datasets described before. However for this dataset, the stopping point is not exactly at the point where the TN instances start. Beyond the stopping point, we obtain some TP instances, nevertheless, these are few in number, and the overall accuracy is not affected significantly.

5.4 Additional Examples

We conclude our experimental results for three additional datasets in Fig. 16. This datasets are not from a medical domain, and are designed to hint at the generality of our ideas.

The result in Fig. 16 *left* corresponds to the SwedishLeaf_TEST dataset [19]. For this dataset, beyond the stopping point, we obtain a tiny number of TP instances. For the Fish_TEST dataset [19], we stop exactly at the moment when the TN instances begin to be included in the labeled set (Fig. 16 *center*). However, beyond this point, we obtain *very* few TP instances, which do not hurt our classifier performance significantly. For the FaceAll_TEST dataset [19], we obtain some TP instances beyond our stopping point (Fig. 16 *right*). This is because we started with a *single* instance to seed in our training set, and this dataset is known to be polymorphic. Given just a single example, it is hard for the classifier to distinguish between the faces of the same person, let us say, *with* and *without glasses*. Nevertheless, our algorithm could still identify the point where the vast majority TP instances were classified already, and the TN instances started to be included in the training set.

5.5 Comparison with Rival Approaches

We perform a comparison between our algorithm and the state-of-the-art algorithm [41] on the Fish_TEST dataset [19]. From Fig. 17, the readers can easily observe that our algorithm suggests a better stopping point (Fig. 17 *bottom*) compared to the too conservative stopping point (Fig. 17 *top*) suggested by the state-of-the-art algorithm.

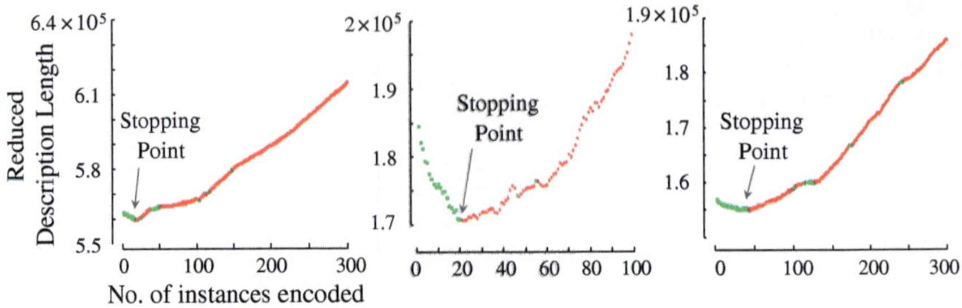

Fig. 16 Reduced description length of the labeled set versus number of encoded time series plots. *left*) SwedishLeaf_TEST Dataset. *center*) Fish_TEST Dataset. *right*) FaceAll_TEST Dataset

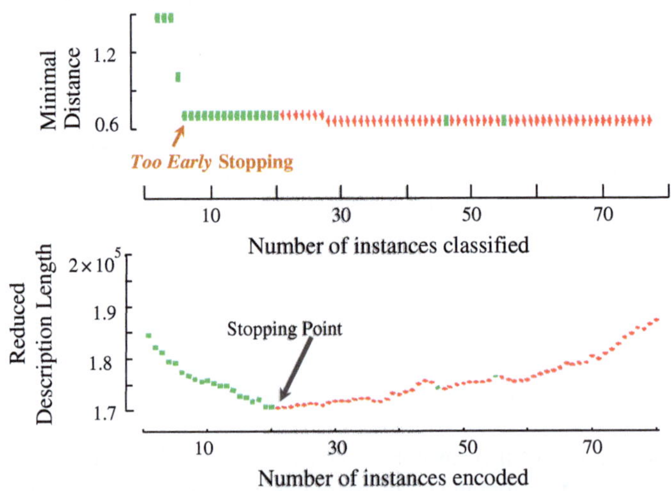

Fig. 17 Comparison of our algorithm (*bottom*) with the state-of-the-art algorithm (*top*) [41] for the Fish_TEST dataset [19]. Results show that the state-of-the-art algorithm stops too early, whereas our algorithm suggests better stopping point

Experiments on other datasets yielded similar results. In Fig. 18, we show the results for SVDB, INCARTDB and SDDB datasets respectively.

From Fig. 18 *top*, we can see that for all the three datasets, the state-of-the-art algorithm [41] is too conservative and therefore, missed too many true positives. In contrast, our algorithm suggested a better stopping point (Fig. 18 *bottom*) with *almost* no prohibitive expansion of false positives or false negatives.

5.6 ROC Curve

We measure the performance of our self-training classifier by varying the discrimination threshold, which is the stopping point. We consider all instances before the stopping point as positive instances and all instances after the point as negative

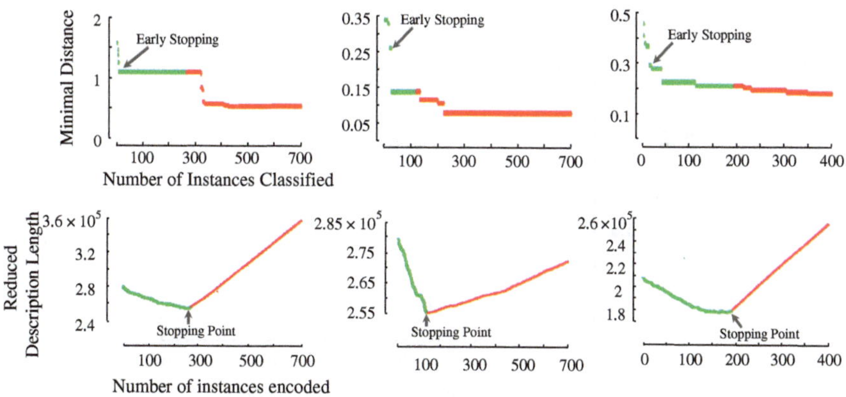

Fig. 18 Comparison of our algorithm (*bottom*) with the state-of-the-art algorithm (*top*) [41] for SVDB (*left*), INCARTDB (*center*) and SDDB (*right*) datasets

Fig. 19 ROC curve for the SVDB dataset. The stopping point denotes a point with majority of the TP instances identified, with almost no FP instances selected

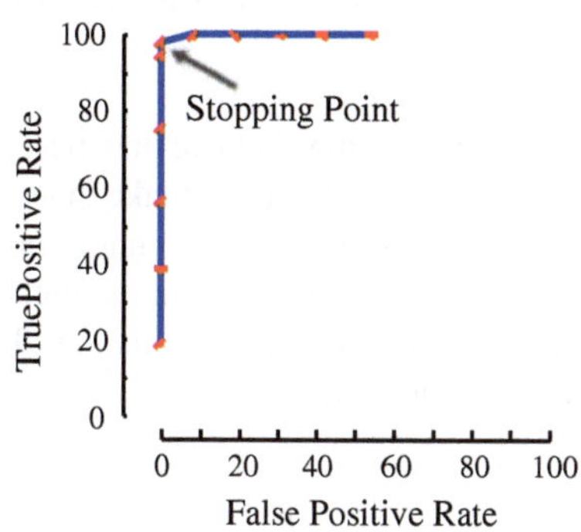

instances. Out of some potential stopping points, we look for the point where our algorithm actually suggests stopping the classification process. Experimental results show that our classifier stops at a point with majority of the TP instances identified with almost no FP instances selected (Fig. 19).

5.7 When does the Algorithm Fail?

Our MDL based algorithm for finding the stopping criterion of a self-training classifier does not perform well when the patterns of the two classes under consideration look very similar. As an example, we show the results for the TwoLeadECG_TEST dataset [19] in Fig. 20.

From Fig. 20, we can see that because the positive and negative instances are visually very similar, therefore, the expected increase in the Reduced Description Length separating the two class boundaries is not obvious in this case.

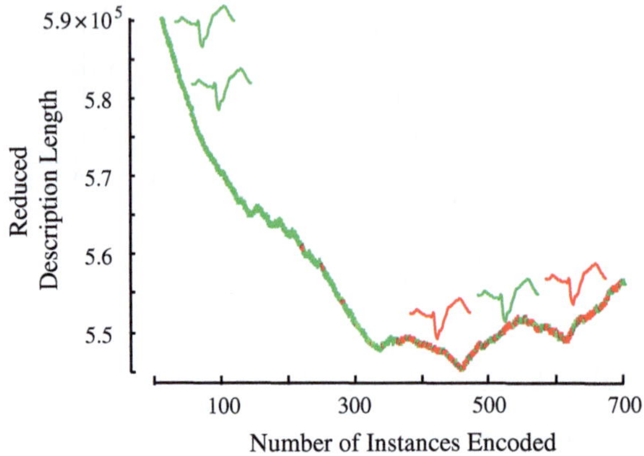

Fig. 20 Reduced description length of the labeled set versus number of encoded time series plot for the TwoLeadECG_TEST dataset. The positive/*green* and negative/*red* instances are very similar looking, and therefore the MDL based approach does not work well

The reader may believe that there must be *some* difference that allowed the original annotator of the data to make class distinctions in the first place. However, it must be noted that the original annotator has access to additional information, including other telemetry recorded in parallel, which is not available to our algorithm. It is not clear if even an expert human could do better than our algorithm, given the same view of the data.

6 Conclusions and Future Work

In this chapter, we proposed a novel method of semi-supervised classification of time series with as few as a single labeled instance. Previous approaches for stopping the classification process required extensive parameter tuning, and hence remain something of a black art. We devised a stopping criterion for the semi-supervised classification based on MDL, which is parameter free, and leverages the intrinsic structure of the data. To our knowledge, this is the *first* work addressing time series SSL using MDL. Extensive experiments demonstrate that our method allows us to classify real-world medical datasets with near perfect accuracy.

Our current approach works only on offline time series data; we plan extending it to perform online semi-supervised time series classification. We also plan to generalize our algorithm to the setting of multiple classes.

Acknowledgments This research was funded by NSF grant IIS—1161997.

References

1. Besemer, J., Lomsadze, A., Borodovsky, M.: GeneMarkS: a self-training method for prediction of gene starts in microbial genomes. Implications for finding sequence motifs in regulatory regions. Nucleic Acids Res. **29**(12), 2607–2618 (2001)
2. Blum, A., Mitchell, T.: Combining labeled and unlabeled data with co-training. In: Proceedings of the 11th ACM Annual Conference on Computational Learning Theory, pp. 92–100 (1998)
3. Bouchard, D., Badler, N.: Semantic segmentation of motion capture using Laban movement analysis. In: Intelligent Virtual Agents, pp. 37–44. Springer, Heidelberg (2007)
4. Chapelle, O., Schölkopf, B., Zien, A.: Semi-Supervised Learning, vol. 2. MIT press, Cambridge (2006)
5. Chazal, P.D., O'Dwyer, M., Reilly, R.B.: Automatic classification of heartbeats using ECG morphology and heartbeat interval features. IEEE Trans. Biomed. Eng. **51**, 1196–1206 (2004)
6. Chen, Y., Hu, B., Keogh, E., Batista, G.E.: DTW-D: time series semi-supervised learning from a single example. In: The 19th ACM SIGKDD, pp. 383–391 (2013)
7. Demšar, J.: Statistical comparisons of classifiers over multiple data sets. J. Mach. Learn. Res. **7**, 1–30 (2006)
8. Druck, G., Pal, C., Zhu, X., McCallum, A.: Semi-supervised classification with hybrid generative/discriminative methods. In: The 13th ACM SIGKDD (2007)
9. Florea, F., Müller, H., Rogozan, A., Geissbuhler, A., Darmoni, S.: Medical image categorization with MedIC and MedGIFT. In: Medical Informatics Europe (MIE) (2006)
10. Geurts, P.: Pattern extraction for time series classification. In: Proceedings of the 5th European Conference on Principles of Data Mining and Knowledge Discovery, pp. 115–127 (2001)
11. Goldberger, A.L., et al.: PhysioBank, PhysioToolkit, and PhysioNet: components of a new research resource for complex physiologic signals. Circulation **101**(23), 215–220 (2000)
12. Greenwald, S.D., Patil, R.S., Mark, R.G.: Improved detection and classification of arrhythmias in noise-corrupted electrocardiograms using contextual information. In: Proceedings of IEEE Conference on Computing in Cardiology (1990)
13. Greenwald, S.D.: The Development and Analysis of a Ventricular Fibrillation Detector. M.S. thesis, MIT Department of Electrical Engineering and Computer Science, Cambridge (1986)
14. Grünwald, P.: A Tutorial Introduction to the Minimum Description Length Principle. MIT Press, Cambridge (2005)
15. Herwig, M.: Google's Total Library: Putting the World's Books on the Web (2007)
16. Hills, J., Lines, J., Baranauskas, E., Mapp, J., Bagnall, A.: Classification of time series by shapelet transformation. Data Min. Knowl. Disc. **2**, 1–31 (2013)
17. Hu, B., Rakthanmanon, T., Hao, Y., Evans, S., Lonardi, S., Keogh, E.: Discovering the intrinsic cardinality and dimensionality of time series using MDL. In: Proceedings of ICDM, pp. 1086–1091 (2011)
18. Jones, P.D., Hulme, M.: Calculating regional climatic time series for temperature and precipitation: methods and illustrations. Int. J. Climatol. **16**(4), 361–377 (1996)
19. Keogh, E., Zhu, Q., Hu, B., Hao, Y., Xi, X., Wei, L., Ratanamahatana, C.A.: The UCR Time Series Classification/Clustering. www.cs.ucr.edu/~eamonn/time_series_data
20. Li, M., Vitanyi, P.: An Introduction to Kolmogorov Complexity and Its Applications, 2nd edn. Springer, New York (1997)
21. Maeireizo, B., Litman, D., Hwa, R.: Co-training for predicting emotions with spoken dialogue data. In: Proceedings of ACL (2004)
22. McClosky, D., Charniak, E., Johnson, M.: Effective self-training for parsing. In: Proceedings of the Main Conference on Human Language Technology and Conference of the North American Chapter of the Association of Computational Linguistics, pp. 152–159 (2006)
23. Nemenyi, P.B.: Distribution-free Multiple Comparisons. PhD Thesis, Princeton University (1963)
24. Nguyen, M.N., Li, X.L., Ng, S.K.: Positive unlabeled learning for time series classification. In: Proceedings of AAAI (2011)

25. Nguyen, M.N., Li, X.L., Ng, S.K.: Ensemble Based Positive Unlabeled Learning for Time Series Classification. Database Systems for Advanced Applications. Springer, Heidelberg (2012)
26. Ordonez, P., Oates, T., Lombardi, M.E., Hernandez, G., Holmes, K.W., Fackler, J., Lehmann, C.U.: Visualization of multivariate time-series data in a Neonatal ICU. IBM J. Res. Dev. **56**(5), 7–1 (2012)
27. Patton, A.J.: Copula-based models for financial time series. In: Andersen, T.G., Davis, R.A., Kreiss, J-P., Mikosch, T. (eds.) Handbook of Financial Time Series, pp. 767–785. Springer, Heidelberg (2009)
28. Philipose, M.: Large-Scale Human Activity Recognition Using Ultra-Dense Sensing. The Bridge, vol. 35, issue 4. National Academy of Engineering, Winter (2005)
29. Radovanovic, M., Nanopoulos, A., Ivanovic, M.: Time-series classification in many intrinsic dimensions. In: Proceedings of SIAM SDM, pp. 677–688 (2010)
30. Rakthanmanon, T., Keogh, E., Lonardi, S., Evans, S.: Time series epenthesis: clustering time series streams requires ignoring some data. In: Proceedings of ICDM (2011)
31. Raptis, M., Wnuk, K., Soatto, S.: Flexible dictionaries for action classification. In: The 1st International Workshop on Machine Learning for Vision-based Motion Analysis (2008)
32. Ratanamahatana, C.A., Keogh, E.: Making time-series classification more accurate using learned constraints. In: Proceedings of SIAM SDM (2004)
33. Ratanamahatana, C.A., Wanichsan, D.: Stopping criterion selection for efficient semi-supervised time series classification. In: Lee, R.Y. (ed.) Software Engineering, Artificial Intelligence, Networking and Parallel/Distributed Computing, Studies in Computational Intelligence, vol. 149, pp. 1–14. Springer (2008)
34. Rosenberg, C., Hebert, M., Schneiderman, H.: Semi-supervised Self-training of Object Detection Models. WACV/MOTION, 29–36 (2005)
35. Simon, B.P., Eswaran, C.: An ECG classifier designed using modified decision based neural networks. Comput. Biomed. Res. **30**(4), 257–272 (1997)
36. Sun, A., Grishman, R.: Semi-supervised semantic pattern discovery with guidance from unsupervised pattern clusters. In: Proceedings of the 23rd International Conference on Computational Linguistics: Posters, pp. 1194–1202 (2010)
37. Sykacek, P., Roberts, S.J.: Bayesian time series classification. In: Jordan, M., Reams, M., Solla, S. (eds.) Advances in Neural Information Processing Systems. MIT Press, Cambridge (2002)
38. Tsumoto, S.: Rule discovery in large time-series medical databases. In: In: Zytkow, J., Rauch, J. (eds.) Principles of Data Mining and Knowledge Discovery, pp. 23–31. Springer, Heidelberg (1999)
39. Veeraraghavan, A., Chellappa, R., Srinivasan, M.: Shape and behavior encoded tracking of bee dances. IEEE Trans. Pattern Anal. Mach. Intell. **30**(3), 463–476 (2008)
40. Wang, X., Mueen, A., Ding, H., Trajcevski, G., Scheuermann, P., Keogh, E.J.: Experimental comparison of representation methods and distance measures for time series data. Data Min. Knowl. Discov. **26**(2), 275–309 (2013)
41. Wei, L., Keogh, E.: Semi-supervised time series classification. In: Proceedings of SIGKDD (2006)
42. Xi, X., Keogh, E., Shelton, C., Wei, L., Ratanamahatana, C.A.: Fast time series classification using numerosity reduction. In: Proceedings of the 23rd ACM International Conference on Machine Learning, pp. 1033–1040 (2006)
43. Yarowsky, D.: Unsupervised word sense disambiguation rivaling supervised methods. In: Proceedings of ACL (1995)
44. Zhu, X.: Semi-supervised Learning Literature Survey. Technical Report No. 1530. Computer Sciences, University of Wisconsin-Madison (2005)
45. http://www.cs.ucr.edu/~nbegu001/SSL_myMDL.htm

Interpreting Random Forest Classification Models Using a Feature Contribution Method

Anna Palczewska, Jan Palczewski, Richard Marchese Robinson and Daniel Neagu

Abstract Model interpretation is one of the key aspects of the model evaluation process. The explanation of the relationship between model variables and outputs is relatively easy for statistical models, such as linear regressions, thanks to the availability of model parameters and their statistical significance. For "black box" models, such as random forest, this information is hidden inside the model structure. This work presents an approach for computing feature contributions for random forest classification models. It allows for the determination of the influence of each variable on the model prediction for an individual instance. By analysing feature contributions for a training dataset, the most significant variables can be determined and their typical contribution towards predictions made for individual classes, i.e., class-specific feature contribution "patterns", are discovered. These patterns represent a standard behaviour of the model and allow for an additional assessment of the model reliability for new data. Interpretation of feature contributions for two UCI benchmark datasets shows the potential of the proposed methodology. The robustness of results is demonstrated through an extensive analysis of feature contributions calculated for a large number of generated random forest models.

A. Palczewska (✉) · D. Neagu
Department of Computing, University of Bradford, Bradford, BD7 1DP, UK
e-mail: a.m.wojak@bradford.ac.uk

D. Neagu
e-mail: d.neagu@bradford.ac.uk

J. Palczewski
School of Mathematics, University of Leeds, Leeds, LS2 9JT, UK
e-mail: j.palczewski@leeds.ac.uk

R. M. Robinson
School of Pharmacy and Biomolecular Sciences, Liverpool John Moores University, Liverpool, L3 3AF, UK
e-mail: R.L.MarcheseRobinson@ljmu.ac.uk

T. Bouabana-Tebibel and S. H. Rubin (eds.), *Integration of Reusable Systems*, Advances in Intelligent Systems and Computing 263, DOI: 10.1007/978-3-319-04717-1_9,

Keywords Random forest · Classification · Variable importance · Feature contribution · Cluster analysis

1 Introduction

Models are used to discover interesting patterns in data or to predict a specific outcome, such as drug toxicity, client shopping purchases, or car insurance premium. They are often used to support human decisions in various business strategies. This is why it is important to ensure model quality and to understand its outcomes. Good practice of model development [1] involves: (1) data analysis (2) feature selection, (3) model building and (4) model evaluation. Implementing these steps together with capturing information on how the data was harvested, how the model was built and how the model was validated, allows us to trust that the model gives reliable predictions. But, how to interpret an existing model? How to analyse the relation between predicted values and the training dataset? Or which features contribute the most to classify a specific instance?

Answers to these questions are considered particularly valuable in such domains as chemoinformatics, bioinformatics or predictive toxicology [2]. Linear models, which assign instance-independent coefficients to all features, are the most easily interpreted. However, in the recent literature, there has been considerable focus on interpreting predictions made by non-linear models which do not render themselves to straightforward methods for the determination of variable/feature influence. In [3], the authors present a method for local interpretation of Support Vector Machine (SVM) and Random Forest models by retrieving the variable corresponding to the largest component of the decision-function gradient at any point in the model. Interpretation of classification models using local gradients is discussed in [4]. A method for visual interpretation of kernel-based prediction models is described in [5]. Another approach, which is presented in detail later, was proposed in [6] and aims at shedding light at decision-making process of regression random forests.

Of interest to this chapter is a popular "black-box" model—the random forest model [7]. Its author suggests two measures of the significance of a particular variable [8]: the variable importance and the Gini importance. The variable importance is derived from the loss of accuracy of model predictions when values of one variable are permuted between instances. Gini importance is calculated from the Gini impurity criterion used in the growing of trees in the random forest. However, in [9], the authors showed that the above measures are biased in favor of continuous variables and variables with many categories. They also demonstrated that the general representation of variable importance is often insufficient for the complete understanding of the relationship between input variables and the predicted value.

Following the above observation, Kuzmin et al. propose in [6] a new technique to calculate a feature contribution, i.e., a contribution of a variable to the prediction, in a random forest model. Their method applies to models generated for data with numerical observed values (the observed value is a real number). Unlike in the variable importance measures [8], feature contributions are computed separately for each instance/record. They provide detailed information about relationships between variables and the predicted value: the extent and the kind of influence (positive/negative) of a given variable. This new approach was positively tested in [6] on a Quantitative Structure-Activity (QSAR) model for chemical compounds. The results were not only informative about the structure of the model but also provided valuable information for the design of new compounds.

The procedure from [6] for the computation of feature contributions applies to random forest models predicting numerical observed values. This chapter aims to extend it to random forest models with categorical predictions, i.e., where the observed value determines one from a finite set of classes. The difficulty of achieving this aim lies in the fact that a discrete set of classes does not have the algebraic structure of real numbers which the approach presented in [6] relies on. Due to the high-dimensionality of the calculated feature contributions, their direct analysis is not easy. We develop three techniques for discovering class-specific feature contribution "patterns" in the decision-making process of random forest models: the analysis of median feature contributions, of clusters and log-likelihoods. This facilitates interpretation of model predictions as well as allows a more detailed analysis of model reliability for unseen data.

The chapter is organised as follows. Section 2 provides a brief description of random forest models. Section 3 presents our approach for calculating feature contributions for binary classifiers, whilst Sect. 4 describes its extension to multi-class classification problems. Section 5 introduces three techniques for finding patterns in feature contributions and linking them to model predictions. Section 6 contains applications of the proposed methodology to two real world datasets from the UCI Machine Learning repository. Section 7 concludes the work presented in this chapter.

2 Random Forest

A random forest (RF) model introduced by Breiman [7] is a collection of tree predictors. Each tree is grown according to the following procedure [8]:

1. the bootstrap phase: select randomly a subset of the training dataset—a local training set for growing the tree. The remaining samples in the training dataset form a so-called out-of-bag (OOB) set and are used to estimate the RF's goodness-of-fit.
2. the growing phase: grow the tree by splitting the local training set at each node according to the value of one variable from a randomly selected subset of variables (the best split) using classification and regression tree (CART) method [10].
3. each tree is grown to the largest extent possible. There is no pruning.

The bootstrap and growing phases require an input of random quantities. It is assumed that these quantities are independent between trees and identically distributed. Consequently, each tree can be viewed as sampled independently from the ensemble of all tree predictors for a given training dataset.

For prediction, an instance is run through each tree in a forest down to a terminal node which assigns it a class. Predictions supplied by the trees undergo a voting process: the forest returns the class with the maximum number of votes. Draws are resolved through a random selection.

To present our feature contribution procedure in the following section, we have to develop a probabilistic interpretation of the forest prediction process. Denote by $C = \{C_1, C_2, \ldots, C_K\}$ the set of classes and by Δ_K the set

$$\Delta_K = \Big\{(p_1, \ldots, p_K) : \sum_{k=1}^{K} p_k = 1 \quad \text{and} \quad p_k \geq 0\Big\}.$$

An element of Δ_K can be interpreted as a probability distribution over C. Let e_k be an element of Δ_K with 1 at position k—a probability distribution concentrated at class C_k. If a tree t predicts that an instance i belongs to a class C_k then we write $\hat{Y}_{i,t} = e_k$. This provides a mapping from predictions of a tree to the set Δ_K of probability measures on C. Let

$$\hat{Y}_i = \frac{1}{T} \sum_{t=1}^{T} \hat{Y}_{i,t}, \tag{1}$$

where T is the overall number of trees in the forest. Then $\hat{Y}_i \in \Delta_K$ and the prediction of the random forest for the instance i coincides with a class C_k for which the k-th coordinate of $\hat{Y}_i$ is maximal.[1]

3 Feature Contributions for Binary Classifiers

The set Δ_K simplifies considerably when there are two classes, $K = 2$. An element $p \in \Delta_K$ is uniquely represented by its first coordinate p_1 ($p_2 = 1 - p_1$). Consequently, the set of probability distributions on C is equivalent to the probability weight assigned to class C_1.

Before we present our method for computing feature contributions, we have to examine the tree growing process. After selecting a training set, it is positioned in the root node. A splitting variable (feature) and a splitting value are selected and the set of instances is split between the left and the right child of the root node. The procedure is repeated until all instances in a node are in the same class or further splitting does not improve prediction. The class that a tree assigns to a terminal node is determined through majority voting between instances in that node.

[1] The distribution $\hat{Y}_i$ is calculated by the function `predict` in the R package `randomForest` [11] when the type of prediction is set to `prob`.

We will refer to instances of the local training set that pass through a given node as the training instances in this node. The fraction of the training instances in a node n belonging to class C_1 will be denoted by Y^n_{mean}. This is the probability that a randomly selected element from the training instances in this node is in the first class. In particular, a terminal node is assigned to class C_1 if $Y^n_{mean} > 0.5$ or $Y^n_{mean} = 0.5$ and the draw is resolved in favor of class C_1.

The feature contribution procedure for a given instance involves two steps: (1) the calculation of local increments of feature contributions for each tree and (2) the aggregation of feature contributions over the forest. A local increment corresponding to a feature f between a parent node (p) and a child node (c) in a tree is defined as follows:

$$LI^c_f = \begin{cases} Y^c_{mean} - Y^p_{mean}, & \text{if the split in the parent is performed over the feature f,} \\ 0, & \text{otherwise.} \end{cases}$$

A local increment for a feature f represents the change of the probability of being in class C_1 between the child node and its parent node provided that f is the splitting feature in the parent node. It is easy to show that the sum of these changes, over all features, along the path followed by an instance from the root node to the terminal node in a tree is equal to the difference between Y_{mean} in the terminal and the root node.

The contribution $FC^f_{i,t}$ of a feature f in a tree t for an instance i is equal to the sum of LI_f over all nodes on the path of instance i from the root node to a terminal node. The contribution of a feature f for an instance i in the forest is then given by

$$FC^f_i = \frac{1}{T} \sum_{t=1}^{T} FC^f_{i,t}. \tag{2}$$

The feature contributions vector for an instance i consists of contributions FC^f_i of all features f.

Notice that if the following condition is satisfied:

(U) for every tree in the forest, local training instances in each terminal node are of the same class

then $\hat{Y}_i$ representing forest's prediction (1) can be written as

$$\hat{Y}_i = \left(Y^r + \sum_f FC^f_i,\ 1 - Y^r - \sum_f FC^f_i\right) \tag{3}$$

where Y^r is the coordinate-wise average of Y_{mean} over all root nodes in the forest. If this unanimity condition (U) holds, feature contributions can be used to retrieve predictions of the forest. Otherwise, they only allow for the interpretation of the model.

Table 1 Selected records from the UCI Iris Dataset. Each record corresponds to a plant. The plants were classified as iris versicolor (class 0) and virginica (class 1)

	iris.row	Sepal.Length (f1)	Sepal.Width (f2)	Petal.Length (f3)	Petal.Width (f4)	Class
x_1	52	6.4	3.2	4.5	1.5	0
x_2	73	6.3	2.5	4.9	1.5	0
x_3	75	6.4	2.9	4.3	1.3	0
x_4	90	5.5	2.5	4.0	1.3	0
x_5	91	5.5	2.6	4.4	1.2	0
x_6	136	7.7	3.0	6.1	2.3	1
x_7	138	6.4	3.1	5.5	1.8	1
x_8	139	6.0	3.0	4.8	1.8	1
x_9	145	6.7	3.3	5.7	2.5	1
x_{10}	148	6.5	3.0	5.2	2.0	1

3.1 Example

We will demonstrate the calculation of feature contributions on a toy example using a subset of the UCI Iris Dataset [12]. From the original dataset, ten records were selected—five for each of two types of the iris plant: versicolor (class 0) and virginica (class 1) (see Table 1). A plant is represented by four attributes: Sepal.Length (f1), Sepal.Width (f2), Petal.Length (f3) and Petal.Width (f4). This dataset was used to generate a random forest model with two trees, see Fig. 1. In each tree, the local training set (LD) in the root node collects those records which were chosen by the random forest algorithm to build that tree. The LD sets in the child nodes correspond to the split of the above set according to the value of a selected feature (it is written between branches). This process is repeated until reaching terminal nodes of the tree. Notice that the condition **(U)** is satisfied—for both trees, each terminal node contains local training instances of the same class: Y_{mean} is either 0 or 1.

The process of calculating feature contributions runs in 2 steps: the determination of local increments for each node in the forest (a preprocessing step) and the calculation of feature contributions for a particular instance. Figure 1 shows Y^n_{mean} and the local increment LI^c_f for a splitting feature f in each node. Having computed these values, we can calculate feature contributions for an instance by running it through both trees and summing local increments of each of the four features. For example, the contribution of a given feature for the instance x_1 is calculated by summing local increments for that feature along the path $p_1 = n_0 \rightarrow n_1$ in tree T_1 and the path $p_2 = n_0 \rightarrow n_1 \rightarrow n_4 \rightarrow n_5$ in tree T_2. According to Formula (2) the contribution of feature f2 is calculated as

$$FC^{f2}_{x_1} = \frac{1}{2}\left(0 + \frac{1}{4}\right) = 0.125$$

and the contribution of feature f3 is

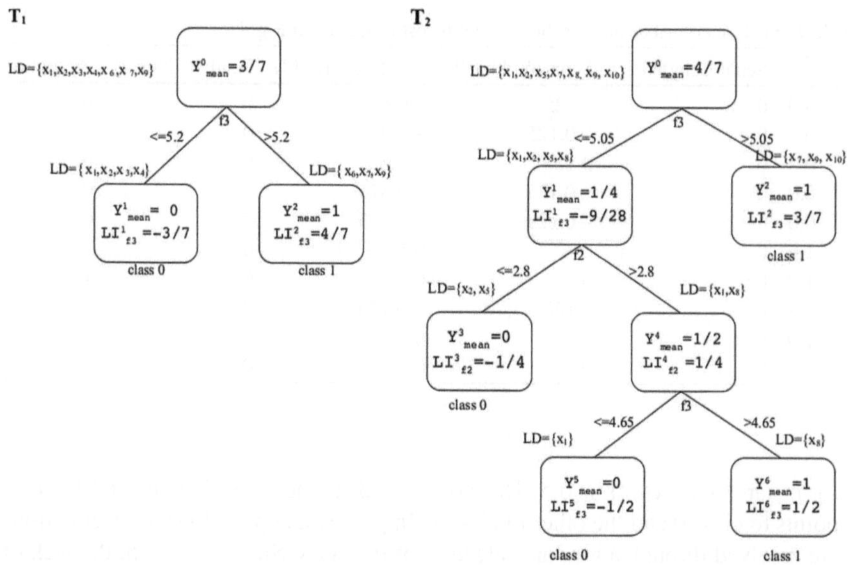

Fig. 1 A random forest model for the dataset from Table 1. The set LD in the root node contains a local training set for the tree. The sets LD in the child nodes correspond to the split of the above set according to the value of selected feature. In each node, Y^n_{mean} denotes the fraction of instances in the LD set in this node belonging to class 1, whilst LI^n_f shows non-zero local increments

$$FC^{f3}_{x_1} = \frac{1}{2}\Big(-\frac{3}{7} - \frac{9}{28} - \frac{1}{2}\Big) = -0.625.$$

The contributions of features f1 and f4 are equal to 0 because these attributes are not used in any decision made by the forest. The predicted probability $\hat{Y}_{x_1}$ that x_1 belongs to class 1 (see Formula (3)) is

$$\hat{Y}_{x_1} = \underbrace{\frac{1}{2}\Big(\frac{3}{7} + \frac{4}{7}\Big)}_{\hat{Y}^r} + \underbrace{(0 + 0.125 - 0.625 + 0)}_{\sum_f FC^f_{x_1}} = 0.0$$

Table 2 collects feature contributions for all 10 records in the example dataset. These results can be interpreted as follows:

- for instances x_1, x_3, the contribution of f2 is positive, i.e., the value of this feature increases the probability of being in class 1 by 0.125. However, the large negative contribution of the feature f3 implies that the value of this feature for instances x_1 and x_3 was decisive in assigning the class 0 by the forest.
- for instances x_6, x_7, x_9, the decision is based only on the feature f3.
- for instances x_2, x_4, x_5, the contribution of both features leads the forest decision towards class 0.

Table 2 Feature contributions for the random forest model from Fig. 1

	$\hat{Y}$	Sepal.Length (f1)	Sepal.Width (f2)	Petal.Length (f3)	Petal.Width (f4)	Prediction
x_1	0.0	0	0.125	−0.625	0	0
x_2	0.0	0	−0.125	−0.375	0	0
x_3	0.0	0	0.125	−0.625	0	0
x_4	0.0	0	−0.125	−0.375	0	0
x_5	0.0	0	−0.125	−0.375	0	0
x_6	1.0	0	0	0.5	0	1
x_7	1.0	0	0	0.5	0	1
x_8	0.5	0	0.125	−0.125	0	?
x_9	1.0	0	0	0.5	0	1
x_{10}	0.5	0	0	0	0	?

- for instances x_8, x_{10}, $\hat{Y}$ is 0.5. This corresponds to the case where one of the trees points to class 0 and the other to class 1. In practical applications, such situations are resolved through a random selection of the class. Since $\hat{Y^r} = 0.5$, the lack of decision of the forest has a clear interpretation in terms of feature contributions: the amount of evidence in favour of one class is counterbalanced by the evidence pointing towards the other.

4 Feature Contributions for General Classifiers

When $K > 2$, the set Δ_K cannot be described by a one-dimensional value as above. We, therefore, generalize the quantities introduced in the previous section to a multi-dimensional case. Y^n_{mean} in a node n is an element of Δ_K, whose k-th coordinate, $k = 1, 2, \ldots, K$, is defined as

$$Y^n_{mean,k} = \frac{|\{i \in TS(n) : i \in C_k\}|}{|TS(n)|}, \tag{4}$$

where $TS(n)$ is the set of training instances in the node n and $|\cdot|$ denotes the number of elements of a set. Hence, if an instance is selected randomly from a local training set in a node n, the probability that this instance is in class C_k is given by the k-th coordinate of the vector Y^n_{mean}. Local increment LI^c_f is analogously generalized to a multidimensional case:

$$LI^c_f = \begin{cases} Y^c_{mean} - Y^p_{mean}, & \text{if the split in the parent is performed over the feature f,} \\ \underbrace{(0, \ldots, 0)}_{K \text{ times}}, & \text{otherwise,} \end{cases}$$

where the difference is computed coordinate-wise. Similarly, $FC_{i,t}^{f}$ and FC_{i}^{f} are extended to vector-valued quantities. Notice that if the condition (U) is satisfied, Eq. (3) holds with Y^{r} being a coordinate-wise average of vectors Y_{mean}^{n} over all root nodes n in the forest.

Take an instance i and let C_k be the class to which the forest assigns this instance. Our aim is to understand which variables/features drove the forest to make that prediction. We argue that the crucial information is the one which explains the value of the k-th coordinate of $\hat{Y}_i$. Hence, we want to study the k-th coordinate of FC_{i}^{f} for all features f.

Pseudo-code to calculate feature contributions for a particular instance towards the class predicted by the random forest is presented in Algorithm 1. Its inputs consist of a random forest model RF and an instance i which is represented as a vector of feature values. In line 1, $k \in \{1, 2, \ldots, K\}$ is assigned the index of a class predicted by the random forest RF for the instance i. The following line creates a vector of real numbers indexed by features and initialized to 0. Then for each tree in the forest RF the instance i is run down the tree and feature contributions are calculated. The quantity $SplitFeature(parent)$ identifies a feature f on which the split is performed in the node $parent$. If the value $i(f)$ of that feature f for the instance i is lower or equal to the threshold $SplitValue(parent)$, the route continues to the left child of the node $parent$. Otherwise, it goes to the right child (each node in the tree has either two children or is a terminal node). A position corresponding to the feature f in the vector FC is updated according to the change of value of $Y_{mean,k}$, i.e., the k-th coordinate of Y_{mean}, between the parent and the child.

Algorithm 1 FC(RF,i)

1: $k \leftarrow forest\_predict(RF, i)$
2: $FC \leftarrow vector(features)$
3: **for** each tree T in forest F **do**
4: $\quad parent \leftarrow root(T)$
5: $\quad$ **while** $parent$!= TERMINAL **do**
6: $\quad\quad f \leftarrow SplitFeature(parent)$
7: $\quad\quad$ **if** $i[f] <= SplitValue(parent)$ **then**
8: $\quad\quad\quad child \leftarrow leftChild(parent)$
9: $\quad\quad$ **else**
10: $\quad\quad\quad child \leftarrow rightChild(parent)$
11: $\quad\quad$ **end if**
12: $\quad\quad FC[f] \leftarrow FC[f] + Y_{mean,k}^{child} - Y_{mean,k}^{parent}$
13: $\quad\quad parent \leftarrow child$
14: $\quad$ **end while**
15: **end for**
16: $FC \leftarrow FC$ / nTrees(F)
17: **return** FC

Algorithm 2 provides a sketch of the preprocessing step to compute Y_{mean}^{n} for all nodes n in the forest. The parameter D denotes the set of instances used for training of the forest RF. In line 2, TS is assigned the set used for growing tree T. This set is further split in nodes according to values of splitting variables. We propose to use DFS (depth first search [13]) to traverse the tree and compute the vector Y^n_{mean} once a training set for a node n is determined. There is no need to store a training set for a node n once Y^n_{mean} has been calculated.

Algorithm 2 $Y_{mean}(RF, D)$

1: **for** each tree T in forest F **do**
2: $TS \leftarrow$ training set for tree T
3: use DFS algorithm to compute training sets in all other nodes n of tree T and compute the vector Y^n_{mean} according to formula (4).
4: **end for**

5 Analysis of Feature Contributions

Feature contributions provide the means to understand mechanisms that lead the model towards particular predictions. This is important in chemical or biological applications where the additional knowledge of the forest's decision-making process can inform the development of new chemical compounds or explain their interactions with living organisms. Feature contributions may also be useful for assessing the reliability of model predictions for unseen instances. They provide complementary information to the forest's voting results. This section proposes three techniques for finding patterns in the way a random forest uses available features and linking these patterns with the forest's predictions.

5.1 Median

The median of a sequence of numbers is such a value that the number of elements bigger than it and the number of elements smaller than it is identical. When the number of elements in the sequence is odd, this is the central elements of the sequence. Otherwise, it is common to take the midpoint between the two most central elements. In statistics, the median is an estimator of the expectation which is less affected by outliers than the sample mean. We will use this property of the median to find a "standard level" of feature contributions for representatives of a particular class. This standard level will facilitate an understanding of which features are instrumental for the classification. It can also be used to judge the reliability of forest's prediction for an unseen instance.

For a given random forest model, we select those instances from the training dataset that are classified correctly. We calculate the medians of contributions of every feature separately for each class. The medians computed for one class are combined into a vector which is interpreted as providing the aforementioned "standard level" for this class. If most of the instances from the training dataset belonging to a particular class are close to the corresponding vector of medians, we may treat this vector justifiably as a standard level. When a prediction is requested for a new instance, we query the random forest model for the fraction of trees voting for each class and calculate feature contributions leading to its final prediction. If a high fraction of trees votes for a given class and the feature contributions are close to the standard level for this class, we may reasonably rely on the prediction. Otherwise we may doubt the random forest model prediction.

It may, however, happen that many instances from the training dataset correctly predicted to belong to a particular class are distant from the corresponding vector of medians. This might suggest that there is more than one standard level, i.e., there are multiple mechanisms relating features to correct classes. The next subsection presents more advanced methods capable of finding a number of standard levels—distinct patterns followed by the random forest model in its prediction process.

5.2 Cluster Analysis

Clustering is an approach for grouping elements/objects according to their similarity [14]. This allows us to discover patterns that are characteristic for a particular group. As we discussed above, feature contributions in one class may have more than one "standard level". When this is discovered, clustering techniques can be employed to find if there is a small number of distinct standard levels, i.e., feature contributions of the instances in the training dataset group around a few points with only a relatively few instances being far away from them. These few instances are then treated as unusual representatives of a given class. We shall refer to clusters of instances around these standard levels as "core clusters".

The analysis of core clusters can be of particular importance for applications. For example, in the classification of chemical compounds, the split into clusters may point to groups of compounds with different mechanisms of activity. We should note that the similarity of feature contributions does not imply that particular features are similar. We examined several examples and noticed that clustering based upon the feature values did not yield useful results whereas the same method applied to feature contributions was able to determine a small number of core clusters.

Figure 2 demonstrates the process of analysis of model reliability for a new instance using cluster analysis. In a preprocessing phase, feature contributions for instances in the training dataset are obtained. The optimal number of clusters for each class can be estimated by using, e.g., the Akaike information criterion (AIC), the Bayesian information criterion (BIC) or the Elbow method [14, 15]. We noticed that these methods should not be rigidly adhered to: their underlying assumption is that the data is clustered and we only have to determine the number of these clusters. As we argued above, we expect feature contributions for various instances to be grouped into a small number of clusters and we accept a reasonable number outliers interpreted as unusual instances for a given class. Clustering algorithms try to push

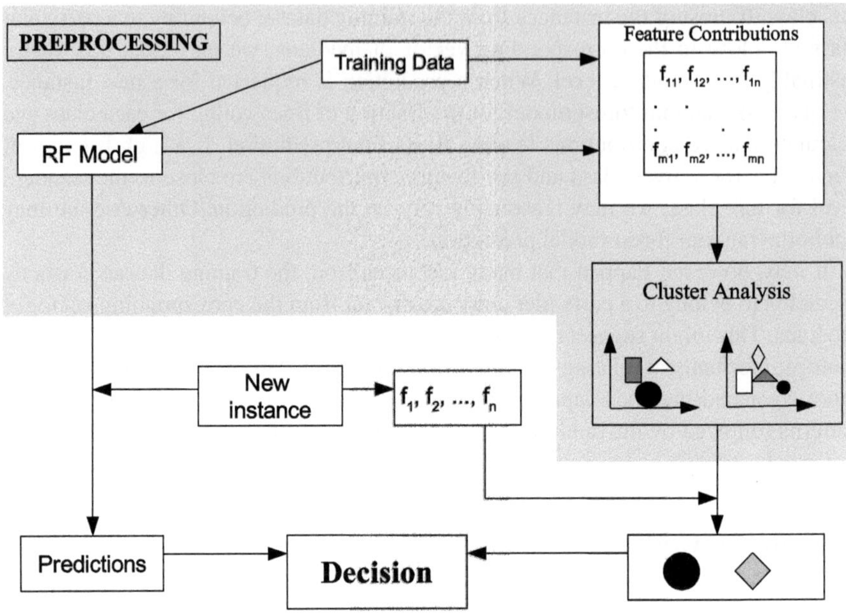

Fig. 2 The workflow for assessing the reliability of the prediction made by a random forest (RF) model

those outliers into clusters, hence increasing their number unnecessarily. We recommend, therefore, to treat the calculated optimal number of clusters as the maximum value and consecutively decrease it looking at the structure and performance of the resulting clusters: for each cluster we assess the average fraction of trees voting for the predicted class across the instances belonging to this cluster as well as the average distance from the centre of the cluster. Relatively large clusters with the former value close to 1 and the latter value small form the group of core clusters.

To assess the reliability of the model prediction for a new instance, we recommend looking at two measures: the fraction of trees voting for the predicted class as well as the cluster to which the instance is assigned based on its feature contributions. If the cluster is one of the core clusters and the distance from its centre is relatively small, the instance is a typical representative of its predicted class. This together with high decisiveness of the forest suggests that the model's prediction should be trusted. Otherwise, we should allow for an increased chance of misclassification.

5.3 Log-likelihood

Feature contributions for a given instance form a vector in a multi-dimensional Euclidean space. Using a popular k-mean clustering method, for each class we divide vectors corresponding to feature contributions of instances in the training dataset into

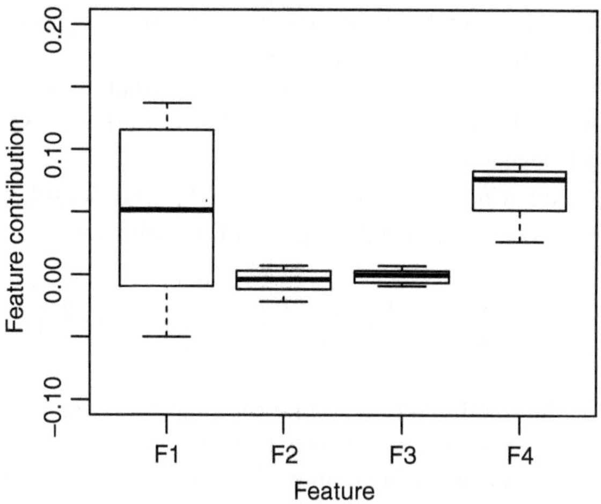

Fig. 3 The box-plot for feature contributions within a core cluster for a hypothetical random forest model

groups minimizing the Euclidean distance from the centre in each group. Figure 3 shows a box-plot of feature contributions for instances in a core cluster in a hypothetical random forest model. Notice that some features are stable within a cluster—the height of the box is small. Others (F1 and F4) display higher variability. One would therefore expect that the same divergence of contributions for features F3 and F4 from their mean value should be treated differently. It is more significant for the feature F3 than for the feature F4. This is unfortunately not taken into account when the Euclidean distance is considered. Here, we propose an alternative method for assessing the distance from the cluster centre which takes into account the variation of feature contributions within a cluster. Our method has probabilistic roots and we shall present it first from a statistical point of view and provide other interpretations afterwards.

We assume that feature contributions for instances within a cluster share the same base values (μ_f)—the centre of the cluster. We attribute all discrepancies between this base value and the actual feature contributions to a random perturbation. These perturbations are assumed to be normally distributed with the mean 0 and the variance σ_f^2, where f denotes the feature. The variance of the perturbation for each feature is selected separately—we use the sample variance computed from feature contributions of instances of the training dataset belonging to this cluster. Although it is clear that perturbations for different features exhibit some dependence, it is impossible to assess it given the number of instances in a cluster and a large number of features typically in use.[2] Therefore, we resort to a common solution: we assume

[2] A covariance matrix of feature contributions has $F(F+1)/2$ distinct entries, where F is the number of features. This value is usually larger than the size of a cluster making it impossible to retrieve useful information about the dependence structure of feature contributions. Application of more advanced methods, such as principal component analysis, is left for future research.

that the dependence between perturbations is small enough to justify treating them as independent. Summarising, our statistical model for the distribution of feature contributions within a cluster is as follows: feature contributions for instances within a cluster are composed of a base value and a random perturbation which is normally distributed and independent between features.

Take an instance i with feature contributions FC_i^f. The log-likelihood of being in a cluster with the centre (μ_f) and variances of perturbations (σ_f^2) is given by

$$LL_i = \sum_f \Big(- \frac{(FC_i^f - \mu_f)^2}{2\sigma_f^2} - \frac{1}{2}\log(2\pi\sigma_f^2)\Big). \tag{5}$$

The higher the log-likelihood the bigger the chance of feature contributions of the instance i to belong to the cluster. Notice that the above sum takes into account the observations we made at the beginning of this subsection. Indeed, as the second term in the sum above is independent of the considered instance, the log-likelihood is equivalent to

$$\sum_f \Big(- \frac{(FC_i^f - \mu_f)^2}{2\sigma_f^2}\Big),$$

which is the negative of the squared weighted Euclidean distance between FC_i^f and μ_f. The weights are inversely proportional to the variability of a given feature contribution in the training instances in the cluster. In our toy example of Fig. 3, this corresponds to penalizing more for discrepancies for features F2 and F3, and significantly less for discrepancies for features F1 and F4.

In the following section, we analyse relations between the log-likelihood and classification for a UCI Breast Cancer Wisconsin Dataset.

6 Applications

In this section, we demonstrate how the techniques from the previous section can be applied to improve understanding of a random forest model. We consider one example of a binary classifier using the UCI Breast Cancer Wisconsin Dataset [16] (BCW Dataset) and one example of a general classifier for the UCI Iris Dataset [12]. We complement our studies with a robustness analysis.

6.1 Breast Cancer Wisconsin Dataset

The UCI Breast Cancer Wisconsin Dataset contains characteristics of cell nuclei for 569 breast tissue samples; 357 are diagnosed as benign and 212 as malignant. The characteristics were captured from a digitized image of a fine needle aspirate

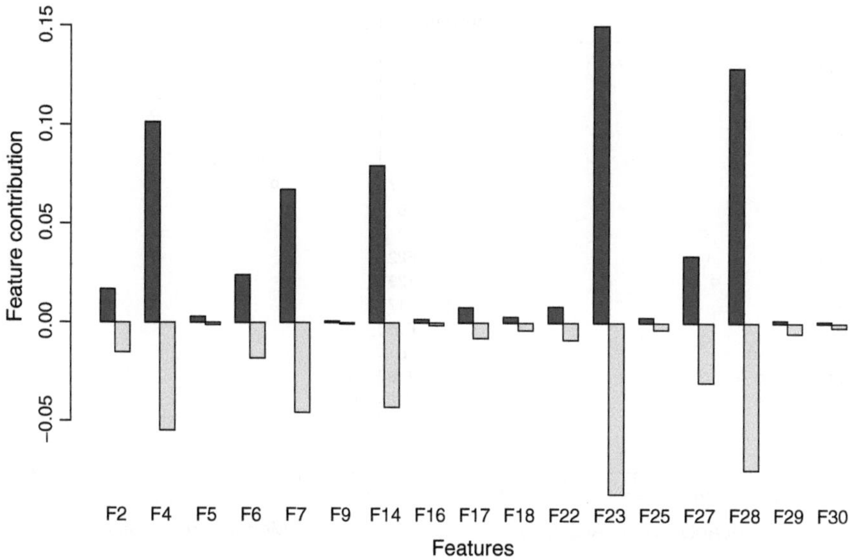

Fig. 4 Medians of feature contributions for each class for the BCW Dataset. The *light grey bars* represent contributions for class 0 and the *black bars* show contributions for class 1

(FNA) of a breast mass. There are 30 features, three (the mean, the standard error and the average of the three largest values) for each of the following 10 characteristics: radius, texture, perimeter, area, smoothness, compactness, concavity, concave points, symmetry and fractal dimension. For brevity, we numbered these features from $F1$ to $F30$ according to their order in the data file.

To reduce correlation between features and facilitate model interpretation, the min-max (minimal-redundancy-maximal-relevance) method was applied and the following features were removed from the dataset: 1, 3, 8, 10, 11, 12, 13, 15, 19, 20, 21, 24, 26. A random forest model was generated on 2/3 randomly selected instances using 500 trees. The other 1/3 of instances formed the testing dataset. The validation showed that the model accuracy was 0.9682 (only 6 instances out of 189 were classified incorrectly); similar accuracy was achieved when the model was generated using all the features.

We applied our feature contribution algorithm to the above random forest binary classifier. To align notation with the rest of the paper, we denote the class "malignant" by 1 and the class "benign" by 0. Aggregate results for the feature contributions for all training instances and both classes are presented in Fig. 4. Light-grey bars show medians of contributions for instances of class 0, whereas black bars show medians of contributions for instances of class 1 (malignant). Notice that there are only a few significant features in the graph: F4—the mean of the cell area, F7—the mean of the cell concavity, F14—the standard deviation of the cell area, F23—the average of three largest measurements of the cell perimeter and F28—the average of three largest

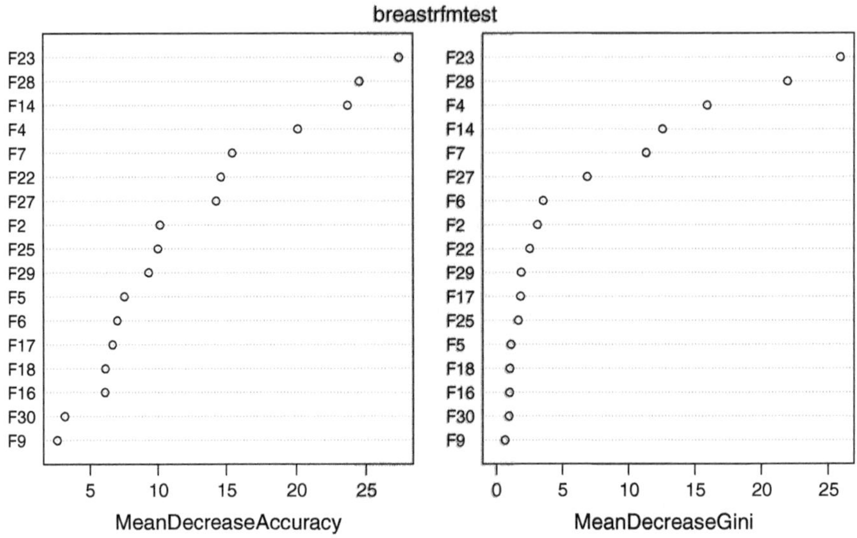

Fig. 5 The *left panel* shows permutation based variable importance and the *right panel* displays Gini importance for a RF binary classification model developed for the BCW Dataset. Graphs generated using randomForest package in R

measurements of concave points. This selection of significant features is perfectly in agreement with the results of the permutation based variable importance (the left panel of Fig. 5) and the Gini importance (the right panel of Fig. 5). Interpreting the size of bars as the level of importance of a feature, our results are in line with those provided by the Gini index. However, the main advantage of the approach presented in this chapter lies in the fact that one can study the reasons for the forest's decision for a *particular instance*.

Comparison of feature contributions for a particular instance with medians of feature contributions for all instances of one class provides valuable information about the forest's prediction. Take an instance predicted to be in class 1. In a typical case when the large majority of trees votes for class 1 the feature contributions for that instance are very close to the median values (see Fig. 6a). This happens for around 80% of all instances from the testing dataset predicted to be in class 1. However, when the decision is less unanimous, the analysis of feature contributions may reveal interesting information. As an example, we have chosen instances 194 and 537 (see Table 3) which were classified correctly as malignant (class 1) by a majority of trees but with a significant number of trees expressing an opposite view. Figure 6b presents feature contributions for these two instances (grey and light grey bars) against the median values for class 1 (black bars). The largest differences can be seen for the contributions of very significant features F23, F4 and F14: it is highly negative for the two instances under consideration compared to a large positive value commonly found in instances of class 1. Recall that a negative value

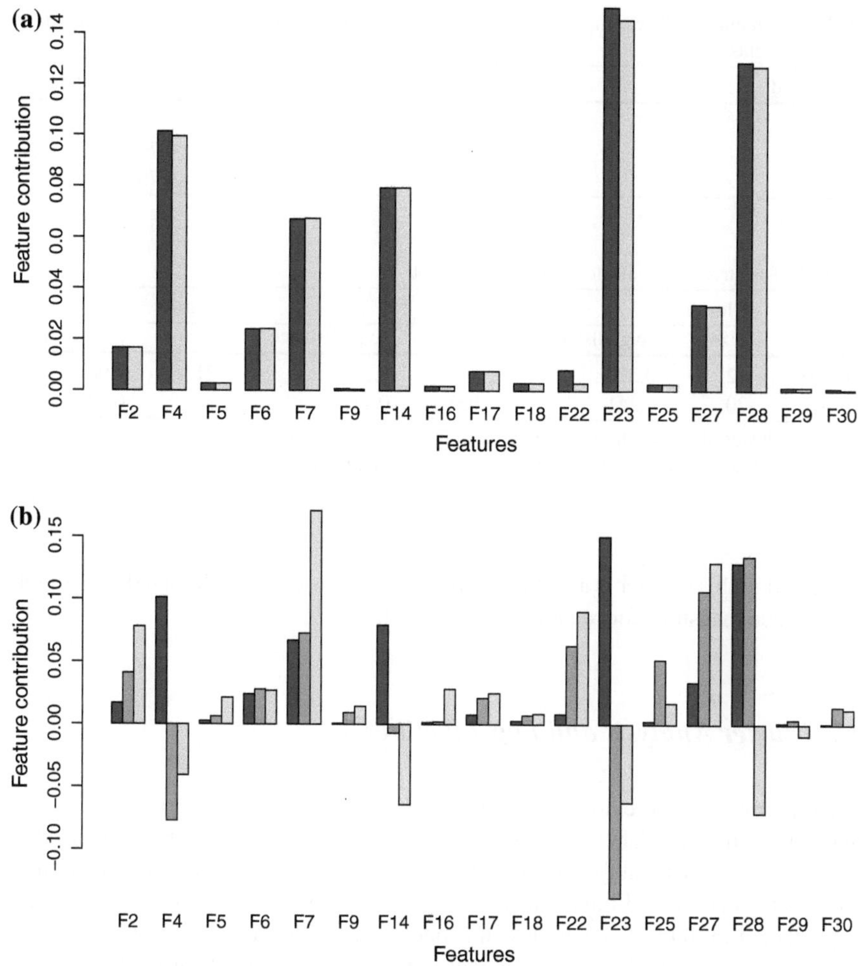

Fig. 6 Comparison of the medians of feature contributions (toward class 1) over all instances of class 1 (*black bars*) with **a** feature contributions for instance number 3 (*light-grey bars*) **b** feature contributions for instances number 194 (*grey bars*) and 537 (*light-grey bars*) from the BCW Dataset. The fractions of trees voting for class 0 and 1 for these three instances are collected in Table 3

contributes towards the classification in class 0. There are also three new significant attributes (F2, F22 and F27) that contribute towards the correct classification as well as unusual contributions for features F7 and F28. These newly significant features are judged as only moderately important by both of the variable importance methods in Fig. 5. It is, therefore, surprising to note that the contribution of these three new features was instrumental in correctly classifying instances 194 and 537 as malignant. This highlights the fact that features which may not generally be important for the model may, nonetheless, be important for classifying specific instances. The approach

Table 3 Percentage of trees that vote for each class in RF model for a selection of instances from the BCW Dataset

Instance Id	Benign (class 0)	Malignant (class 1)
3	0	1
194	0.298	0.702
537	0.234	0.766

Table 4 The structure of clusters for BCW Dataset

	Cluster 1		Cluster 2		Cluster 3	
	Size	Avg. distance	Size	Avg. distance	Size	Avg. distance
Class 0	12	0.220	16	0.262	213	0.068
Class 1	20	0.241	109	0.111	10	0.336

For each cluster, the size (the number of training instances) is reported in the *left column* and the average Euclidean distance from the cluster centre among the training dataset instances belonging to this cluster is displayed in the *right column*

presented in this chapter is able to identify such features, whilst the standard variable importance measures for random forest cannot.

6.2 Cluster Analysis and Log-Likelihood

The training dataset previously derived for the BCW Dataset was partitioned according to the true class labels. A clustering algorithm implemented in the R package kmeans was run separately for each class. This resulted in the determination of three clusters for class 0 and three clusters for class 1. The structure and size of clusters is presented in Table 4. Each class has one large cluster: cluster 3 for class 0 and cluster 2 for class 1. Both have a bigger concentration of points around the cluster centre (small average distance) than the remaining clusters. This suggests that there is exactly one core cluster corresponding to a class. This explains the success of the analysis based on the median as the vectors of medians are close to the centres of unique core clusters.

Figure 7 lends support to our interpretation of core clusters. The left panel shows the box-plot of the fraction of trees voting for class 0 among training instances belonging to each of the three clusters. A value close to one represents predictions for which the forest is nearly unanimous. This is the case for cluster 3. Two other clusters comprise around 10 % of the training instances for which the random forest model happened to be less decisive. A similar pattern can be observed in the case of class 1, see the right panel of the same figure. The unanimity of the forest is observed for the most numerous cluster 2 with other clusters showing lower decisiveness. The reason for this becomes clear once one looks at the variability of feature contributions within each cluster, see Fig. 8. The upper and lower ends of the box correspond to the

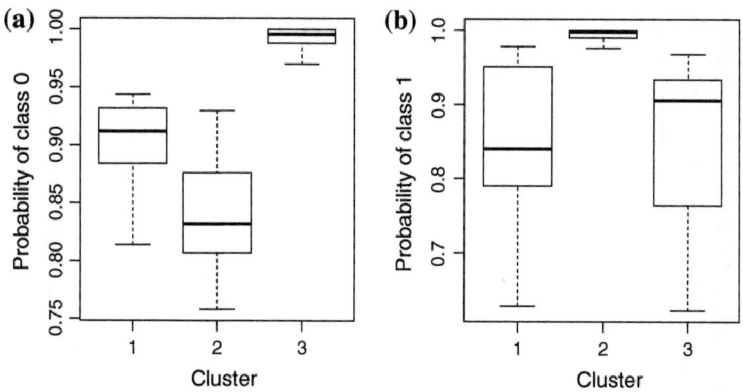

Fig. 7 Fraction of forest trees voting for the correct class in each cluster for training part of the BCW Dataset

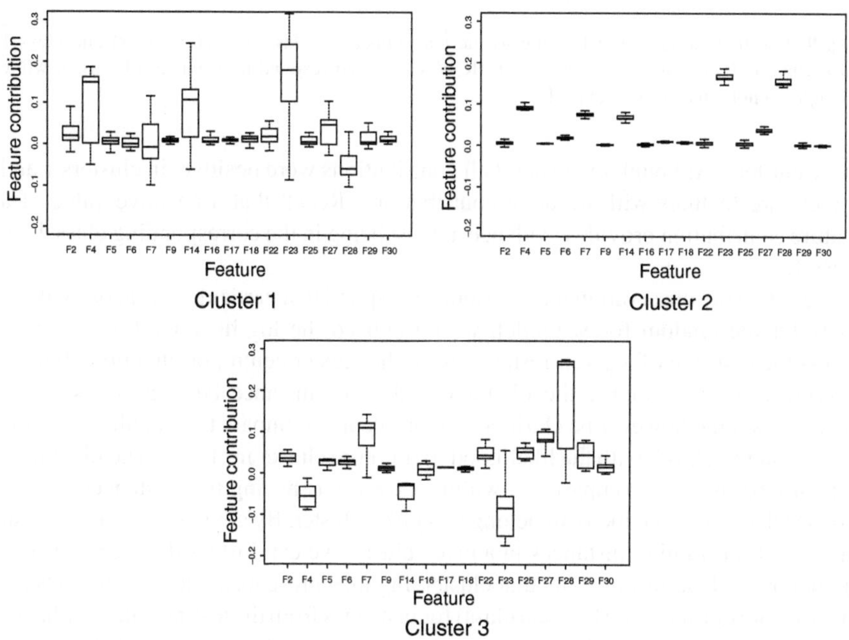

Fig. 8 Boxplot of feature contributions (towards class 1) for training instances in each of the three clusters obtained for class 1

75 % and 25 % quantiles, whereas the whiskers show the full range of the data. Cluster 2 enjoys a minor variability of all the contributions which supports our earlier claims regarding the similarity of instances (in terms of their feature contributions) in the core class. One can see much higher variability in two remaining clusters showing that the forest used different features as evidence to classify instances in each of

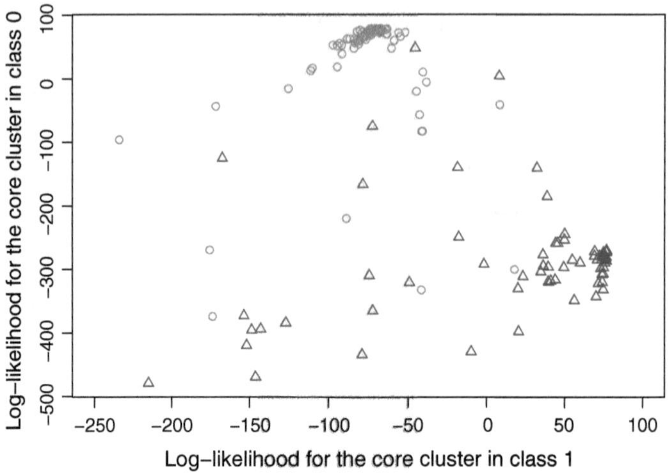

Fig. 9 Log-likelihoods for belonging to the core cluster in class 0 (*vertical axis*) and class 1 (*horizontal axis*) for the testing dataset in BCW. *Circles* correspond to instances of class 0 while *triangles* denote instances of class 1

these clusters. Although in cluster 2 all contributions were positive, in clusters 1 and 3 there are features with negative contributions. Recall that a negative value of a feature contribution provides evidence against being in the corresponding class, here class 1.

Based on the observation that clusters correspond to a particular decision-making route for the random forest model, we introduced the log-likelihood as a way to assess the distance of a given instance from the cluster centre, or, in a probabilistic interpretation, to compute the likelihood[3] that the instance belongs to the given cluster. It should however be clarified that one cannot compare the likelihood for the core cluster in class 0 with the likelihood for the core cluster in class 1. The likelihood can only be used for comparisons within one cluster: having two instances we can say which one is more likely to belong to a given cluster. By comparing it to a typical likelihood for training instances in a given cluster we can further draw conclusions about how well an instance fits that cluster. Figure 9 presents the log-likelihoods for the two core clusters (one for each class) for instances from the testing dataset. Shapes are used to mark instances belonging to each class: circles for class 0 and triangles for class 1. Notice that likelihoods provide a very good split between classes: instances belonging to class 0 have a high log-likelihood for the core cluster of class 0 and rather low log-likelihood for the core cluster of class 1. And vice-versa for instances of class 1.

[3] The likelihood is obtained by applying the exponential function to the log-likelihood.

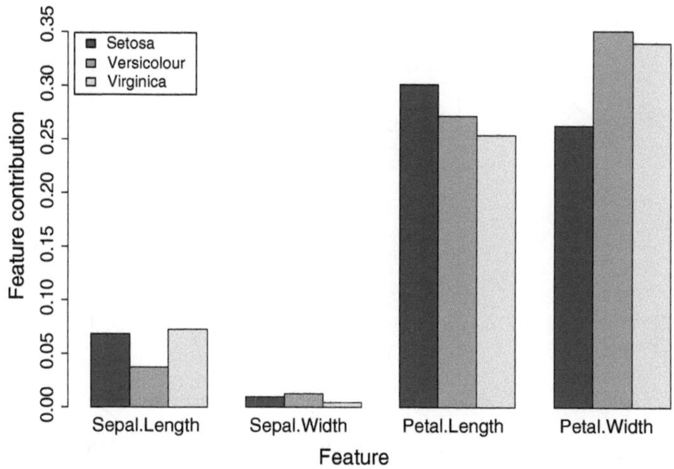

Fig. 10 Medians of feature contributions for each class for the UCI Iris Dataset

Table 5 Feature contributions towards predicted classes for selected instances from the UCI Iris Dataset

Instance	Sepal		Petal	
	Length	Width	Length	Width
120	0.059	0.014	0.053	0.448
150	−0.097	0.035	0.259	0.339

6.3 Iris Dataset

In this section we use the UCI Iris Dataset [12] to demonstrate interpretability of feature contributions for classification models. We generated a random forest model on 100 randomly selected instances. The remaining 50 instances were used to assess the accuracy of the model: 47 out of 50 instances were correctly classified. Then we applied our approach for determining the feature contributions for the generated model. Figure 10 presents medians of feature contributions for each of the three classes. In contrast to the binary classification case, the medians are positive for all classes. A positive feature contribution for a given class means that the value of this feature directs the forest towards assigning this class. A negative value points towards the other classes.

Feature contributions provide valuable information about the reliability of random forest predictions for a particular instance. It is commonly assumed that the more trees voting for a particular class, the higher the chance that the forest decision is correct. We argue that the analysis of feature contributions offers a more refined picture. As an example, take two instances: 120 and 150. The first one was classified in class Versicolour (88% of trees voted for this class). The second one was assigned class Virginica with 86% of trees voting for this class. We are, therefore, tempted to trust both of these predictions to the same extent. Table 5 collects feature contributions for

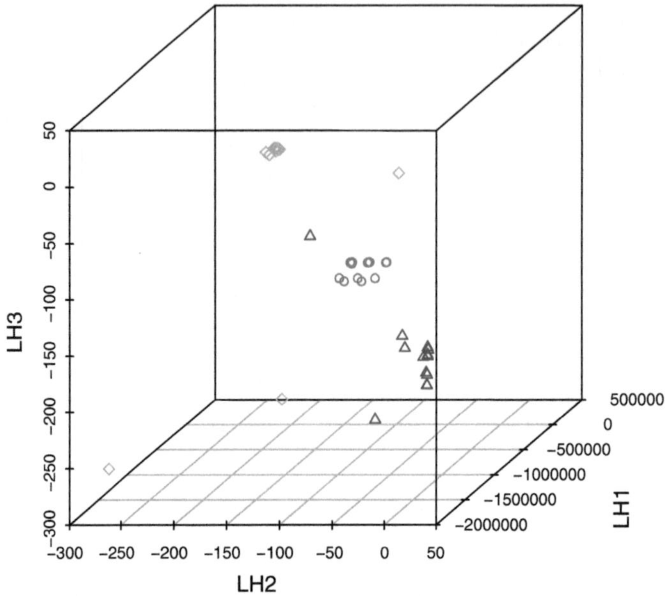

Fig. 11 Log-likelihoods for all instances in UCI Iris Dataset towards core clusters for each class. *Circles* represent the Setosa class, *triangles* represent Versicolour and *diamonds* represent the Virginica class. Points corresponding to the same class tend to group together and there are only a few instances that are far from their cores

these instances towards their predicted classes. Recall that the highest contribution to the decision is commonly attributed to features 3 (Petal.Length) and 4 (Petal Width), see Fig. 10. These features also make the highest contributions to the predicted class for instance 150. The indecisiveness of the forest may stem from an unusual value for the feature 1 (Sepal.Length) which points towards a different class. In contrast, the instance 120 shows standard (low) contributions of the first two features and unusual contributions of the last two features: very low for feature 3 and high for feature 4. Recall that features 3 and 4 tend to contribute most to the forest's decision (see Fig. 10) with values between 0.25 and 0.35. The low value for feature 3 is non-standard for its predicted class, which increases the chance of it being wrongly classified. Indeed, both instances belong to class Virginica while the forest classified the instance 120 wrongly as class Versicolour and the instance 150 correctly as class Virginica.

The cluster analysis of feature contributions for the UCI Iris Dataset revealed that it is sufficient to consider only two clusters for each class. Cluster sizes are 4 and 38 for class Setosa, 2 and 25 for class Versicolour and 3 and 28 for class Virginica. Core clusters were straightforward to determine: for each class, the largest of the two clusters was selected as the core cluster. Figure 11 displays an analysis of log-likelihoods for all instances in the dataset. For every instance, we computed feature contributions towards each class and calculated log-likelihoods of being in the core clusters of the respective classes. On the graph, each point represents one instance. The coordinate LH1 is the log-likelihood for the core cluster of class Setosa, the coordinate LH2 is the log-likelihood for the core cluster of class Versicolour and the coordinate LH3 is the log-likelihood for the core cluster of class Virginica. Shapes of points show the true classification: class Setosa is represented by circles, Versicolour by triangles and Virginica by diamonds. Notice that points corresponding to instances of the same class tend to group together. This can be interpreted as the existence of coherent patterns in the reasoning of the random forest model.

6.4 Robustness Analysis

For the validity of the study of feature contributions, it is crucial that the results are not artefacts of one particular realization of a random forest model but that they convey actual information held by the data. We therefore propose a method for robustness analysis of feature contributions. We will use the UCI Breast Cancer Wisconsin Dataset studied in Sect. 6.1 as an example.

We removed instance number 3 from the original dataset to allow us to perform tests with an unseen instance. We generated 100 random forest models with 500 trees with each model built using an independent randomly generated training set with $379 \approx 2/3 \cdot 568$ instances. The rest of the dataset for each model was used for its validation. The average model accuracy was 0.963. For each generated model, we collected medians of feature contributions separately for training and testing datasets and each class. The variation of these quantities over models for class 1 and the training dataset are presented using a box plot in Fig. 12a. The top of the box is the 75 % quantile, the bottom is the 25 % quantile, while the bold line in the middle is the median (recalling that this is the median of the median feature contributions across multiple models). Whiskers show the extent of minimal and maximal values for each feature contribution. Notice that the variation between simulations is moderate and conclusions drawn for one realization of the random forest model in Sect. 6.1 would hold for each of the generated 100 random forest models.

A testing dataset contains those instances that do not take part in the model generation. One can, therefore, expect more errors in the classification of the forest, which, in effect, should imply lower stability of the calculated feature contributions. Indeed, the box plot presented in Fig. 12b shows a slight tendency towards increased variability of the feature contributions when compared to Fig. 12a. However, these results are qualitatively on a par with those obtained on the training datasets. We can, therefore, conclude that feature contributions computed for a new (unseen) instance provide reliable information. We further tested this hypothesis by computing feature contributions for instance number 3 that did not take part in the generation of models. The statistics for feature contributions for this instance over 100 random forest models are shown in Fig. 12c. Similar results were obtained for other instances.

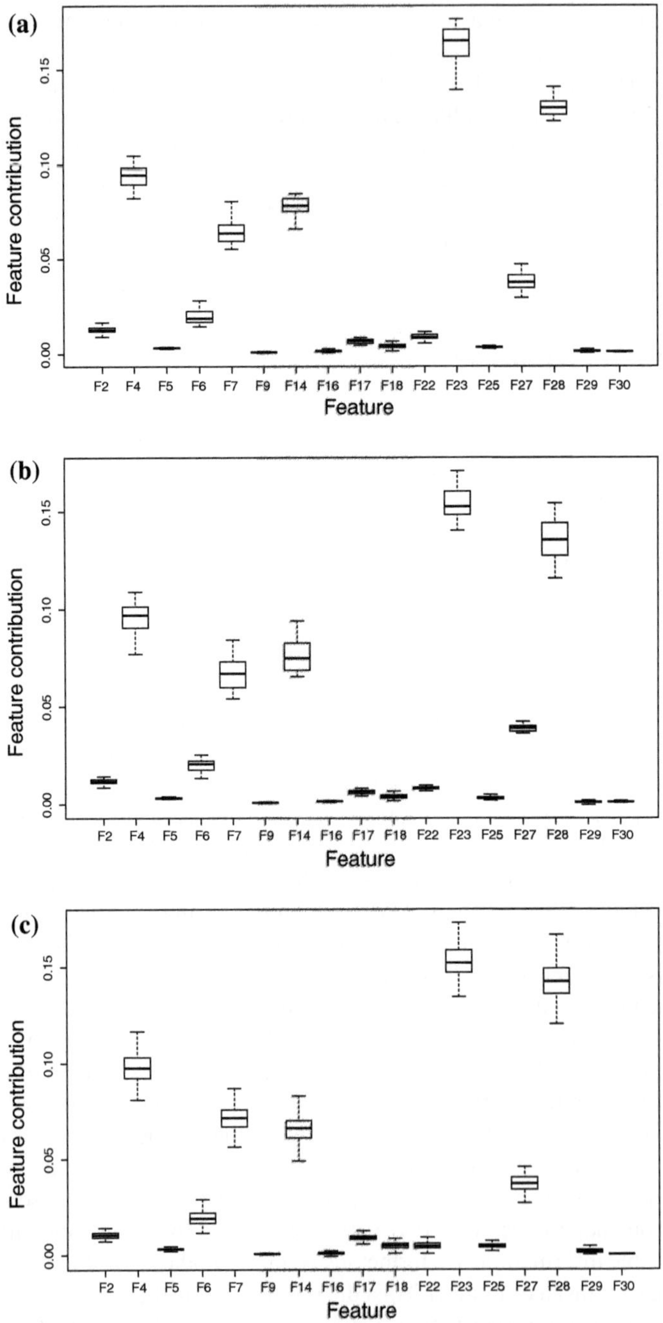

Fig. 12 Feature contributions towards class 1 for 100 random forest models for the BCW dataset, **a** Medians of feature contributions for training datasets, **b** Medians of feature contributions for testing datasets, **c** Feature contributions for an unseen instance

7 Conclusions

Feature contributions provide a novel approach towards black-box model interpretation. They measure the influence of variables/features on the prediction outcome and provide explanations as to why a model makes a particular decision. In this work, we extended the feature contribution method of [6] to random forest classification models and we proposed three techniques (median, cluster analysis and log-likelihood) for finding patterns in the random forest's use of available features. Using UCI benchmark datasets we showed the robustness of the proposed methodology. We also demonstrated how feature contributions can be applied to understand the dependence between instance characteristics and their predicted classification and to assess the reliability of the prediction. The relation between feature contributions and standard variable importance measures was also investigated. The software used in the empirical analysis was implemented in R as an add-on for the `randomForest` package and is currently being prepared for submission to CRAN [17] under the name `rfFC`.

Acknowledgments This work is partially supported by BBSRC and Syngenta Ltd through the Industrial CASE Studentship Grant (No. BB/H530854/1).

References

1. Tropsha, A.: Best practices for QSAR model development, validation, and exploitation. Mol. Inform. **29**(6–7), 476–488 (2010)
2. Rosenbaum, L., Hinselmann, G., Jahn, A., Zell, A.: Interpreting linear support vector machine models with heat map molecule coloring. J. Cheminf. **3**(1), 11 (2011)
3. Carlsson, L., Helgee, E.A., Boyer, S.: Interpretation of nonlinear QSAR models applied to ames mutagenicity data. J. Chem. Inf. Model. **49**(11), 2551–2558 (2009)
4. Baehrens, D., Schroeter, T., Harmeling, S., Kawanabe, M., Hansen, K., Muller, K.R.: How to explain individual classification decisions. J. Mach. Learn. Res. **11**, 1803–1831 (2010)
5. Hansen, K., Baehrens, D., Schroeter, T., Rupp, M., Muller, K.R.: Visual interpretation of kernel-based prediction models. Mol. Inform. **30**(9), 817–826 (2011)
6. Kuz'min, V.E., Polishchuk, P.G., Artemenko, A.G., Andronati, S.A.: Interpretation of QSAR models based on random forest methods. Mol. Inform. **30**(6–7), 593–603 (2011)
7. Breiman, L.: Random forests. Mach. Learn. **45**(1), 5–32 (2001)
8. Breiman, L., Cutler, A.: Random forests. http://www.stat.berkeley.edu/~breiman/RandomForests (2008)
9. Strobl, C., Boulesteix, A.-L., Zeileis, A., Hothorn, T.: Bias in random forest variable importance measures: Illustrations, sources and a solution. BMC Bioinf. **8**(1), 25 (2007)
10. Breiman, L., Friedman, J.H., Olshen, R.A., Stone, C.J.: Classification and Regression Trees. Wadsworth & Brooks/Cole Advanced Books & Software, Monterey (1984)
11. Liaw, A., Wiener, M.: Classification and regression by randomforest. R News **2**(3), 18–22 (2002)
12. Iris dataset. http://archive.ics.uci.edu/ml/datasets/Iris
13. Cormen, T.H., Stein, C., Rivest, R.L., Leiserson, C.E.: Introduction to Algorithms. 2nd edn. McGraw-Hill Higher Education, New York (2001)
14. Hand, D.J., Smyth, P., Mannila, H.: Principles of Data Mining. MIT Press, Cambridge (2001)

15. Rajaraman, A., Ullman, J.D.: Mining of Massive Datasets. Cambridge University Press, Cambridge (2012)
16. Breast Cancer Wisconsin Diagnostic dataset. http://archive.ics.uci.edu/ml/datasets/Breast+Cancer+Wisconsin+%28Diagnostic%29
17. CRAN—The Comprehensive R Archive Network. http://cran.r-project.org/

Towards a High Level Language for Reuse and Integration

Thouraya Bouabana-Tebibel, Stuart H. Rubin, Kadaouia Habib, Asmaa Chebba, Sofia Mellah and Lynda Allata

Abstract The modeling and design of complex systems continues to face grand challenges in feedback and control. Existing languages and tools, either textual or graphical, bring some improvement for such purposes, but much remains to be done in order to readily insure scalability. In this chapter, we propose a language which gathers specialization and composition properties. It is our belief that the latter properties bear the necessary capabilities to overcome the difficulties raised when developing these systems. The language is designed in a way to be specific to complex system domains. It supports a component-based structure that conforms to a user-friendly component assembly. The proposed structure is based on static, dynamic, functional and parametric parts. It is conceived in the spirit of SysML concepts. The language also supports textual and graphical specification. The specified models generate Internal Block Diagrams. A modeling tool is built on the basis of the Eclipse framework.

Keywords Complex systems · Component-based language · Domain-specific language · SysML

1 Introduction

Customized Domain Specific Languages (DSL) have gained a lot of popularity in the last few decades. There are many DSLs with textual or graphical/visual concrete syntax to support creative thinking, communication, and organization of the work

T. Bouabana-Tebibel (✉) · K. Habib · A. Chebba · S. Mellah · L. Allata
Laboratoire de Communication dans les Systèmes Informatiques - LCSI, Ecole Nationale Supérieure d'Informatique, Algiers, Algeria
e-mail: t_tebibel@esi.dz

S. H. Rubin
SPAWAR Systems Center Pacific, San Diego, USA
e-mail: stuart.rubin@navy.mil

T. Bouabana-Tebibel and S. H. Rubin (eds.), *Integration of Reusable Systems*, Advances in Intelligent Systems and Computing 263, DOI: 10.1007/978-3-319-04717-1_10,

to be performed [10]. In contrast to General Purpose Languages (GPL) like C, Java, and UML, DSLs enable more concise and precise specifications, which even non-programmer domain experts can understand. This is due to the domain concepts which are directly represented by syntactic constructs [11]. A DSL may also enable optimizing sentences by restricting what the user can do. This is typically not possible with a GPL. Besides, a sentence expressed in a DSL usually makes use of higher-level constructs than an equivalent sentence in a GPL [22].

On the other hand, component-based development is an attractive approach that receives much interest from both academia and industry. In recent years, component models, such as Enterprise Java Beans (EJB) of Sun [21], Component Object Model (COM) of Microsoft [13], and CORBA Component Model (CCM) of the OMG [14] have been well developed. In component-based systems, the dependencies between components are controlled through well-defined interfaces, as communication ports. Furthermore, components are more independent from each other compared with procedures in procedure-based systems and with classes/objects in object-oriented systems. At runtime, these components also have their own life cycles and runtime environment.

The general idea of our work falls under the context of complex systems modeling using a component-based language. Complex systems are systems with interdependent parts [3]. Interdependence means that we cannot identify the system behavior by just considering each of the parts and combining them. Instead, we must consider how the relationship between the parts affects the behavior of the whole. Several features of complex systems makes them difficult for learners to understand, among which are the large number of entities and interactions between them, non-linearity of changes, self-organization mechanisms, and exchanges with the environment.

Our principal goal is to facilitate the complex systems specification phase by offering the developer a language which is simple enough to be quickly learned and easily handled. The proposed language should take into account two main features. It will be domain specific to the modeling of complex systems, thus allowing the specification step to be performed in a way that hides all details that may bother the developer and complicate the development process. It will, furthermore, be designed based on component assembly, thus providing independent deployment and component reuse.

Systems are mainly characterized by their structural, behavioral, and functional aspects. Structures refer to the elements of a system. Behaviors refer to the mechanisms within a system. Finally functions refer to outcomes or roles in a system [3]. Systems Modeling Language (SysML) Internal Block Diagrams (IBD) [15], a standard for complex systems modeling widely used in engineering, support most of these aspects. However, they sometimes lack flexibility, relative to their high abstraction level, which may involve difficulties in handling. They, furthermore, present an exclusive graphical notation that makes it difficult to specify and visualize, in a readable way, scalable model components. Formal verification of such a visual specification proves, furthermore, to be unrealizable. For all these reasons, we propose, on one hand, a DSL that integrates both textual and graphical components. The main advantage behind the textual specification is to provide the beginnings of a high-level language supporting possible formal verification of the models. We based, on the other hand, our DSL features on IBD concepts and provided mechanisms to systematically generate IBD models. The whole of the transformation rules are formally written.

The remainder of this chapter starts with a brief presentation of the system used to illustrate the proposed DSL. In Sects. 3, 4 and 5 the DSL is presented and the techniques on which it rests are developed. Section 6 deals with its implementation. We discuss, in Sect. 7, some related works. We finally conclude, in Sect. 8, with some observations on the results obtained.

2 Case Study

To illustrate our study, we select a refrigeration system, which is commonly recognized as a benchmark for the representation of complex systems. The system uses a circulating fluid refrigerant to absorb and remove heat from the space to be cooled, and radiates that heat of condensation. It is mainly composed of:

- Compressor: transforms the low pressure vapor drawn from the evaporator to high pressure vapor.
- Condenser: transforms the high pressure refrigerant vapor to liquid state by radiating the heat of condensation.
- Expander: reduces the liquid refrigerant pressure sufficiently to allow the liquid vaporization (heat absorption).
- Evaporator: transforms the liquid refrigerant drawn from the valve to low pressure vapor.

Multistage refrigeration systems are widely used where ultra low temperatures are required, but cannot be obtained economically through the use of a single-stage system. They use two or more condensers connected in series in the same refrigerator—albeit with different refrigerants. Figure 1 depicts a part of such systems.

3 SSL Abstract Syntax

The language we propose will be termed SSL for System Specification Language. The abstract syntax of a given DSL represents the domain concepts. In our case, we consider the abstract syntax of SysML to extract complex systems concepts.

The block is the basic unit in SysML structure. Modeling with IBD consists in representing a given system by assembling blocks connected via ports. Ports are a special class of property used to specify allowable types of interactions between blocks. Blocks may specify operations or other features that describe the behavior of a system. They also support multi-level nesting of connector ends. A block can

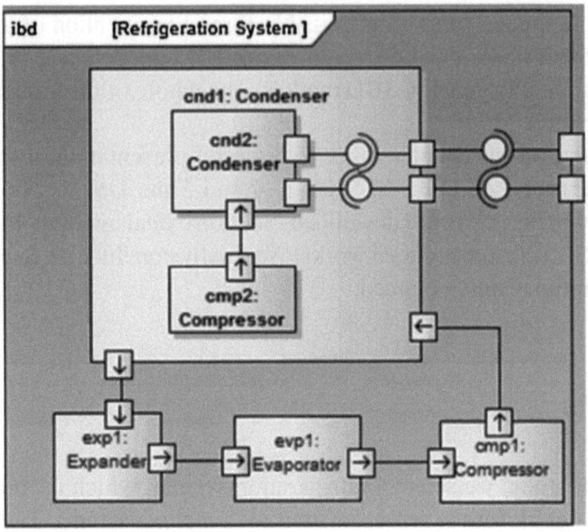

Fig. 1 IBD of the refrigeration system

be simple or composed of several sub-blocks. Composition is a property of encapsulation.

Like SysML, the SSL architecture will be based on blocks. Such a structure would be interesting for users familiar with SysML. It is composed of two types of components:

- Simple components which are elementary components that do not have sub-components, for example, *Expander*, *Evaporator*, and *Compressor* (see Fig. 1).
- Complex components that encapsulate sub-components, whether simple or complex, for example, the Condenser which encapsulates a Condenser and a Compressor.

The component architecture, which we propose is based on the principle of reuse. In order to favor component reuse, we suggest an appropriate architecture where the components include distinct static (invariant), dynamic (variable/scalable), functional (role/result) and parametric (parameters/constraints) parts, called compartments. Figure 2 depicts that architecture. The static and functional compartments assure the reuse property whereas the dynamic part serves to express the composite property of blocks, which is omnipresent in complex systems. As for the parametric compartment, it assures the control of physical properties of the system. Both simple and complex blocks include the static, functional and parametric parts; whereas, the dynamic part is present only in complex blocks. The latter specificity is closely connected to the idea according to which a composite block is necessarily composed of instantiated elements; thus, it includes information of a dynamic nature. That fragmentary architecture provides much ease in reusing predefined blocks by separate specifications and instantiated ones within the same specification.

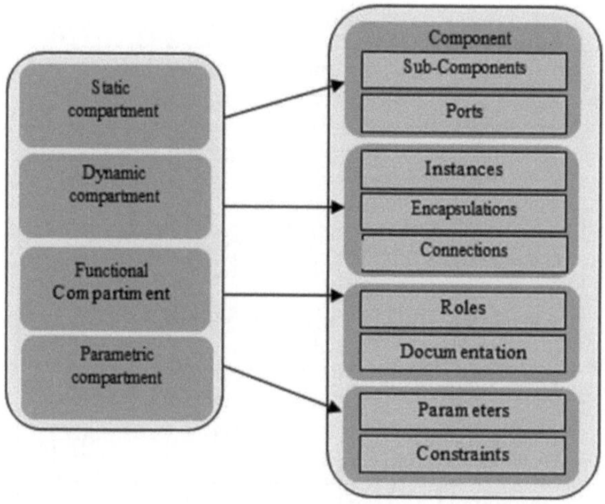

Fig. 2 Component structure

3.1 The Static Aspect

Sub-blocks of a composite block constitute fixed information. They are included in the static compartment of the block. Communication ports are also considered as invariant information, and integrated in the static compartment. More specifically, the static component comprises the following elements. Figure 3 shows the structure of the Condenser component.

- Component name, which usually represents the component type. It is of the form BlockName.
- Sub-components, which are the list of the blocks encapsulated in a complex component. They are listed using the component name form.
- Ports, which represent the access points of the blocks and allow the latter assembly. According to SysML IBDs, we distinguish between two port types, namely, standard and flow ports. They may be named using the service or flow name.
  - **Standard ports** deliver required and provided services between a block and its environment. This type of port allows data circulation from the provided ports (which specify the provided services) towards the required ports (which specify the required services). A block can call all services specified in its required ports and it must, at the same time, assure all services specified in its provided ports. For example, Penv1 and Penv2 are the *Condenser* standard ports.
  - **Flow ports** allow the circulation of any energy or material between the block and its environment. A flow port is atomic if it allows the circulation of only one type of flow and it is non-atomic if it allows the circulation of several types of flow. For example, Pexp and Pcmp are the *Condenser* flow ports.

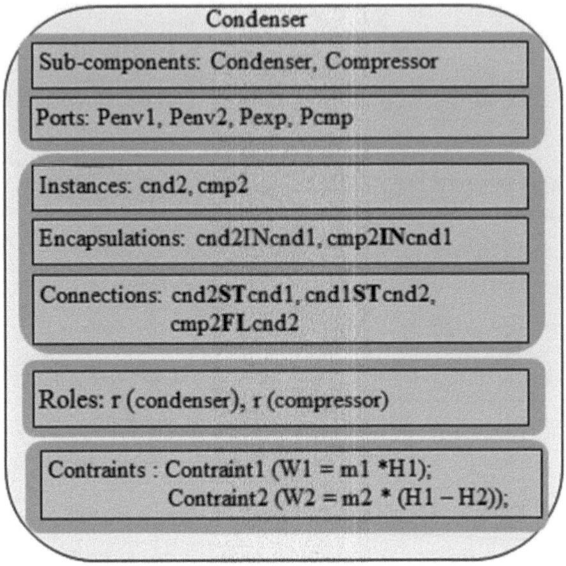

Fig. 3 Structure of the condenser component

3.2 The Dynamic Aspect

Blocks are instantiated. The resulting instances provide information, which varies. They are thus part of the component dynamic compartment. Communication between blocks is another part of the dynamic compartment. This communication is modeled using connections between blocks via their ports. More specifically, the dynamic compartment comprises the following elements.

- **Instances** derived from the sub-components already declared in the components list. We can see, in Fig. 3, two declared instances for the block Condenser. They are of the form instance: Block.
- **Encapsulations** provide all instances encapsulated within other instances using the symbol IN.
- **Connections** are all links between components. A connection is defined by the name of the two connected instances separated by a keyword indicating the type of the connection, which may be ST for standard services and FL for flows. For each of these connections, the direction of the communication is specified by placing the sender ahead followed by the keyword and the receiver. We can see the declared connections for the Condenser block in Fig. 3.

3.3 The Functional Aspect

Every function assured by a system component is described on the functional compartment of the block. In SSL, such a function expresses the role of the block. Roles—also called association-ends in object-oriented languages, such as SysML, are a key idea in modeling the functionalities that a given object may render. They enable us to distinguish among the numerous roles an object plays, based on the set of the operations it performs. By analogy to the component-based structure, which we propose, roles will characterize an instantiated block on the basis of the connections the latter has with its surroundings. Connections are supported by the ports associated with each instance. Thus, the roles a block instance plays are merely provided by the ports it supports. We thus propose to express a role of a block instance in a function of the block instance to which it is connected. More formally, we can write:

- Inst = {inst} is a set of block instances
- Rol = {rol} is a set of roles played by block instances
- r : Inst → $\mathcal{P}$(Rol) is a function that returns the role that a block instance plays.

For example, in Fig. 3, roles are expressed using the function r.

The role statement that we assign to block functions provides a formal aspect to the concept. Practically, a block function (i.e. role) is described in a paraphrased way—as documentation of the instantiated block. Thus, in the functional compartment, the user defines the functional specification by answering questions such as: What is expected from the block instance? How can the services be used? etc. A source programming code may also accompany the textual description. Besides, global documentation, describing the functioning of the whole block, is provided in order to be of assistance in reusing block documentation. In this manner, future users can thoroughly understand the function realized by the reused component.

One might question the significance of such a concept within SSL. How could roles be handled to improve systems specification? As we claimed above, roles bring a specific touch to block instances according to their connections. Identifying and consequently gathering block instances according to that property may be of interest. Among other advantages, it provides collections of objects grouped according to identical characteristics. Languages based on set theory, like OCL and Z, may complete SSL for system properties expression.

3.4 The Parametric Aspect

To enrich the proposed component structure, we add to the previous compartments another one that allows the integration of mechanisms for system analysis in terms of performance and reliability. It is a parametric compartment inspired from SysML parametric diagram, which allows users to specify constraints that control the physical properties of a system.

The parametric compartment is composed of two parts. The first part concerns the constraint parameters. The second part contains the constraints which must be verified to ensure that the system is functioning correctly. More specifically, the parametric compartment is composed of the following elements:

- Parameters: they represent the physical properties of a system, such as temperature, pressure, etc. In the parametric compartment each parameter is defined by its name, its components or source instance that represents the source of the physical property.
- Constraints: they express physical controls on a component and define relationships between the physical properties of a system. These constraints are expressed using mathematical or logical expressions composed of parameters. A constraint is defined by its name and its parameters that can be of two types:
  - In parameters: represent physical properties from other components.
  - Out parameters: represent physical properties of the component.

4 SSL Grammar and Concrete Syntax

To develop a new language one can proceed either by building its metamodel or establishing its grammar. The method differs depending on the language form. In our case, we mainly aim at creating a textual language that integrates graphics handling. We, therefore, opt for the grammatical approach.

4.1 Concrete Syntax

SSL concrete syntax is represented by two different sets of notations, either textual or graphical, respectively supported by textual and graphical editors. The latter may be indifferently handled by the user, according to his/her preferences, to facilitate modeling. Any specification introduced by a user in a given form, either textual or graphical, is systematically transformed and displayed in the other form. This approach allows users to check their model validity. Besides, the two forms of notation are checked for all types of lexical and syntax errors while typing the specification.

To facilitate SSL learning and manipulation, we thought to use some keywords used in IBD, such as: Block to designate a component, Port, and Flow. We also select keywords very close to the natural language used to describe complex systems in terms of components, such as SimpleBlock, ComplexBlock, and Connection.

The program "DiagramIbd SYSTEM" illustrates some visual features of the SSL text editor. It shows a part of the program specifying the design of the refrigeration system.

The SSL graphical editor includes a palette that contains all of the icons that the designer can use to model his system, namely, complex blocks, simple blocks, standard and flow ports, and standard and flow connections. Figure 4 shows the SSL graphical editor with a part of the refrigerator system.

```
DiagramIbd  SYSTEM
SimpleBlock   Compressor
SimpleBlock   Condenser
SimpleBlock   Expander
SimpleBlock   Evaporator
ComplexBlockcnd1: Condenser {
  StaticPart {
  Ports{StandardPortID: Penv1 {};StandardPortID: Penv2 {}
FlowPortID: Pexp{}; FlowPortID:Pcmp{}}EndPorts
  SubBlocks{
  SimpleSubBlock:Compressor; Condensser}
  EndSubBlocks } EndStaticPart
  DynamicPart{
      Instances {
    SimpleInstance
InstanceID:cmp2: Compressor; InstanceID: cnd2:Condenser
{Ports{StandardPortID: Pcnd1In {Is Behavioral};
StandardPortID: Pcnd1Out {Is Behavioral}
FlowPortID:Pcmp2{IsBehavioral;IsAtomic}}
EndPorts}}
EndInstances
     Connections {
StandardConnection{ConnectionID:cnd1STcnd2 ,
               Source: cnd1.Penv1, Target: cnd2.Pcnd1In}
StandardConnection{ConnectionID:cnd1STcnd2 ,
               Source: cnd2.Pcnd1Out,Target: cnd1.Penv2}

FlowConnection {ConnectionID: cmp2FLcnd2,
Source:cmp2.Pcnd2,Target: cnd2.Pcmp2}}
EndConnections }EndDynamicPart
FunctionalPart{
         Role: r (cnd2), r(cmp2)}EndFunctionalPart
PrametricPart {
          Parameters
{W1:"energy1",W2:"energy2"H1:"fluid_enthalpy1
",H2:"fluid_enthalpy2", m1:"fluid_masse1 ",m2 "flu-
id_masse2"}
EndParameters
Constraints {Contraint1: W1= m1 * H1;
                      Contraint2: W2= m2 * (H1- H2) ;}
EndConstraints}
EndPrametricPart}
EndBlock
```

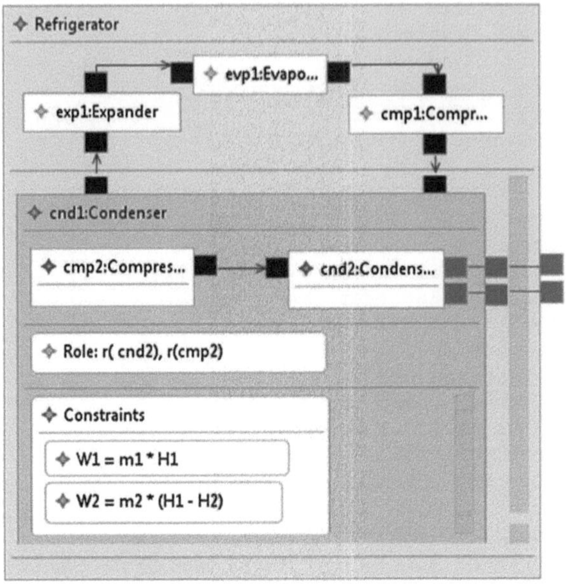

Fig. 4 Refrigerator system scheme using SSL

4.2 The Grammar

SSL is characterized by: (1) a vocabulary inspired from SysML constructs, which are well recognized to be basic, thus facilitating the user programming, and (2) a very simple syntax, which respects the component architecture that we proposed. It is based on a context-free grammar. The general form of the proposed grammar is presented in what follows:

- R1: <Axiome>→ "DiagramIbd" < DiagramName >< Model > "EndDiagram"
- R2: <Model>→ < SimpleBlock >+< ComplexBlock >
- R3: <SimpleBlock>→ "SimpleBlock"<BlockName>< SimpleStaticPart >"EndBlock"
- R3: <SimpleStaticPart>→"Ports"<StandardPorts>*<FlowPorts >* "EndPorts"
- R4: <ComplexBlock>→"ComplexBlock"<BlockName><ComplexStaticPart> <Comple- x DynamicPart> <ComplexFunctionalPart> <ComplexParametricPart>"EndBloc"
- R5: <ComplexStaticPart>→"Ports"<StandardPort >* <FlowPort >* "EndPorts""Sub-Blocks" <Blocks> + "EndSubBlocks"
- R6: <ComplexDynamicPart>→"Instances"<BlockInstance> + "EndInstances"Connections" <StandardConnection> * <FlowConnection>* "EndConnections"
- R7: <ComplexFunctionalPart >→ "FunctionalPart"<Role> ∗"EndFunctionalPart"
- R8: <ComplexParametricPart>→ "ParametricPart"<Parameters>*<Constraints> ∗"End-ParametricPart"
- R9: <Standard Ports >→"StandardPort"<PortName>"Behavioral"<Service> +|"Standard-Port"<PortName><Service> +
- R10: <Service>→"Service"<ServiceName>

- R11: <Flow*Ports*>→"Flow*Ports*"<*PortName*>"Behavioral" "Conjugated" <Atomic> *| <PortName>"Behavioral""Conjugated"|"FlowPorts"<PortName>"Behavioral" <Atomic> *| <PortName>"Behavioral" <Atomic>*
- R12: <Atomic>→"ItemFlow:" <ItemFlowName>
- R13: *<StandardConnection>*→"StandardConnection"<ConnectionName>"Source:"*<StandardPort1Name>* "Target :"*<StandardPort2Name>*
- R14: *<FlowConnection>*→"FlowConnection"<ConnectionName>"Source:"<Flow*Port1-Name*>"Target :"<Flow*Port2Name*>
- R15: *<Blocks>*→< BlockName> "; "
- R16: *<BlockInstance>*→<InstanceName> ": " < BlockName>
- *<Role>*→"Role:" < RoleName>"("<InstanceName>")"
- R17: *<Parameters>*→"*Parameters*:" <ParameterName">"*EndParameters*"
- R18: *<Constraints>*→"*Constraint*"<ConstraintName><Expression>"*EndConstraint*"
- R19: < BlockName>→ String
- R20: *<PortName>*→ String
- R21: <InstanceName>→ String

The production rule R1 is the axiom of the grammar. It states that each program must begin with the keyword DiagramIbd, followed by its name, and ending with the keyword EndDiagram. The body of the diagram is represented by the non-terminal <Model>. Rules R3 and R4 express simple and complex block types.

5 SSL Semantics

Besides being a dedicated language that simplifies complex systems modeling, SSL guarantees syntactical and functional coherence, while modeling validity thanks to its semantics that we describe as SSL features in what follows.

5.1 Component Assembly Control

Component assembly counts among the most significant functionalities of the SSL language. To preserve the semantics consistency of that assembly, we must assure that components are assembled in accordance with their respective type, which must fit. According to the SSL structure and concept, component assembly may only be performed through block encapsulation or port connection.

In reference to block encapsulation, at the static structure level—defined by the static compartment, blocks may be assembled without any restriction on their type. Formal verification does not make sense. A simple validation, which is left to the modeler's appreciation, or based on logical languages like the Object Constraint Language (OCL) [16], which is used to express properties of systems modeled with UML [17], or temporal logics, may be performed. At the dynamic structure level—defined by the dynamic compartment, a rigorous control must be performed on the instantiated embedded blocks. The latter requires the same type as that of the blocks previously declared within the static structure.

Ports connection is declared only at the dynamic structure level, where every composite block describes: (1) connections between its sub-components and (2) its own connections with the sub-components. On the other hand, ports convey services through standard ports and flows through flow ports. They may support several sub-types given as port names. We state that a connection between two ports is allowed if and only if:

- **Constraint 1:** both ports concerned with the connection are of the same type, which may be standard or flow.

This constraint can be realized syntactically using grammatical rules, as expressed in what follows with EBNF:

```
FlowConnection:
'FlowConnection'{''ConnectionID'':'(NameConnection= ID)
  ','  ('Source' ':' Port1=[FlowPort])
  ','  ('Target' ':' Port2=[FlowPort]) '}'  ;

StandardConnection:
'StandardConnection''{''ConnectionID'':'  (NameConnection=ID)
',' ( 'Source'   ':' Port1=[StandardPort])
  ',' ( 'Target'   ':' Port2=[StandardPort])
'}' ;
```

- **Constraint 2:** a standard port whose service is provided may only be connected with a standard port whose service is required.

To realize this constraint, we use an OCL invariant which checks that a port whose service is not "Provided and Required" at the same time, is connected to a port of a dual service. If the invariant is false then the message 'SSL Error: Incorrect Provided/Required Service' is displayed.
Constraint 2 expression in OCL

```
Context StandardConnection
inv:
ifself.Service1.type<> 'ProvidedRequired'
thenself.Service1.type<>self.Service2.type
elseself.Service2.type = 'Provided'
orself.Service2.type = 'Required'
orself.Service2.type = 'ProvidedRequired'
endif;
```

- **Constraint 3:** an output flow port may only be connected with an input flow port.

To realize this constraint, we use an OCL invariant which checks that if the direction of a flow port is not "InOut" then the direction of the flow port to which it is connected is opposite. If the checking fails then the message 'SSL Error: Incorrect Flow Direction' is shown.
Constraint 3 expression in OCL

```
Context FlowConnection
inv:
ifself.flowdirection1<> 'InOut'
thenself.flowdirection1<>self.flowdirection2
elseself.flowdirection2 = 'In'
orself.flowdirection2 = 'Out'
orself.flowdirection2 = 'InOut'
endif;
```

Constraint 4: connected required and provided services—between two standard ports, must have the same sub-type.

To realize this constraint, we use an OCL invariant which checks that the service name of the first port is the same as that of the second one. If the invariant returns false then the message 'SSL Error: Incorrect Service Name' is displayed.
Constraint 4 expression in OCL

```
Context StandardConnection
inv:
self.Service1 = self.Service2
```

- **Constraint 5:** flow circulating between two flow ports must be of the same sub-type; for example: gas, water, etc.

To realize this constraint, we use an OCL invariant. If the invariant returns false then the message 'SSL Error : Incorrect Flow Type' is displayed.
Constraint 5 expression in OCL

```
Context FlowConnection
inv:
self.Flow1 = self.Flow2
```

The conditions above constitute formal constraints. SSL is based on a grammar that allows expression of such system constraints through syntactical controls. We used production rules, expressed in EBNF, to carry out these controls.

5.2 Sensitivity Study

Language efficiency may be measured according to the model sensitivity, which consists in a systematic propagation of changes throughout the model. In regard to SSL, the proposed architecture allows all modifications, undertaken on a given block, to be automatically propagated in all its instances. This is realized thanks to the block identifier, which also identifies block instances. Besides, changes between the textual and graphical models are also automatically propagated.

The grammar rule R5, expressed in EBNF format, shows the relationship between a block and its instances, where the cross-reference Component = [componentID] assures that an instance of a block inherits the latter features including its static part. This inheritance makes it possible to propagate, in all instances of a given block, any modification carried out on the latter.

- R5 Instance:

  name=ID ":" "(" Component = [*componentID*] ")"

6 Transformation of the Specification

Our approach involves two principal transformation processes. The first one concerns IBD generation and the second one is related to the textual and graphical editors.

6.1 From SSL Program to IBD Diagram

To generate IBD diagrams, from the proposed grammar, we resort to a processing method using metamodels and transformation rules. The ATLAS transformation (ATL) Eclipse Plugin [4] supports such processing (Fig. 5).

It requires, as inputs, the source metamodel, the target one, the program to transform, and the transformation rules. In our case, the source and target metamodels will be, respectively, those corresponding to the proposed grammar and the IBD one. As for the program to transform, it simply corresponds to the program written in SSL. As output, ATL provides the IBD model in XMI format. This file can be interpreted to generate the graphical representation, or for any other transformations.

To build the metamodel of the proposed grammar, we used the Xtext Eclipse Plugin, which systematically derives a metamodel from an EBNF grammar. On the other hand, our numerous investigations, starting from the OMG document, reveal the absence of a metamodel describing the IBD specification. Thus, we had to build it ourselves.

We edited it in the Eclipse Modeling Framework (EMF) Eclipse Plugin, for this purpose. Figure 6 shows this metamodel. The program to transform to IBD must be in XMI format. It is generated from the program written in SSL.

Finally, we defined the transformation rules between the two metamodels. The rule below shows the beginning of the transformation. All transformation rules are included in a file .atl. They result in an IBD instance expressed in XMI format (see Fig. 7).

```
Rule Diagram2Ibd {
/This Rule transforms the SSL diagram class attributes
to the IBD class attributes
From s: MyDsl!Diagram /The Source metamodel class
To t: Ibd!Diagram /The target metamodel class
Head <- s.name, /Transform the given name of the Diagram
to the head name of the IBD
HasBloc<- s.complexe_block,
HasBloc<- s.simple_block ) }
```

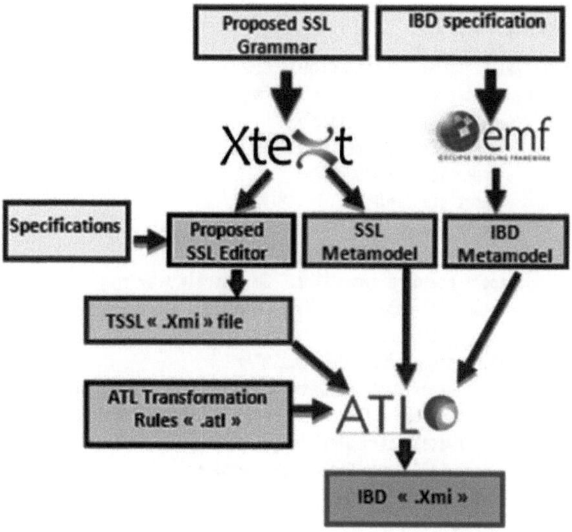

Fig. 5 Transformation process to IBD

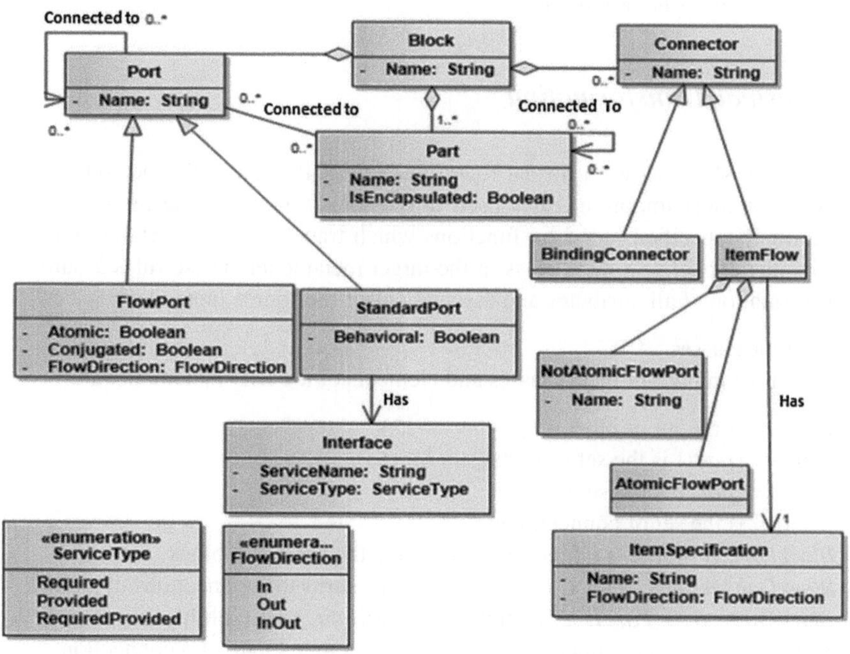

Fig. 6 The IBD metamodel

```
resultat.xmi
<?xml version="1.0" encoding="ASCII"?>
<myDsl:Diagram xmi:version="2.0" xmlns:xmi="http://www.omg.org/
  <simple_block name="Simp1">
    <port_standard name="Port1" Behavioral="true">
      <service name="Ser1"/>
    </port_standard>
  </simple_block>
  <complexe_block name="Comp1">
    <port_standard name="Port2" Behavioral="true">
      <service name="Ser2" Type="Required"/>
    </port_standard>
    <bloc name="//@simple_block.0"/>
    <case name="Var1" NomComposantSimple="//@simple_block.0"/>
    <standard_connexion name="conn1" Instance1="//@simple_block
      <service1 name="Ser1"/>
      <service2 name="Ser2" Type="Required"/>
    </standard_connexion>
  </complexe_block>
</myDsl:Diagram>
```

Fig. 7 IBD resulting in XMI format

6.2 Formal Transformation

We give in what follows a formal representation of IBD and SSL meta-classes as well as the most important rules used in the transformation. The mapping rules are formally specified based on functions which transform each meta-class in the source metamodel to a meta-class in the target metamodel. These rules assure the transformation of all attributes and associations of the source meta-class.

IBD Meta-classes

The most important meta-classes and elements of the IBD metamodel are:

- $B = \{b\}$ is the set of blocks.
- $Part = \{part\}$ is the set of parts (sub-blocks).
- $Port = \{port\}$ is the set of ports.
- $C = \{c\}$ is the set of connections.
- *BlockPort:* $B \rightarrow PR$ *is* a function that returns the ports of a block.
- *BlockConnections:* $B \rightarrow C$ *is* a function that returns the connections of a block.
- *BlockParts:* $B \rightarrow Part$ *is* a function that returns the parts (sub-blocks) of a block.
- *SourceB:* $C \rightarrow B$ *is* a function that returns the source block of a connection.
- *TargetB:* $C \rightarrow B$ *is* a function that returns the target block of a connection.
- *SourceP:* $C \rightarrow Port$ *is* a function that returns the source port of a connection.
- *TargetP:* $C \rightarrow Port$ *is* a function that returns the target port of a connection.

SSL Meta-classes

The most important meta-classes and elements of the SSL metamodel are:

- $CC = \{cc\}$ is the set of complex components.
- $SC = \{sc\}$ is the set of simple components.
- $I = \{i\}$ is the set of instances.
- $SubC = \{subc\}$ is the set of sub-components of a complex block.
- $SP = \{sp\}$ is the set of standard ports.
- $FP = \{fp\}$ is the set of flow ports.
- $StC = \{stc\}$ is the set of standard connections.
- $FlC = \{flc\}$ is the set of flow connections.
- $CompPorts : CC \cup SC \rightarrow P(SP \cup FP)\, is$ a function that returns the standard ports and the flow ports of a complex component or a simple component.
- $CompInstances$: $CC \rightarrow P(I)$ is a function that returns the encapsulated instances of a complex block.
- CompConnections: CC ∪ SC → P(StC ∪ FlC) is a function that returns the connections of a complex component or a simple component.
- SourceIns: SC ∪ FC → I is a function that returns the source instance of a standard connection or a flow connection.
- TargetIns: SC ∪ FC → I is a function that returns the target instance of a standard connection or a flow connection
- SourceSP: StC → SP is a function that returns the source standard port of a sstandard connection.
- TargetSP: StC → SP is a function that returns the target standard port of a standard connection.
- SourceFP: FlC → FP is a function that returns the source flow port of a flow connection.
- TargetFP: FlC → FP is a function that returns the target flow port of a flow connection.
- InstanceComp: I → CC ∪ SC is a function that returns the simple block or the complex block of an instance.

The transformation rules

To generate IBD models from SSL programs, we define the flowing functions which represent our transformation rules:

- Function ComplexComponent ( ):

```
For each cc ∈ CC
    Create b ∈ B; BlockParts (b) = InstanceComp (CompInstances (cc));
                  BlockPorts (b) = CompPorts (cc);
                  BlockConnetions (b) = CompConnections (cc);
EndFor
```

- Function SimpleComponent ( ):

```
For each sc ∈ SC
        Create b ∈ B; BlockPorts (b) = CompPorts (sc);
                      BlockConnetions (b) = CompConnections (sc);
EndFor
```

- Function StandardConnection ( ):

```
For each stc ∈ StC
        Create c ∈ C; SourceB (c) = InstanceComp (SourceIns(stc));
                      TargetB (c) = InstanceComp (TargetIns(stc));
                      SourceP(c) = SourceSP(stc); TargetP(c)= TargetFP(stc);
EndFor
```

- Function FlowConnection ( ):

```
For each flc ∈ FlC
        Create c ∈ C; SourceB (c) = InstanceComp (SourceIns(flc));
                      TargetB (c) = InstanceComp (TargetIns(flc));
                      SourceP(c) = SourceSP(flc); TargetP(c) = TargetFP(flc);
EndFor
```

6.3 The SSL Textual and Graphical Editors

We used Graphical Modeling Framwork (GMF) Eclipse plugins to generate a graphical specification from a textual one and vice versa. GMF yields a set of generative components and runtime infrastructures to develop graphical editors for instances of a given metamodel. Thus, to get a graphical editor dedicated to the modeling of programs written with SSL, we provided GMF with the metamodel derived from our grammar using Xtext. Once the two editors were created, changes between the two forms of respectively supported specifications are systematically undertaken.

7 Related Works

Various DSLs are widespread in academia and industry domains. HUNT [18] is a well-known DSL proposed by the OMG to remedy the XMI [19] syntax, which is rather verbose and not easily readable. HUNT is used for storing models in a human understandable format. However, it imposes very strict constraints on the notation. In [8] Mandeep Gill et al. present Ode, a DSL for modeling biological systems based on mathematical concepts. Ode is inspired by functional programming languages (e.g., Common LISP).

Other studies have focused on a particular type of DSLs, called Domain-Specific Modeling (DSMs), which are graphical representations used to develop systems in a more expressive way. They are usually developed using metamodels. However, when looking to their development process, there is little to distinguish them from DSLs [12]. In [10] P. Laforcade provides end-users, acting as both teachers and designers, with dedicated Visual Instructional Design Languages and authoring tools. These tools are built to help practitioners in specifying learning scenarios with some specific terminology and graphical formalism so that the produced artifacts will be machine-readable. In [9] a DSM called the Sequencer, integrated with the measuring equipment DEWESoft, enables domain experts to model their own data acquisitions.

A further category of languages combines graphical and textual specifications. Heidenreich et al. mentioned in [6] that graphical and textual modeling are not alternatives, but rather complementary. However, few researchers investigated this field. KIEL [20] is a DSL that aims at enhancing the graphical model-based design of complex systems. Among other facilities, it enables editing the abstract syntax of the model as well as modifying the latter via its textual representation.

On the other hand, component-based specification is a privileged approach for language design due to the modeling flexibility, which it provides. In the academic arena, Fabresse proposed SCL [7], a minimal language that provides abstractions and mechanisms suitable for component programming. Elizondo and Ndjatchi, present in [5] FSCC, a components approach based on composite components. They focus on deriving the specification of functional properties using a set of connector-specific functions, which allow deriving functional specifications in a systematic and consistent manner.

In the industry domain, CCM (CORBA Component Model) [14] introduces the component concept on the famous CORBA language, thus providing a standard component model - defined by the OMG, enabling the development of distributed applications. Valentina and Vlado propose in [23] a component-based approach for modeling Multi-Processor Systems-on-Chip (MPSoC). This approach uses the technology SaveComp Component Model (SaveCCM) [1], which facilitates the specification of non-functional properties inherent to embedded systems. Another well-known language is ArchJava [2], an extension of Java, where objects and components support different constraints. However, ArchJava does not support behavioral aspect descriptions; and, the integrity control of component communication constrains the programmer to the use of complex annotations.

Thus, despite the wide adoption of based-component and domain-specific concepts by the community, we still need languages, which integrate both specificities. SSL is intended to explicitly integrate these features for the modeling of complex systems.

8 Conclusion

In this chapter, we proposed SSL, a component-based domain specific language supporting textual and graphical specification of complex systems. In SSL, the model structure is represented by means of connected blocks, its behavior is described by using the block roles and system properties are expressed as constraints. SSL provides semantic links between the structural and dynamic concepts and assures a high level of consistency for the global model. A sensitivity analysis shows that any change introduced on the static part is automatically reproduced on the dynamic one; and, a systematic and bidirectional transformation of changes between the textual and graphical parts is also supported. These textual and graphical specifications are automatically translated into IBD models. The transformation rules have been formally specified.

For future work, we aim to integrate our work with the reuse process, based on an appropriate repository accompanied with an efficient retrieval mechanism.

References

1. Akerholm, M., Moller, A., Hansson, H., Nolin, M.: Towards a dependable component technology for embedded system applications. In: 10th IEEE International Workshop on Object-Oriented Real-Time Dependable Systems—WORDS'05, pp. 320–32. Washington, USA (2005)
2. Aldrich, J.C. Chambers, Notkin, D.: ArchJava: connecting software architecture to implementation. In: ICSE ACM, pp. 187–197 (2002)
3. Bar-Yam, Y.: When systems engineering fails toward complex systems engineering. In: IEEE International Conference on Systems, Man and Cybernetics, vol. 2, pp. 2021–2028 (2003)
4. Eclipse Project on http://www.eclipse.org
5. Elizondo, P.V., Ndjatch, M.K.C.: Deriving functional interface specifications for composite components. In: 10th International Conference on Software Composition—SC'11, pp. 1–17. Springer (2011)
6. Fabresse, L., Dony, C., Huchard, M.: SCL: a simple, uniform and operational language for component-oriented programming in smalltalk. In: LNCS, vol. 4406, pp. 91–110. Springer (2007)
7. Heidenreich, F., Johannes, J., Karol, S., Seifert, M., Wende, C.: Derivation and refinement of textual syntax for models. In: Model Driven Architecture-Foundations and Applications, pp. 114–129. Springer, Enschede, The Netherlands (2009)
8. Gill, M., McKeever, S., Gavaghan, D.: Modular mathematical modelling of biological systems. In: 2013 Symposium on Theory of Modeling and Simulation—DEVS, pp. 36:1–36:8. San Diego, USA (2013)
9. Kos, T., Kosar, T., Mernik, M.: Development of data acquisition systems by using a domain-specific modeling language. Comput. Ind. **63**, 181–192 (2012)
10. Laforcade, P.: A domain-specific modeling approach for supporting the specification of visual instructional design languages and the building of dedicated editors. J. Vis. Lang. Comput. **21**(6), 347–358 (2010)
11. Mannadiar, R., Vangheluwe, H.: Domain-specific engineering of domain-specific languages. In: 10th Workshop on Domain-Specific Modeling, p. 11 (2010)
12. Mernik, M., Heering, J., Sloane, A.: When and how to develop domain-specific languages. ACM Comput. Surv. **37**(4), 316–344 (2005)
13. Microsoft: The Component Object Model Specification (1995)

14. Object Management Group: CORBA Component Model (CCM) 3.0 (2004)
15. Object Management Group: Systems Modeling Language, SyML v. 1.3 (2012)
16. Object Management Group: Object Constraint Language, OCL v. 2.3.1 (2012)
17. Object Management Group: Unified Modeling Language UML v. 2.4.1 (2011)
18. Object Management Group: Human-Usable Textual Notation, v1.0 (2004)
19. Object Management Group: MOF 2.0/XMI Mapping Specification, v 2.4.1 (2011)
20. Prochnow, S., von Hanxleden, R.: Statechart development beyond WYSIWYG. In: The ACM/IEEE 10th International Conference on Model Driven Engineering Languages and Systems—MODELS'07, LNCS 4735, pp. 635–649 (2007)
21. Sun Microsystems: Enterprise JavaBeans (TM) Specification, Version2.1 (2003)
22. Taha, W.M.: Domain-Specific Languages. In: LNCS, vol. 5658, pp. 148–169. Springer (2009)
23. Zadrija, V., Sruk, V.: Component-based specification for multi-processor system-on chip design. In: 15th IEEE Mediterranean Electrotechnical Conference, pp. 1044–1049 (2010)

An Exploratory Case Study on Exploiting Aspect Orientation in Mobile Game Porting

Tanmay Bhowmik, Vander Alves and Nan Niu

Abstract Portability is a crucial requirement in the mobile game domain. Aspect-oriented programming has been shown to be a promising solution to implement the portability concerns, and more generally, to be a key technical enabler to transition mobile application development toward systematic software reuse. In this chapter, we report an exploratory case study that critically examines how aspect orientation is practiced in industrial-strength mobile game applications. Our analysis takes into account technical artifacts, organizational structures, and their relationships. Altogether these complementary and synergistic viewpoints allow us to formulate a set of hypotheses and to offer some concrete insights into developing information reuse and integration strategies in the rapidly changing landscape of mobile software development.

Keywords Software reuse · Aspect-oriented software development · Mobile game development · Porting · Software ecosystem · Exploratory case study

1 Introduction

Mobile game is a booming industry whose development challenges can be addressed by leveraging software engineering principles and sound practices [19]. For many game vendors, deploying their mobile games to various platforms is key to their business successes. A critical enabler for realizing the multiple-platform deployment

T. Bhowmik (✉) · N. Niu
Department of Computer Science and Engineering, Mississippi State University, Starkville, MS, USA
e-mail: tb394@msstate.edu

N. Niu
e-mail: niu@cse.msstate.edu

V. Alves
Computer Science Department, University of Brasilia, Brasilia, Brazil
e-mail: valves@unb.br

T. Bouabana-Tebibel and S. H. Rubin (eds.), *Integration of Reusable Systems*, Advances in Intelligent Systems and Computing 263, DOI: 10.1007/978-3-319-04717-1_11,

vision is *porting*, which is aimed at reusing the existing code instead of creating new code when moving software from one environment to another [16]. In other words, porting helps reduce development cost by ensuring that the same game application can be executed properly across different operational contexts.

Porting can be achieved in practice by several means, such as abstract specification, program transformation, and preprocessing [2]. A novel way to implement portability is *aspect-oriented programming* [20] in which code modules (known as aspects) are created to handle a mobile game's porting features. These aspects define *where* porting takes place, *when* porting needs to be triggered, and *how* porting should be carried out [27]. The aspects are then weaved with base modules to form an integrated system.

Our earlier work [3] has demonstrated the technical feasibility of using aspects to implement a mobile game's portability. By localizing porting-related features in code aspects, not only a modular separation of concern is devised, but the software development of mobile games can also be incrementally transitioned to product line engineering where orders of magnitude improvements in productivity and time-to-market are expected [3].

While we reported positive experience of refactoring portability concerns into code aspects (e.g., [1]), we also observed practitioner's hesitation of adopting aspect-orientation (AO) in mobile game development. The main obstacle seemed to be organizational rather than technical, e.g., some mobile game vendors enjoyed the culture of small and highly-coupled project teams and suspected that the newly created (porting) aspects would introduce unnecessary latency in their business processes [26]. These mixed findings motivate us to investigate further about the strategies of coping with AO-enabled portability in the rapidly changing landscape of mobile software development.

The goal of our work is to gain an in-depth understanding of the pros and cons when exploiting AO to address the mobile game's porting needs. To that end, we report an industrial case study by examining the software practices in two mobile game companies, one in Brazil and the other in Canada. We analyze two units (one game from each company) and study them from both technical and organizational perspectives. From a technical standpoint, we compare the implementations with and without the AO support, and find that AO reduces code coupling but causes increase in size. From an organizational standpoint, we analyze the project teams' communications and collaborations during game development, and find that AO could alleviate role coupling but increase business process latency.

The contributions of our work lie in the critical evaluation of AO as a porting strategy for mobile game development. Our study goes beyond the source code artifacts and expands to the project team's organizational characteristics. This novel synergy between technical and organizational levels is in line with the emerging trend of studying AO from social perspectives [24, 31]. Our findings therefore offer some concrete insights into developing AO-enabled reuse strategies in practice. Extending upon our recent work [4], the study presented here carries out further quantitative data analyses and explores how porting could be handled from the software ecosystem perspective [17, 23]. In what follows, we present background information and related work in Sect. 2. We then detail our research methodology in Sect. 3. Section 4 analyzes the results, describes the study implications, and discusses the threats to validity. Section 5 concludes the chapter.

2 Background and Related Work

Mobile game is a fast growing domain. According to Juniper Research [18], the mobile game market had global revenues of $2.1bn in 2011 and it is estimated to reach $17.6bn by 2016. Several characteristics of mobile game development are worth mentioning. The development of mobile games is greatly affected by time-to-market constraints [2]. Typically, the mobile game development cycle is very short, lasting between 4–6 months and is often rigid with respect to deadlines, e.g., a game based on a movie to be released or covering a major sports event could take more advantage if launched at a specific time frame. Such games are mainly developed by small companies. The development process is considered especially creative but still rarely follows a well-defined software engineering process. Thus, the quality of mobile games has been greatly dependent on the ability of game designers and developers rather than on good practices from other software domains [1, 19].

It has become increasingly recognized that one of the critical requirements in the mobile game domain is *portability*. It was estimated that, in 2011, the cost of porting represented approximately a third, i.e., $1bn, of total development costs in the mobile game domain [18]. The need of porting stems from a combination of technical and business constraints. Manufactures release in ever-shortening time periods different devices targeting diverse customer profiles. As a result, a large number of mobile devices coexist to serve different segments of the market with distinct needs and financial resources. In order to maximize revenue over the heterogeneous space of devices, mobile operators and publishers are required to develop games for the greatest possible number of users. This leads the game vendors to provide multiple versions and variants of the application, each optimized to a specific device [2].

The portability requirement is often stated in the business contract explicitly: games must be deployed in various devices, each with specific capacities, functionalities, and retail prices. Take a demanding case as an example, a single game was required to be ported to 69 different devices [2]. Due to the rapidly changing and increasingly diverse nature of the mobile game industry, the main challenges involved in fulfilling the portability requirement include:

- Different features of the devices regarding user interface, such as screen size, number of colors, screen resolution, sounds, and keyboard layout;
- Total heap capacity and maximum application size;
- Different behavior of the same API (application programming interface), due to unclear or flexible specification;
- Proprietary APIs and internationalization;

- Different software development environments, such as J2ME [28] and Brew [29]; and
- Different profiles in the same environment (e.g., MIDP 1.0 and MIDP 2.0 in J2ME) and optional packages.

Prior research has supported porting in various ways, including preprocessing tools [12], practice guidelines [13], program transformations [35], and incremental approaches [36]. In [2], we compared three porting mechanisms, namely incremental approach, program transformation, and preprocessing, based on granularity, cohesion, coupling, maintainability, and tool support. Our comparative analysis, together with those conducted by other researchers (e.g., [22]), shows that portability is a highly crosscutting concern in software development. In other words, platform variations affect a large number of modules. For example, memory constraints might demand omitting some game features in certain devices. Thus, these features become optional, whose implementation is often tangled with the implementation of other features. Another example is that device-specific APIs' invocations tend to be scattered throughout the code base, making it difficult to modularize by using traditional object-oriented (OO) techniques.

As porting-related concerns tend to cut across technological and organizational boundaries, researchers offer the new *software ecosystem* perspective to understanding these broadly scoped phenomena [17]. Messerschmitt and Szyperski [23] gave one of the earliest definitions of software ecosystem which is composed of a set of businesses functioning as a unit and interacting with a shared market for software and services, together with the relationships among them. Bosch [5] argued that reuse across traditional organizational boundaries was the main reason for software vendors to become networked in an ecosystem, and posited that software product line companies, such as mobile game vendors, were likely to transition to an ecosystem approach. The main strategic rationale was that software companies could gain dominance and sustainability by forging and enriching an ecosystem around their core applications underpinned by a common technological platform, such as Web and mobile [5]. We therefore expand our work from technical and organizational dimensions [4, 25] towards the entire software ecosystem.

A promising technique to facilitate porting is aspect-oriented programming. The basic idea, as illustrated in Fig. 1, is to modularize features into aspects and to use the weaving mechanism to bind features into different instances of the product line. Young [35] was among the first to integrate AspectJ with J2ME and demonstrated the feasibility of AO implementation on 18 device emulators from different platform providers including Nokia, Siemens, Blackberry, and Motorola. We advanced the literature by creating aspects in AspectJ to handle *portability* concerns [3], as well as extracting and evolving aspect code to reduce the substantial upfront effort associated with engineering a mobile game product line from scratch [1]. Despite these recent endeavors, little is known about how AO is practiced in industrial-strength mobile game applications. Next, we present a case study to fill the knowledge gap.

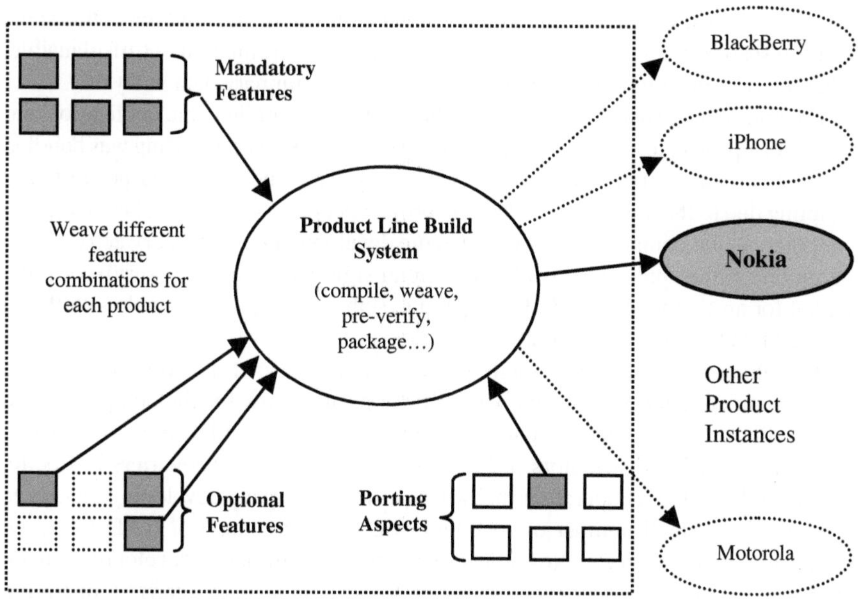

Fig. 1 Using aspects to build a mobile produce line (adapted from [35])

3 Research Methodology

We report in this section an exploratory case study [34] by collaborating with two companies on their mobile game projects within the projects' real-life contexts. Our overall goal is to gain concrete insights into how AO can help port mobile games in practice. The main rationale for adopting case study as the basis of our research design is that the investigation of a contemporary phenomenon is suitable for addressing the 'how' and 'why' questions that can otherwise be difficult to be answered through controlled experiments [11]. Essentially, the benefits and obstacles of using AO to tackle porting are only likely to be evident for the ongoing real-world projects, under conditions that cannot be replicated in the lab. In particular, the study of applying AO in mobile game development cannot be separated from the organizational context and the effects may take weeks or months to appear.

We therefore designed an *exploratory* case study to deepen our understanding about the practical impacts of AO on porting mobile games to multiple platforms. According to Yin [34], an exploratory case study is appropriate for preliminary inquiries and is ideal for analyzing what is common and different across cases that share some key criteria. The intention of adopting an exploratory case study as the basis for our research design is to gather data with the aim of deriving specific hypotheses that characterize the benefits and obstacles faced by mobile game development organizations; rather than attempting to confirming or refuting any preconceived hypotheses. We expect follow-up studies, including those of our own, to test the hypotheses in a more quantitative and definite manner. However, at this stage of the research, we feel that it is more important to formulate a set of plausible, coherent, or even competing hypotheses based on preliminary data analysis.

The unit of analysis in our study was the mixture of mobile game application and its vendor's project team. This choice allowed us to focus on how porting was handled within a software vendor in an autonomous way. The study had a purposeful case sampling due to the mingled product, and project perspectives. Our inclusion criteria were that: (i) the company does mobile game development as a primary activity; (ii) the mobile game application is of commercial significance and has survived in the market for an extended period of time; and (iii) the development team has explored the use of AO to fulfill the portability requirement.

We collaborated with two mobile game vendors that met the above criteria. To honor confidentiality agreements, we will use pseudonyms for the companies and their products throughout this chapter. Meantime is one of the leading mobile game developer-publisher-distributors in Latin America, with its headquarters in Recife, Brazil [1]. It started its operation in 2001 as a small company developing prototype of games for a small number of specific devices in the context of R&D (research and development) projects. Nowadays, the company's in-house development team has about 30 people, and develops games for several wireless carriers operating worldwide. Meantime has produced more than 60 games, both proprietary and third-party titles, in many game genres such as racing games, shooting games, puzzle games, etc.

We chose Rain of Fire [2], a shooting game developed at Meantime, to be a unit of analysis in our study. Figure 2 displays a screenshot of the game. In Rain of Fire, the player is the guard master of a city and controls ballistas and catapults to defend his or her town from several types of flying dragons with different speeds, powers, and attack patterns. It is not necessary to kill every dragon, but to destroy as many as possible in order to prevent the main city buildings from being destroyed.

FC—the other company that we collaborated with—is a small mobile game company located in Toronto, Canada. FC started in 1998 with six people specializing in real-time video game development. The company has approximately 50 employees as of 2009. We chose one of FC's proprietary games, Genuine Soccer, in our study. Two screenshots of the game are shown in Fig. 3. Genuine Soccer was first released in 2002 as an OEM (original equipment manufacturer) game bundled with a specific mobile device provider to exploit the business opportunities offered by the FIFA World Cup. After serially building different versions and variants of the product, FC began adopting software product line technologies to manage Genuine Soccer in 2005 [26]. In Genuine Soccer, the player can play a single game or in a tournament. The player can set line up, make trades, and control soccer player's moves like bicycle kick, elastico, slide tackle, penetrative pass, etc.

Table 1 summarizes some basic characteristics of the selected mobile games. Both games were developed by small teams within short cycles. The games were rigid with respect to deadlines, e.g., FC had taken more advantage by launching some versions of Genuine Soccer over the period of a major event like the FIFA World Cup. Both games were built upon the J2ME tenet [28].

Fig. 2 Screenshot of "Rain of Fire"

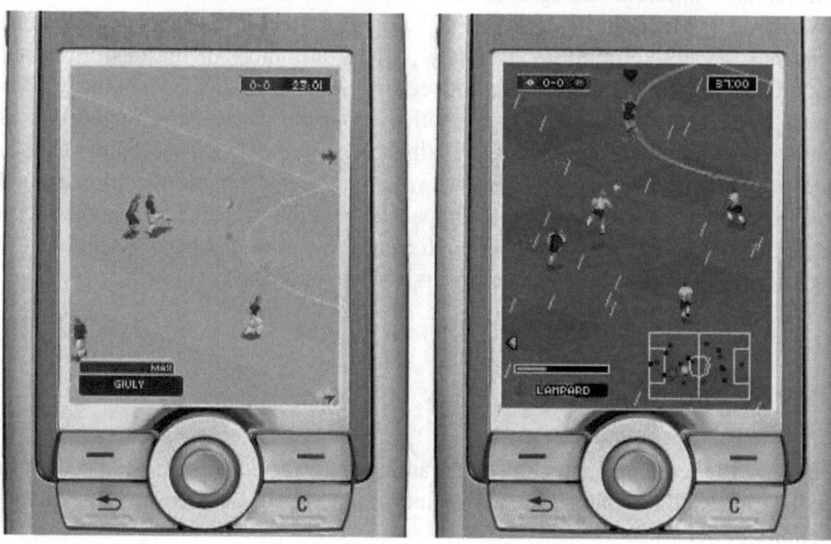

Fig. 3 Screenshots of "Genuine Soccer"

Table 1 Characteristics of the two units of analysis

	Rain of Fire	Genuine Soccer
Company size[1]	30	50
Project team Size[2]	10	9
Product longevity	8 years	9 years
Release cycle	~4 months	~4 months
Development environment	J2ME, Eclipse	J2ME, Eclipse
Platforms[3]	Motorola T720, Nokia series 40, Nokia series 60	BlackBerry OS, Nokia/Symbian S60, Google android
Scope & Maturity of using aspects	Organization-level and repeatable practice (mature)	Project-level and preliminary tryouts (immature)

[1] Company sizes are approximate as both companies are currently experiencing acquisition or hiring new staff
[2] See Sect. 4 for detailed discussion on the roles within the project team
[3] These platforms are what the companies have shared with the authors of this chapter, and are only representative of the diverse range of mobile platforms actually supported

It is worth pointing out that, though the companies shared with us only a partial list, there was a wide range of mobile platforms supported by the games. This reflects portability challenges posed by diverse mobile devices, as discussed in Sect. 2. Due to our inclusion criteria, both companies have explored AO in game development. While Meantime has accumulated extensive experience at the organizational level [1–3], FC restricted the use of aspects within some selected projects. This illustrates mobile game vendors' different strategies in adopting AO in practice.

We selected Meantime's Rain of Fire and FC's Genuine Soccer as ideal units of analysis in our exploratory case study [34] for a number of reasons. First, our selected games are representative in both companies. This indicates that the lessons learned from our case study are informative about the experiences of the typical situation. Second, the development of both games represents a longitudinal case in that the companies are able to assess the role of aspects from an evolutionary and retrospective perspective. Third, the companies and their project teams under study were highly cooperative and generous with regards to our research, so we anticipated a high degree of access to key stakeholders and projects' data.

4 Results and Analysis

We analyze the results from technical, organizational, and ecosystem perspectives in Sect. 4.1, present the implications of our findings in Sect. 4.2, and discuss the threats to validity in Sect. 4.3.

4.1 AO-Enabled Portability in Practice

Technical perspective analysis. In order to inquire about the role of AO in implementing portability and other mobile game features, we collected source code artifacts and related documentations. Note that FC shared with us only the OO code, mainly due to the Genuine Soccer's incomplete trial-and-error AO implementation. In contrast, Meantime shared both the OO and AO implementations, which allowed us to assess the effect of aspects on implementation by measuring metrics like size and coupling. Our analysis from the technical perspective therefore hinges on Rain of Fire's OO and AO code.

The OO and AO implementations Meantime shared with us are developed to allow Rain of Fire to be deployed into three mobile platforms: Motorola T720, Nokia S40, and Nokia S60. The game is developed using J2ME's MIDP profile [28]. For every platform, the OO implementation has 18 different Java class files with the code concerning platform dependencies and portability issues distributed among them. The AO implementation has those 18 Java class files with the same name as they are in OO, and 3 aspects for the different platforms to localize the portability concerns.

To illustrate the program transformation, Fig. 4a shows an OO code pattern which is refactored using aspect. Fig. 4b provides the AO counterpart that guarantees the correctness of the program transformation [3]. As can be seen from this example, AO causes an increase in size due to aspect refactoring and new declarations. Such an increase can be directly measured by the number of LOC (lines of code). Table 2 shows the comparison result. In the AO implementation, "Core" represents the LOC common to all the three platforms, whereas S40, S60, and T720 present the LOC for specific platforms. It can be noted from Table 2 that LOC is slightly higher when comparing each AO instance with the corresponding platform in the OO implementation. The main reason, as noted in [3], is due to aspect refactoring and new declarations.

The increase in the number of LOC further leads to an increase in the size of the packaged game application—in Rain of Fire's case, the increase is reflected in the jar file size for the application developed on J2ME [28]. Table 3 compares the jar file size between the OO and AO implementations of Rain of Fire. In this analysis, the packaged application includes the bytecode files along with all the resources, such as images and sound files, required to execute the game. The "reduced size" in Table 3 represents the resulting jar file size obtained by using a bytecode optimization tool [3]. While the reductions of OO and AO are comparable, AO does result in a greater size of the packaged application. In the mobile game domain, such an increase may not be neglected since the embedded software runs in platforms with limited storage capacity and processing power. Nevertheless, we expect that the rapid advancement of mobile hardware technology will make more room for the AO implementation. The above analyses suggest the following hypothesis:

Hypothesis 1: AO causes increase in size.

(a)

```
ts
class C {
  T f
  fs
  ms
  T1 m(ps) {
    body
    this.f = exp;
    body1
  }
}
```

(b)

```
ts
class C {
  fs
  ms
  T1 m(ps) {
    body
    body1
  }
}
privileged aspect A {
  T C.f;
  after(C cthis, ps) ret(T1 t):
    exec(T1 C.m(T2(ps)))
    && this(cthis)
    && args(aps) {
      cthis.f = exp[cthis/this];
    }
}
```

Fig. 4 **a** OO code snippet. **b** AO code snippet transforming the above OO code. Illustrations of Rain of Fire's OO and AO implementations

Table 2 LOC of the OO and AO implementations of Rain of Fire

OO implementation				AO implementation		
S40	S60	T720	Core	S40	S60	T720
2965	2968	3143	2549	3042	3047	3210

One of the major benefits of using AO is the ability to modularize crosscutting concerns like portability [20]. In this way, AO is aimed at reducing duplicates that would otherwise cause considerable maintenance problems. We use the CBO (coupling between objects) metric [8, 37] to assess the coupling levels in the OO and AO implementations. The basic idea of CBO is that two objects are coupled if the methods or instance variables belonging to one object are used by the methods of the other.

Table 3 Jar file size (kB) of Rain of Fire's OO and AO implementations

	OO implementation		AO implementation	
	Size	Reduced size	Size	Reduced size
S40	61.9	58.5	97.0	67.9
S60	61.7	57.3	97.6	61.8
T720	56.1	52.4	93.5	56.7
Total	179.8	168.2	288.2	186.3

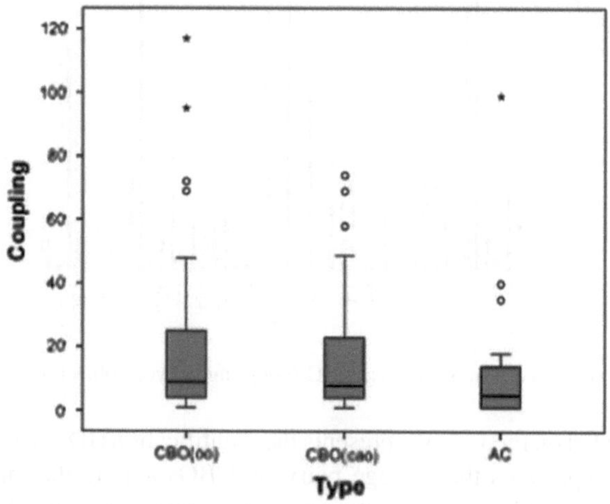

Fig. 5 Box plots of Rain of Fire's CBO (coupling between objects)

Rain of Fire's AO implementation has both Java classes and AspectJ classes (aspects) that address the crosscutting concerns. We calculate CBO for the Java classes in the traditional OO way [8]. However, coupling calculation between the classes and the aspects is different. An aspect may contain several types of attributes and modules: advice, intertype declaration, pointcut, and method. We therefore follow the procedures proposed by Zhao [37] to calculate the coupling in the AO code.

Let us consider a class C and an aspect A. If a module in A (i.e., advice, intertype declaration, pointcut, or method) uses some methods or instance variables defined in C, then the number of such instance variables and/or methods is the measure of dependency between C and A [37]. Note that CBO for two OO classes C_1 and C_2 is the sum of two parts: CBO from C_1 to C_2 and CBO from C_2 to C_1, whereas CBO for a class C and an aspect A is calculated in only one direction: CBO from A to C.

Figure 5 shows the box plots [33] of Rain of Fire's CBO where CBO(oo) represents the CBO for the 18 classes in the OO implementation, CBO(cao) represents the CBO for the 18 classes in the AO implementation, and AC represents the CBO for aspects and classes in the AO implementation. According to Fig. 5, the median of CBO(cao) is slightly less than that of CBO(oo), and the median of AC is considerably less than that of the others. Figure 6 shows further descriptive statistics about the CBO levels.

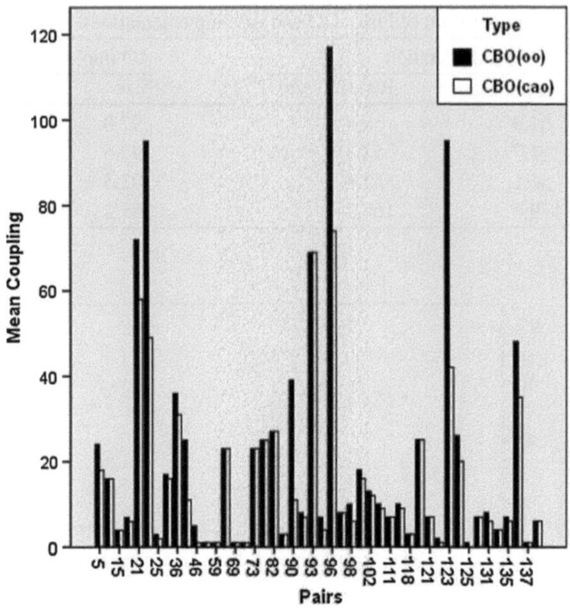

Fig. 6 Bar charts of Rain of Fire's pairwise CBO (coupling between objects)

In Fig. 6, each label on the x-axis represents the identification (ID) of a pair of classes, and the y-axis represents the average pairwise CBOs among the three platforms: T720, S40, and S60. Please note that using class names along x-axis to identify a pair of classes is not feasible because of space constraints. Therefore, we use distinct identification numbers to represent class pairs. It is noticeable from Fig. 6 that in most cases CBO(cao) is less than or equal to CBO(oo). In no case do we find CBO(cao) was greater than CBO(oo). The findings therefore show that the coupling level is indeed reduced in AO compared to its OO counterpart.

To test whether the difference is statistically significant, we perform a one-way ANOVA (analysis of variance) over the collected CBO measures. Table 4 lists the inferential statistics where no statistically significant difference between AO's CBO and OO's CBO is detected at the α=0.05 level ($F = 1.128$, $p = 0.291$). Based on the descriptive and inferential statistical analyses, we formulate the following hypothesis:

Hypothesis 2: AO reduces code coupling.

Despite the positive effect of AO on CBO, our preliminary analysis could not render statistically significant results. Fulfilling AO's fundamental promise of reducing coupling [20] in mobile game development thus needs improved techniques and further evaluations, e.g., testing code coupling level by using larger datasets, devising improved metrics targeting at commercial mobile games that support many platforms, and corroborating findings from other investigations like [14, 15].

Table 4 ANOVA results for CBO (coupling between objects)

Source	Type III sum of squares	df	Mean square	F	Sig.
Corrected model	604.730[a]	1	604.730	1.128	0.291
Intercept	30806.048	1	30806.048	57.467	0.000
Type	604.730	1	604.730	1.128	0.291
Error	47709.556	89	536.062		
Total	79219.000	91			
Corrected total	48314.286	90			

[a] R squared = 0.013 (Adjusted R squared = 0 .001)

Table 5 Adjacency matrix of meantime's roles in developing Rain of Fire

	A	B	C	D	E	F	G	H
A	0	4	4	3	4	4	3	4
B	4	0	2	0	1	1	4	3
C	4	2	0	4	4	4	2	4
D	3	0	4	0	1	1	0	1
E	4	1	4	1	2	4	1	4
F	4	1	4	1	4	2	1	2
G	3	4	2	0	1	1	0	2
H	4	3	4	1	4	2	2	2

A, Project manager; *B*, Product manager; *C*, Architect; *D*, Administrator; *E*, Programmer; *F*, Non-programming expert; *G*, User; *H*, Quality assurance (QA) engineer

Table 6 Adjacency matrix of FC's roles in developing Genuine Soccer

	I	J	K	L	M	N
I	0	3	2	2	4	2
J	3	0	4	4	4	3
K	2	4	3	3	2	3
L	2	4	3	2	1	2
M	2	2	2	2	0	1
N	3	3	4	2	1	2

I, Project manager; *J*, Architect; *K*, Programmer; *L*, Non-programming expert; *M*, User; *N*, QA engineer

Organizational perspective analysis. We now turn the focus of our analysis from the technical side to the organizational side. Inspired by the research on sustainable patterns within software organizations [10], we surveyed the roles and their communication paths in the mobile game projects. Our goal was to identify the social styles of mobile game development, discover the organization patterns of the project team, and analyze AO's impacts beyond the source code level. To that end, we collected two completed questionnaires from each company's project team. We then built organizational models based on the data, drawing on social network analysis [32].

The survey results show that the mobile game team had only a few roles: 8 in Meantime and 6 in FC (cf. Tables 5 and 6). According to both companies, these roles

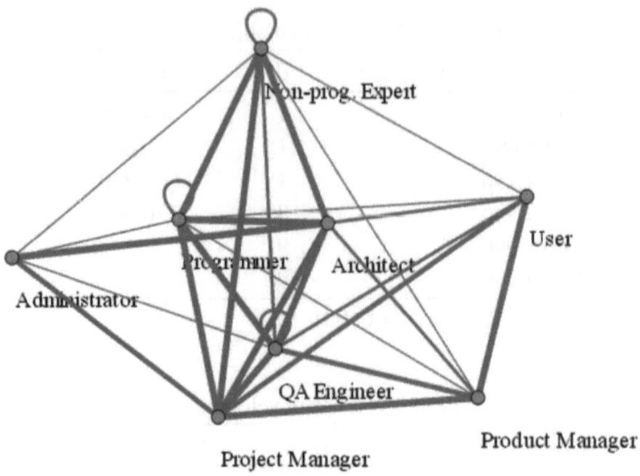

Fig. 7 Sociogram for Meantime's Rain of Fire team

became stable over time, more so than processes and even personnel. Keeping the number of roles low and their responsibilities explicit made it easier for new people to assimilate the organizational culture and become part of it.

For each pair of roles, we asked the respondent to use an ordinal scale to rate their communication: 0—no communication, 1—weak communication, and 2—strong communication. The ratings were intentionally designed to be coarse grained because a finer grained (interval) scale might be difficult to apply at the current exploratory stage. For each company, we aggregated the two responses by adding their individual ratings without altering the original inputs. The aggregated adjacency matrices are shown in Tables 5 and 6. Note that, in general, an adjacency matrix should not be assumed to be symmetric along the diagonal [10]. The rating at cell (p, q) indicates the strength of the interaction p initiates with q. In FC's development team, for instance, the project manger (***I***) initiates stronger interactions with the user (***M***) than the other way around.

We used the SocNetV tool (http://socnetv.sourceforge.net) to perform social network analysis based on the adjacency matrix. In particular, we generated sociograms [32] via SocNetV's force-based placement technique. The algorithm first assigns all nodes with random coordinates, then sets up a repelling force between all pairs of nodes following an inverse square law [32]. Arcs exert an attracting force between the nodes they connect according to the values in the adjacency matrix. The graph reaches a stable state when all of the nodes migrate to positions where their forces balance. The algorithm creates a spatial representation of an organization's interaction graph that offers quick intuitive insights into the coupling between roles [10].

The sociogram for Meantime's team and FC's team is shown in Figs. 7 and 8 respectively. Each node corresponds to an organizational role. Each arc corresponds to a communication path between roles: the stronger the communication

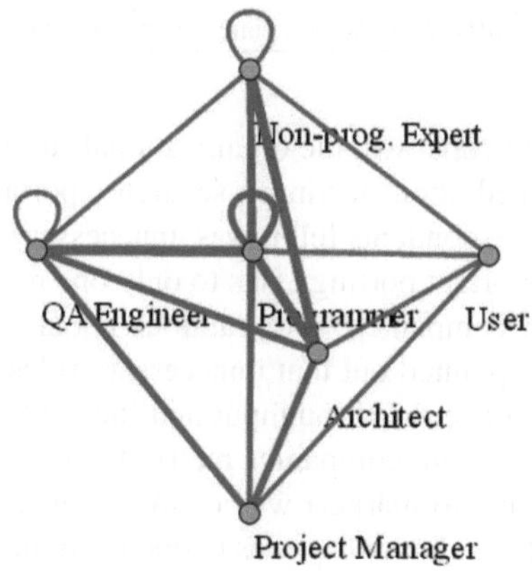

Fig. 8 Sociogram for FC's Genuine Soccer team

and interaction between a pair of roles, the thicker the arc. The self-loop indicates that interactions exist between the people who serve in the same role. The mobile game teams in both companies were highly coupled, as depicted by the dense graphs in Figs. 7 and 8. The density of Meantime's sociogram is 0.98 and that of FC's is 1. As responded by one member of FC, "*in our team, everybody talks to everybody*".

The force-based algorithm places the *programmer* role at the center of both projects. This producers-in-the-middle pattern [10] implies that programmers contribute directly to the end product and company revenue. One way to gain insights into the modularity of the organizational structures is by grouping roles into clusters so that each cluster is characterized by an internal coherence and/or an external isolation.

The *architect* role, according to the sociograms, became a strong candidate to be clustered into a coherent organizational module as the middle programmer role. In both projects, architect worked very closely with programmers. In fact, beyond advising and communicating with programmers, the architects in both companies also participated in implementation, including the implementation of porting strategies. In this way, the organizations perceived buy-in from the guiding architects, and that perception directly availed itself of architectural expertise.

The clustering analysis also helped detect support roles that were not directly involved in software development or porting. For example, Meantime's administrator was not an expert on mobile porting. Due to communication overhead, there might be opportunities to gain efficiency by combining support roles. In FC, for instance, the project manager also acted as product manager and administrator.

The key findings from the social network analysis are that: (1) mobile game development teams have a few number of roles that are highly coupled, (2) these roles become stable in the organization, (3) programmers play a central role, and (4) architects also implement. Some respondents felt that the coupling level was too high in the project team: "*almost all tasks affect the whole team,*" "*every role has similar importance and (porting) failures can easily lead the project to a challenged state,*" and "*a further division might be helpful.*" Accordingly, we derive the following hypothesis:

Hypothesis 3: AO alleviates team coupling.

While extending AO to cope with the organizational structure may help alleviate team coupling, when asked about having a separate "portability engineer" in the development team, most respondents felt it was unnecessary. On one hand, it would be too risky to assign important porting tasks to only one role or one team member. On the other hand, both companies were cautious about creating any new roles. One respondent from FC pointed out that unnecessary roles or artificially complex communications could reduce the throughput and increase the latency of business process. Respondents from both companies realized that, in mobile game business processes, speed (e.g., time-to-market) was of the essence. FC's experience suggested that taking advantage of coupling was to open communication paths between roles to increase the overall coupling/role ratio, particularly between central process roles [26]. We therefore speculate that:

Hypothesis 4: AO increases business process latency.

Similar to the two hypotheses formulated from technical perspective, our investigation into the organizational structures reveals both positive and negative impacts of AO. To some extent, the two hypotheses formed here can be thought of competing. Specifically, Hypothesis 3 is derived from AO's premise that localizing crosscutting concerns and decreasing coupling would be beneficial. On the other hand, Hypothesis 4 argues that leveraging coupling or even furthering coupling would be beneficial, e.g., occasional close coupling between programmers and testers (QA engineers) could reduce administrative overhead and latency [10]. We will provide more discussions on these views in Sect. 4.2.

Software ecosystem perspective analysis. Our final analysis takes the software ecosystem standpoint on porting. The companies we studied both fit into two of the taxonomic characteristics described by Bosch [5]: (1) operating system-centric software ecosystem, and (2) application-centric software ecosystem. In the former, platform provider examples are J2ME [28] and Brew [29], whereas external developers are the mobile game companies. In the latter, the platform providers are the mobile game companies, whereas external developers can be companies specialized in porting. According to Bosch [5], the application-centric software ecosystem type is the most viable option for most companies employing a software product line, since the myriad of product line members attracts a customer base that is potentially attractive to third-party developers.

Based on a survey on how the mobile game development teams might handle portability issues in ongoing and upcoming projects, the respondents from both organizations expressed strong consensus that porting could be addressed by a third-party company. It was raised that some organizations could become specialized to work on GUI components, others on specific platform bugs and details, and still others work only with mobile tests (with mobile phones for a particular country and corresponding regulations). Indeed, the high customization requirement by the mobile game domain exceeds the development capacity within the closed mobile game company's boundary, thus demanding a software ecosystem approach.

However, it was also raised that the high coupling between the teams, in terms of dense communication pattern described earlier, was an issue to make porting to be addressed completely independently by third-party developers. Another key point raised by the respondents was the need for a well-defined process and some robust tools for specification and communication. This is consistent with the software engineering challenges required for transitioning to a software ecosystem, since the architecture becomes a key artifact in the incurred distributed coordination. We therefore formulate our final hypothesis as follows:

> ***Hypothesis 5***: Mobile game porting is such a broadly scoped concern that it can be better handled over the software ecosystem than within a software organization's product line engineering.

As we have discussed earlier, due to the high team coupling, some mobile game development companies are not comfortable involving third-party vendors and handle portability issues themselves. However, the diverse customization required to handle portability exceeds the development capacity of some mobile game vendors. These vendors outsource portability handling to third-party developers or vendors, thereby taking the software ecosystem approach to address portability. In our future works, we plan to study these two ways of handling portability side by side in order to obtain further insights about Hypothesis 5.

4.2 Study Implications

In summary, our case study shows that AO caused the increase in size mainly due to the introduction of programming constructs like pointcut and advice. While this may not be neglected for memory-sensitive applications and platforms, the size increase could be offset by the reduced code coupling due to aspect's ability to localize the crosscutting concerns like portability that would otherwise become tangled and scattered. In another word, having a more AO-enabled modular decomposition could contribute positively to the system's adaptability and maintainability. However, AO's coupling reduction was found to be positive in our study, but not statistically significant. This implies that more efficient methods (e.g., devising programming constructs for embedded systems [6] and domain-specific instrumentations [21]) are needed to best realize AO's fundamental promise of reducing coupling [20] for mobile game development.

The organizational side of our inquiry is novel and can be positioned in the contexts of Conway's law [9] and socio-technical congruence [7]. Specifically, if the mobile game vendor creates porting aspects and weaves them into base modules in the code base, then the software's interface structure should be congruent with the vendor's social structure. This indicates that, in Meantime, aspect programmers already acted as "portability engineers" though such a role or title has yet to appear.

Making crosscutting concerns explicit and using advanced modularity constructs to model these concerns are what AO is about. Following Conway's law [9], AO-embraced organizations would establish a distinct role that handles crosscutting responsibilities. For example, a porting role could be made explicit to take charge of porting games (e.g., racing, puzzle, and arcade) to specific devices. Note that organizational change is risky, so one should add or refactor roles between releases to minimize any turmoil that might confuse work in progress.

Relating to our findings over the technical side, AO's main contributions were positive but not significant. This helps to explain why FC's AO investment stayed at an experimental level, as well as why both companies hesitated to create or designate roles like "portability engineer". In fact, our study reveals that the tension exists between creating any new roles and raising the communication overheads. We therefore speculate that, for small-sized mobile teams, new titles seem unnecessary or even counterproductive.

Extending our analysis toward an ecosystem dimension is also fruitful in that viewing mobile game development as an application-centric software ecosystem enables new strategies of handling porting. In particular, delegating porting tasks to some specialized third-party developers and/or companies might help achieve a sustainable balance over the entire mobile game supply network. In fact, building on their rich experiences in porting, Meantime has begun offering porting-related services like game publishing and distribution. Testing our ecosystem-based hypothesis (Hypothesis 5) can be done in directed or undirected ways [5]. In the directed approach, the mobile game vendor selects partners that are able to provide the porting service and negotiates an agreement with the partners typically involving a form of revenue sharing. This is partially done in the publishing service of the mobile game domain, but porting is still limited. In the undirected approach, the platform is opened without constraining who accesses it to do the porting. This is still largely unexplored in the mobile game domain, but is relevant since it provides a higher potential of scalable porting and innovation.

4.3 Threats to Validity

Several factors can affect the validity of our exploratory case study. Construct validity concerns establishing correct operational measures for the concepts being studied [34]. The main construct in our case study is 'AO-enabled portability' in mobile game development. We intentionally operationalized the main construct in a broad way by considering technical and organizational viewpoints and their relationships.

Regarding internal validity [34], our major sources of data for this study were the OO and AO code and the surveys that we conducted with the two game vendors. The sources of code files should be reliable. For the CBO analysis, we followed the procedures proposed by Zhao [37] in order to calculate the coupling in the AO code. However, our reliance on questionnaires and interviews meant that we needed to trust each participant's description of their own experiences. Participants may have omitted important facts, or we may have misinterpreted them. To address the former threat, we used a coarse grained rating scale for rating the roles' communications. To address the latter threat, we plan to carry out some field observations in the short term as our case study progresses.

The results of our study may not generalize beyond Meantime's and FC's organizational conditions and the Rain of Fire's and Genuine Soccer's project characteristics, a threat to external validity [34]. Both game vendors in our study were headquartered in wireless-carrier-rich areas and survived in the mobile game business for about a decade. This might have generated geographical and industrial-segment biases. We therefore are not certain if our findings apply to mobile game development companies around the world, or those that came into operation for a short period of time. Nevertheless, our investigation of the contemporary projects within their real-life contexts, together with the validation carried out in real industry settings, provides a firm footing for exploiting AO to handle portability concerns.

Finally, in terms of reliability [34], we expect that replications of our study should offer results similar to ours. Of course, the characteristics of each mobile game and its development team under study will differ from our current study, but the underlying trends and implications should remain unchanged.

5 Conclusions

Portability is a broadly scoped concern in mobile software development in that it has an implicit or explicit impact on more than one module. Treating portability as an aspect in mobile game domains is similar to considering persistence as an aspect in data-dominant domains [30]. However, the challenges presented by portability have moved beyond the source code or any other technical artifact's level.

In this chapter, we have conducted an exploratory case study to investigate the state of practice in porting mobile games via aspect orientation. Not only were the AO and OO implementations compared and contrasted, but the organizational patterns and the ecosystem-wide considerations were also analyzed and explored. Overall, we found that modular code was enabled by AO but at the cost of the increase in size. Although a less coupled code base is considered better from the implementation point of view, the mobile game development teams in our study enjoyed high coupling among a few number of stable roles. Such an incongruence, in our opinion, is a fundamental factor that affects the wide adoption of AO in the mobile game domain. At the same time, the incongruence opens up the opportunity of exposing porting-related issues toward the software ecosystem by delegating certain tasks to highly specialized companies over the supply chain.

Our future work includes carrying out further empirical studies to lend strength to the exploratory findings reported here. For example, games targeted at both mobile and tablet platforms can be investigated. We also plan to examine more thoroughly the AO usage barriers and find novel ways to overcome these barriers. Finally, we would like to conduct systematic mobile game ecosystem studies, in both directed and undirected fashions [5, 17, 23, 25], to discover effective reuse strategies across organizational boundaries.

Acknowledgments We thank the partner companies for the generous support throughout our study, especially for sharing their data, time, and expertise. The work is in part supported by the U.S. NSF (National Science Foundation) Grant CCF-1238336.

References

1. Alves, V., Camara, T., Alves, C.: Experiences with mobile games product line development at meantime. In: International Conference on Software Product Lines (SPLC), pp. 287–296. IEEE Press, New York (2008)
2. Alves, V., Cardim, I., Vital, H., Sampaio, P.H.M., Damasceno, A.L.G., Borba, P., Ramalho, G.: Comparative analysis of porting strategies in J2ME games. In: International Conference on Software Maintenance (ICSM), pp. 123–132. IEEE Press, New York (2005)
3. Alves, V., Matos, P., Cole, L., Vasconcelos, A., Borba, P., Ramalho, G.: Extracting and evolving code in product lines with aspect-oriented programming. Trans. Aspect-Oriented Softw. Dev. **4**, 117–142 (2007)
4. Bhowmik, T., Alves, V., Niu, N.: Porting mobile games in an aspect-oriented way: an industrial case study. In: International Conference on Information Reuse and Integration (IRI), pp. 458–465. IEEE Press, New York (2013)
5. Bosch, J.: From software product lines to software ecosystems. In: International Conference on Software Product Lines (SPLC), pp. 111–119. Carnegie Mellon University, Pennsylvania (2009)
6. Cardoso, J.M.P., Carvalho, T., Coutinho, J.G.F., Luk, W., Nobre, R., Diniz, P.C., Petrov, Z.: LARA: an aspect-oriented programming language for embedded systems. In: International Conference on Aspect-Oriented Software Development (AOSD), pp. 179–190. ACM Press, New York (2012)
7. Cataldo, M., Herbsleb, J.D.: Socio-technical congruence: a framework for assessing the impact of technical and work dependencies on software development productivity. In: International Symposium on Empirical Software Engineering and Measurement (ESEM), pp. 2–11. ACM Press, New York (2008)
8. Chidamber, S.R., Kemerer, C.F.: A metrics suite for object oriented design. IEEE Trans. Software Eng. **20**, 476–493 (1994)
9. Conway, M.: How do committees invent? Datamation **14**, 476–493 (1968)
10. Coplien, J.O., Harrison, N.B.: Organizational Patterns of Agile Software Development. Prentice Hall, New Jersey (2004)
11. Easterbrook, S., Singer, J., Storey, M.-A., Damian, D.: Selecting empirical methods for software engineering research. In: Shull, F., Singer, J., Sjøberg, D.I.K. (eds.) Guide to Advanced Empirical Software Engineering, pp. 285–311. Springer, London (2008)
12. Enough Software: J2ME Polish. http://www.j2mepolish.org
13. Facon, X.: Porting your MIDlets to new devices. http://www.webpronews.com/porting-your-midlets-to-new-devices-2003-05
14. Figueiredo, E., Cacho, N., Sant'Anna, C., Monteiro, M., Kulesza, U., Garcia, A., Soares, S., Ferrari, F.C., Khan, S.S., Filho, F.C., Dantas, F.: Evolving software product lines with aspects: an empirical study on design stability. In: International Conference on Software Engineering (ICSE), pp. 261–270. IEEE Press, New York (2008)

15. Figueiredo, E., Sant'Anna, C., Garcia, A., Bartolomei, T.T., Cazzola, W., Marchetto, A.: On the maintainability of aspect-oriented software: a concern-oriented measurement framework. In: European Conference on Software Maintenance and Reengineering (CSMR), pp. 183–192. IEEE Press, New York (2008)
16. Frakes, W.B., Fox, C.J.: Sixteen questions about software reuse. Commun. ACM **38**, 75–87 (1995)
17. Jansen, S., Finkelstein, A., Brinkkemper, S.: A sense of community: a research agenda for software ecosystems (NIER Track). In: International Conference on Software Engineering (ICSE), pp. 187–190. IEEE Press, New York (2009)
18. Juniper Research: Looking at the future of mobile games. http://www.juniperresearch.com
19. Kanode, C.M., Haddad, H.M.: Software engineering challenges in game development. In: International Conference on Information Information Technology: New Generations (ITNG), pp. 260–265. IEEE Press, New York (2009)
20. Kiczales, G., Lamping, J., Mendhekar, A., Maeda, C., Lopes, C.V., Loingtier, J.-M., Irwin, J.: Aspect-oriented programming. In: European Conference on Object-Oriented Programming (ECOOP), pp. 220–242. IEEE, Finland (1997)
21. Marek, L., Villazón, A., Zheng, Y., Ansaloni, D., Binder, W., Qi, Z.: DiSL: A domain-specific language for bytecode instrumentation. In: International Conference on Aspect-Oriented Software Development (AOSD), pp. 239–250. ACM Press, New York (2012)
22. Menon, H.: Portability analysis in mobile gaming using J2ME. Master's Thesis, West Virginia University, USA (2006)
23. Messerschmitt, D.G., Szyperski, C.: Software Ecosystem: Understanding an Indispensable Technology and Industry. The MIT Press, Cambridge (2005)
24. Murphy, G.C.: Everyday aspects (keynote presentation). In: International Conference on Aspect-Oriented Software Development (AOSD), pp. 229–230. ACM Press, New York (2009)
25. Niu, N., Alves, V., Bhowmik, T.: Portability as an aspect: rethinking modularity in mobile game development. In: International Conference on Aspect-Oriented Software Development (AOSD), pp. 3–4. ACM Press, New York (2011)
26. Niu, N., Easterbrook, S.: Concept analysis for product line requirements. In: International Conference on Aspect-Oriented Software Development (AOSD), pp. 137–148. ACM Press, New York (2009)
27. Niu, N., Easterbrook, S., Yu, Y.: A taxonomy of asymmetric requirements aspects. In: International Workshop on Early Aspects (EA), pp. 1–18. Springer, London (2007)
28. Oracle: Jave 2 Platform Micro Edition. http://www.oracle.com/technetwork/java/javame
29. Qualcomm: Brew. http://www.brewmp.com
30. Rashid, A., Chitchyan, R.: Persistence as an aspect. In: International Conference on Aspect-Oriented Software Development (AOSD), pp. 120–129. ACM Press, New York (2003)
31. Rinard, M.: Sociological aspects of aspect-oriented programming (keynote presentation). In: International Conference on Aspect-Oriented Software Development (AOSD). ACM Press, New York (2010)
32. Scott, J.: Social Network Analysis: A Handbook. Sage Publications, London (2000)
33. Tukey, J.: Exploratory Data Analysis. Addison-Wesley, Boston (1977)
34. Yin, R.K.: Case Study Research: Design and Methods. Sage Publications, London (2003)
35. Young, T.J.: Using aspectJ to build a software product line for mobile devices. Master's Thesis, University of British Columbia, Canada (2005)
36. Zhang, W., Jarzabek, S., Loughran, N., Rashid, A.: Reengineering a PC-based system into the mobile device product line. In: International Workshop on Principles of Software Evolution (IWPSE), pp. 149–160. IEEE Press, New York (2003)
37. Zhao, J.: Measuring coupling in aspect-oriented systems. In: International Software Metrics Symposium (METRICS), pp. 14–16. Chicago (2004)

Developing Frameworks from Extended Feature Models

Matheus Viana, Rosângela Penteado, Antônio do Prado and Rafael Durelli

Abstract Frameworks are composed of concrete and abstract classes implementing the functionality of a domain. Applications can reuse framework design and code to improve their quality and be developed more efficiently. However, framework development is a complex task, since it must be adaptable enough to be reused by several applications. In this chapter we present the From Features to Framework (F3) approach, which aims to facilitate the development of frameworks. This approach is divided in two steps: Domain Modeling, in which framework domain is defined in a extended feature model; and Framework Construction, in which the framework is designed and implemented by following a set of patterns from its feature model. Since these steps can be systematically applied, we also present the design of a tool that supports the use of the F3 approach on framework development. Moreover, we performed an experiment that showed that the F3 approach makes framework development easier and more efficient.

Keywords Software reuse · Domain modeling · Patterns · Framework · Domain-specific modeling languages

M. Viana (✉) · R. Penteado · A. do Prado
Department of Computing, Federal University of São Carlos, São Carlos, SP 13565-905, Brazil
e-mail: matheus_viana@dc.ufscar.br

R. Penteado
e-mail: rosangela@dc.ufscar.br

A. do Prado
e-mail: prado@dc.ufscar.br

R. Durelli
Institute of Mathematical and Computer Sciences, University of São Paulo, São Carlos, SP 13566-590, Brazil
e-mail: rdurelli@icmc.usp.br

T. Bouabana-Tebibel and S. H. Rubin (eds.), *Integration of Reusable Systems*, Advances in Intelligent Systems and Computing 263, DOI: 10.1007/978-3-319-04717-1_12,

1 Introduction

Software engineering focuses on more efficient ways to develop quality software [1]. One of the most common practices to achieve this goal consists on the reuse of artifacts. Through reuse, the time spent to develop software is reduced because it is not developed from scratch and its quality is improved because the reused artifacts were previously tested [2]. Programming languages offer different ways of reuse, such as subprograms and inheritance, but all of them are limited to code-level. Therefore, some solutions, such as frameworks, have emerged to provide reuse other than code.

Frameworks are composed of concrete and abstract classes that implement the functionality of a domain [3]. Applications reuse the design and code of a framework by adding their specific characteristics to its functionality [4]. Hence, a framework can be seen as a skeleton that provides a basic structure for applications, which customize this structure according to their requirements [5].

Despite the advantages frameworks offer, they are more complex to develop than applications [6]. Frameworks demand an adaptable design. Their classes will be reused by applications that are unknown during framework development, thereby frameworks need mechanisms to identify and to access application-specific classes. Thus, design patterns and advanced resources of programing languages, such as abstract classes, interfaces, polymorphism, generics and reflection, are often used in framework development. In addition to design and implementation complexities, it is also necessary to determine the domain of applications of the framework, the features that compose this domain and the rules that constraint these features [7].

In a previous chapter we presented an approach for building Domain-Specific Modeling Languages (DSML) to facilitate framework reuse [8]. In that approach a DSML could be built by identifying the features of the framework domain and the information required to instantiate them. Then application models created with the DSML could be used as input for an application generator to transform them into application code. The experiment showed in Viana et al. [8] indicated that, besides the gain of efficiency obtained from code generation, the use of DSML protects developers from framework complexities.

In this chapter we present the From Features to Framework (F3) approach, in which framework domain is defined in an extended version of feature model. Then a set of patterns guides the developers to design and implement the framework according to its domain.

As the F3 approach defines specific steps and rules for designing and implementing the framework, it turns framework development into a systematical process and, therefore, it can be automatized. Thus, in this chapter we also present the design of a tool, called F3T, that supports the use of the F3 approach, making the process of framework development even more efficient.

We also have carried out an experiment in order to verify whether the F3 approach leads the developer to devise frameworks better than the adhoc one. The experiment showed that the F3 approach reduced the problems of incoherence, structure, bad smells and interface found in the outcome frameworks and, consequently, reduced the time spent to develop these frameworks.

The remainder of this chapter is organized as follows: the background concepts applied in this research are discussed in Sect. 2; the F3 approach is presented in Sect. 3; the design of the F3T is presented in Sect. 4; an experiment to evaluate the F3 approach is show in Sect. 5; some related works are discussed in Sect. 6; and conclusions and further works are presented in Sect. 7.

2 Patterns, Frameworks and Domain Engineering

Patterns are successful solutions that can be reapplied to different contexts. They provide reuse of experience to help developers to solve common problems [3]. The documentation of a pattern usually contains its name, the context it can be applied, the problem it is intended to solve, the solution it proposes, illustrative class models and examples of use [9].

Frameworks are reusable software that can be instantiated into applications [3]. According to the way they are reused, frameworks can be classified as: white box, which are reused by class specialization; black box, which work like a set of components; and gray box, which are reused by the two previous ways. Unlike library classes, whose execution flux is controlled by applications, frameworks control the execution flux accessing the application-specific code [5]. Frameworks are composed of two main parts: frozen spots, which implement the common functionality of the framework domain, regardless the application reusing the framework; and the hot spots, which change according to the specifications of the application reusing the framework [4].

Domain engineering represents software development related not to a specific application, but to a domain of applications that share common features [10, 11]. A feature is a distinguishing characteristic that aggregates value to applications. Thus, feature models are often used to model domains, illustrating the features that mandatory or optional, variations and require or exclude other features [12].

Different domain engineering approaches can be found in the literature [11, 13, 14]. Although there are differences between them, the basic idea of these approaches is to identify the features of a domain and to develop the artifacts that implement these features and are reused in application engineering.

Domains can also be modeled with metamodel languages, which are used to create Domain-Specific Languages (DSL) [15]. Metamodels are similar to class models, which makes them more appropriate to developers accustomed to the UML. While in feature models, only features and their constraints are defined, metaclasses in the metamodels can contain attributes and operations. On the other hand, feature models can define dependencies between features, while metamodels depend on declarative languages to do it [15].

3 From Features to Frameworks Approach

The F3 approach allows the definition of a domain at a high level of abstraction and to systematically construct a framework which implements this domain. Thereby, framework development is divided into two steps: (A) Domain Modeling, in which a domain is defined and modeled; and (B) Framework Construction, in which the framework is designed and implemented according to its domain.

3.1 Domain Modeling

The domain of applications that can be developed with the framework is defined and modeled in this step. Usually, a domain is defined by analyzing applications that belong to the desired domain or consulting an specialist in this domain [12–14]. The first features to be identified are the mandatory ones, because they represent the code asset of the domain and the minimum necessary to develop an application. Then optional features and variants can be added and the dependencies between them can be specified. It is possible to develop a small domain framework at first and keep increasing it as more features are required.

In the F3 approach, domains are modeled in F3 models, which are feature models that includes some elements of metamodels, such as attributes, operations and multiplicity, so that frameworks can be developed from these models. As in conventional feature models, features in F3 models must be arranged in a tree-view, in which the main feature is decomposed in others. However, F3 models do not necessarily form a tree, since a feature can have a relationship targeting a sibling or even itself. Moreover, the graphical notation of F3 models is similar to the one of UML class models. This notation has been adopted because it allows that F3 models can be created by using any UML tool. The elements and relationships in F3 models are:

- **Feature**: graphically represented by a rounded square, it must have a name and it can contain any number of attributes and operations;
- **Decomposition**: relationship that indicates that a feature is composed of another feature. Its minimum multiplicity indicates whether the target feature is optional (`0`) or mandatory (`1`). Its maximum multiplicity indicates how many instances of the target feature can be associated to each instance of the source feature. In white box frameworks an instance of a feature is an application class that extends the framework class of this feature. The maximum multiplicity can assume the following values: `1` (simple), for a single feature instance; `*` (multiple), for a list of a single feature instance; and `**` (variant), for a list of different feature instances.
- **Generalization**: relationship that indicates that a feature is a variation and it can be generalized by another feature.
- **Dependency**: relationship that define constraints for feature instantiation. There are two types of dependency: `requires`, when the A feature requires the B feature, an application that contains the A feature has to include the B feature as

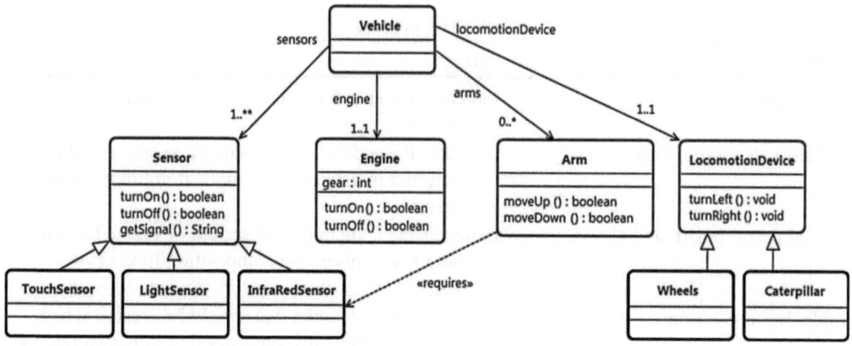

Fig. 1 F3 model for the domain of automated vehicles

well; and `excludes`, when the A feature excludes the B feature, no application can include both features at the same time.

A simplified F3 model for the domain of automated vehicles is shown in Fig. 1. This domain is based on Lego Mindstorms NXT 2.0,[1] whose hardware can be controlled by Lejos Java API.[2] The requirements of this domain are:

1. An automated vehicle is composed of one engine, one locomotion device, zero or more arms and one or more sensors.
2. Engine provides power for the vehicle. It can be turned on and off and movement sense (forward/reverse) is controlled by its gear.
3. Locomotion device uses the power generated by the engine to move the vehicle and to change its direction. There are two types: wheels or caterpillar.
4. Sensors collect information from environment and return a signal. How the vehicle interprets this signal depends on the purpose of the vehicle. There are three types of sensor: infra-red, light and touch.
5. Arms can be used to grab objects. They can only move up and down and require a infrared sensor.

As there are different types of sensor, the relationship between `Vehicle` and `Sensor` is a variant decomposition. A vehicle must have at least one sensor of any type. However, when it has arms, touch sensor becomes necessary, hence the `requires` dependency between `Arm` and `TouchSensor`.

3.2 *Framework Construction*

The F3 approach define a set of patterns to assist developers to design and implement a framework from the domain model. These patterns solves problems that go from

[1] http://mindstorms.lego.com/en-us/products/default.aspx#t

[2] http://lejos.sourceforge.net/#t

Table 1 The F3 patterns that are most commonly applied

Pattern	Purpose
Domain feature	Indicates the structures that should be created for a feature
Mandatory decomposition	Indicates the code units that should be created when there is a mandatory decomposition linking two features
Optional decomposition	Indicates the code units that should be created when there is an optional decomposition linking two features
Simple decomposition	Indicates the code units that should be created when there is a simple decomposition linking two features
Multiple decomposition	Indicates the code units that should be created when there is a multiple decomposition linking two features
Variant decomposition	Indicates the code units that should be created when there is a variant decomposition linking two features
Requiring dependency	Indicates the code units that should be created when a feature requires another one
Excluding dependency	Indicates the code units that should be created when a feature excludes another one

the creation of classes for the features to the definition of the framework interface. Some of the F3 patterns are presented in Table 1.

The documentation of the F3 patterns is organized into topics to help developers to identify when a certain pattern should be used. This documentation is described as follows:

- **Name**: identifies each pattern and summarizes its purpose;
- **Context**: describes a desired behavior for the framework/domain;
- **Scenario/Problem**: describes the arrangement of features and relationships in F3 models that can imply the use of the pattern;
- **Solution**: indicates the code units that should be created to implement the scenario identified by the pattern;
- **Model**: shows a generic graphical representation of the scenario/problem and the solution;
- **Implementation**: displays a fragment of code, in a programming language, that illustrates how the solution can be implemented.

For example, the third pattern listed in Table 1, Optional Decomposition, suggests the creation of an operation that must be overridden in the instances of the source feature to specify which class is an instance of the target feature. The documentation of this pattern is:

- **Name**: Optional Decomposition.
- **Context**: when a target feature is optional to a source feature, every instance of the source feature may be associated with a instance of the target feature.

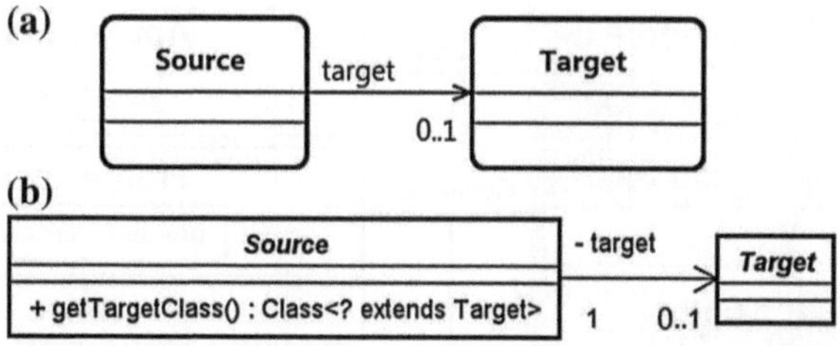

Fig. 2 The (**a**) pattern scenario and (**b**) its design solution

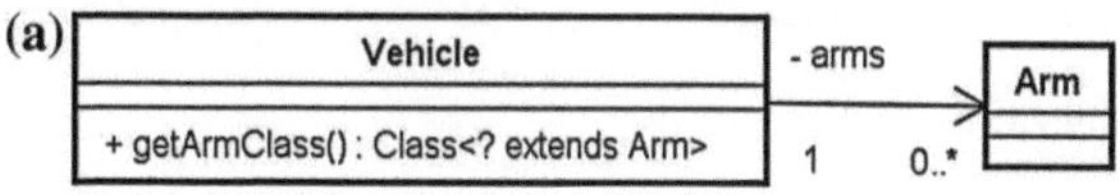

(**b**)
```
public abstract class Vehicle {
    public Class<? extends Arm> getArmClass() {
        return null;
    }
}
```

Fig. 3 Solution applied to the relationship between `Vehicle` and `Arm`

- **Scenario/Problem**: a feature has a decomposition relationship with minimum multiplicity equals `0`.
- **Solution**: the class implementing the source feature must have an operation to indicate what class implements the target feature in the applications. By default, this operation returns null as the target feature is optional.
- **Model**: the (a) scenario and the (b) design solution are shown in Fig. 2.
- **Implementation**: the solution in Java language is:

```
public abstract class Source {
  public Class<? extends Target> getTargetClass() {
     return null;
  }
}
```

Considering the F3 model in Fig. 1, the Optional Decomposition pattern should be applied to the decomposition between `Vehicle` (source) and `Arm` (target). The (a) design and (b) implementation solutions are shown in Fig. 3.

4 Tool Design

In this section we show the design of a tool, F3T, that assists developers to apply the F3 approach in the development of frameworks and reuse these frameworks through their DSMLs [8]. The F3T is a plug-in for Eclipse IDE, so developers can use the resources of this tool as well those provided by the IDE. The F3T is composed of

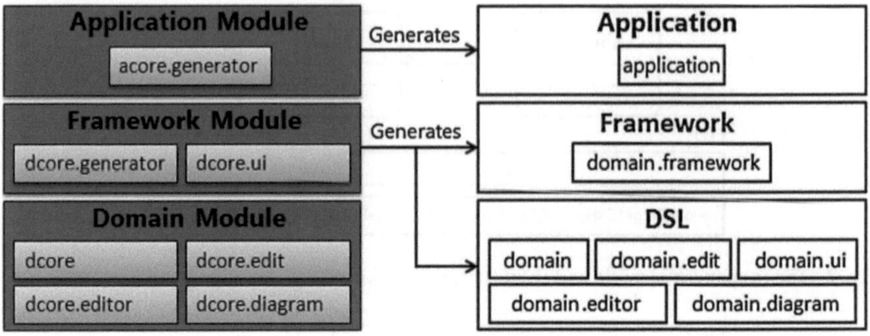

Fig. 4 Modules of the F3T

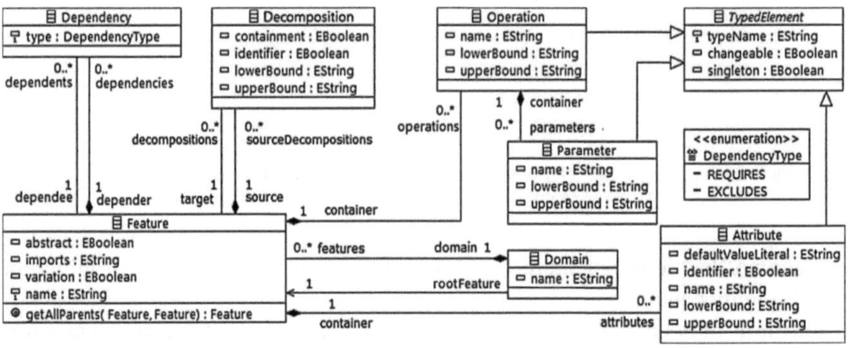

Fig. 5 Metamodel for F3 models

three modules, as seen in Fig. 4: (1) Domain Module; (2) Framework Module; and (3) Application Module.

4.1 Domain Module

The Domain Module (DM) provides a F3 model editor for developers to define domain features. This module was designed by using the Eclipse Modeling Framework (EMF) and the Graphical Modeling Framework (GMF) [15]. The EMF was used to create a metamodel, shown in Fig. 5, in which the elements, relationships and rules of the F3 models were defined as described in the Sect. 3.1.

The GMF has been used to define the graphical notation of the F3 model Editor. This graphical notation is defined through three kind of models provided by the GMF: (1) graphical model, in which the graphical figures representing the editor elements are defined; (2) tool model, in which the menu bar of the editor is created; (3) map model, in which the correspondent elements of the metamodel, graphical model and tool model are linked to each other. For example, in Fig. 6 it is shown how the F3

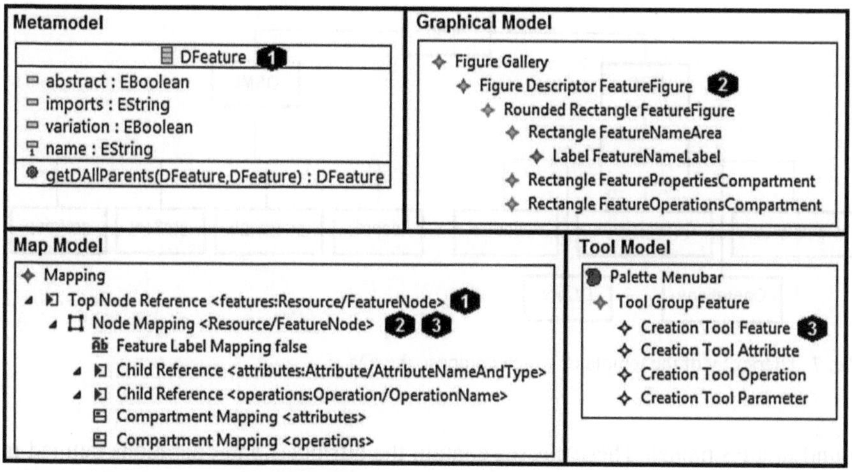

Fig. 6 Defining the Feature element for the F3 model Editor

model feature element was defined in the GMF models. Each number indicate a link between the models.

4.2 Framework Module

Despite their graphical notation, F3 models actually are XML files. It makes them more accessible to other tools, such as a generator. Therefore, the Framework Module (FM) was designed to be a Model-to-Text (M2T) generator that transforms F3 models into framework source-code and DSML.

The FM was developed with the support of the Java Emitter Templates (JET) in the Eclipse IDE [16]. The JET is composed of a framework that is a generic generator and a compiler that translates templates into Java files. These templates are XML files, in which tags are instructions to generate output based on information extracted from the models and text is a fixed content inserted in the output independently of input.

The templates of the FM are organized in two groups: one related to framework source-code (DSC); and another related to framework DSML. Both groups are invoked from the Main template of the FM. The DSC template invokes the templates which originate different parts of the framework classes, such as constructors, attributes, getters, setters, operations and other structures related to the F3 patterns (DCore). The DSML template invokes templates which originates the GMF models of the framework DSML. The hierarchy of the FM templates is shown in Fig. 7.

The framework source-code is generated by the FM according to the patterns defined by the F3 approach. For example, the FM generates a class for each feature

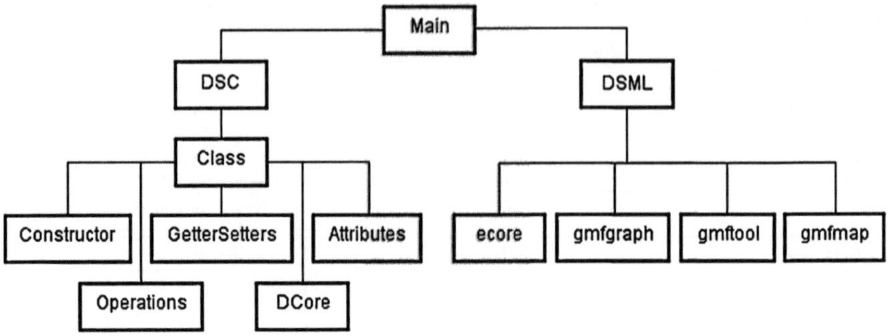

Fig. 7 Hierarchy of the templates which compose the FM

found in a F3 model. These classes contain the attributes and operations defined in its original feature. Generalization relationships result in inheritances and decomposition relationships result in associations between the involving classes. Additional operations are included in the framework classes to treat feature variations and constraints of the domains defined in the F3 models.

For example, from decomposition relationships found in F3 models, the DCore template generates operations to identify applications classes, as determined by the Mandatory Decomposition and the Optional Decomposition F3 patterns (see Sect. 3.2). In this case, the template verifies the minimum multiplicity to determine whether the decomposition relationship is a optional (0) or a mandatory (1). These patterns are implemented in the DCore template as the following:

```
public <c:if test="$decomposition/@lowerBound = '1'">abstract </c:if>
Class<? extends <c:get select="$decomposition/target/@name"/>>
get<c:get select="uppercaseFirst($decomposition/target/@name)"/>Class()
<c:choose select="$decomposition/@lowerBound">
  <c:when test="'1'">;</c:when>
  <c:otherwise> {
  return null;
}</c:otherwise>
</c:choose>
```

Besides the framework source-code, the FM also generates a DSML from a F3 model. This DSML is generated as EMF/GMF models, similar to those created to design the F3 model Editor (Figs. 5 and 6), but each DSML element correspond to a feature of the framework domain.

4.3 Application Module

The Application Module (AM) is an extension to the F3T that has been designed to generate application source-code from an application models based on a framework DSML, as defined in Viana et al. [8]. The AM was also created with the support of

JET. The AM templates generate classes that extend framework classes and override operations (hot spots) indicated by the F3 patterns.

Application source-code is generated in the source folder of the project where the application model is located. The AM generates a class for each feature instantiated in the application model. Since the framework is white box, the application classes extend the framework classes created from the features defined in the F3 models. For example, an application reusing the framework for the domain of automated vehicles (Figs. 1 and 3) contains the classes `MyVehicle` and `MyArm`, respectively extending the classes `Vehicle` and `Arm` of this framework. Part of the `MyVehicle` class is presented as follows:

```
public class MyVehicle extends Vehicle {

  /** @generated */
  public Class<? extends Arm> getArmClass() {
     return MyArm.class;
  }
}
```

5 Evaluation

We have carried out an experiment to compare the F3 approach with an adhoc one. This experiment followed all steps described by Wohlin et al. [17] and it can be defined as: (i) **Analyse** the F3 approach, described in Sect. 3, (ii) **for the purpose of** evaluation, (iii) **with respect to** efficiency (time) and easiness (problems), (iv) **from the point of view of** the developer, and (v) **in the context of** MSc and PhD students of Computer Science.

The context of the experiment corresponds to multi-test within object study [17], since the experiment consisted of experimental tests executed by a group of subjects to study a single approach, which is the F3 approach.

5.1 Planning

The planning phase was divided into the 6 steps, as follows:

1. Context Selection

The experiment has been performed in laboratory of Computer Science at an university environment. It involved the participation of MSc and PhD students of Computer Science with prior experience in software development using Java language, design patterns and frameworks.

2. Formulation of Hypotheses

The first question the experiment had to answer was: **RQ$_1$**: **"Which approach takes to a more efficient framework development in terms of time?"**. In order to answer this question, the subjects had to measure the time spent (τ) to develop each framework. According to this, the following hypotheses were elaborated:

- **RQ$_1$**: **Null hypothesis, H$_0$**: The F3 approach is not more efficient than the adhoc one in terms of time spent in framework development. It can be formalized as: **RQ$_1$H$_0$**: $\tau_{F3} \geq \tau_{adhoc}$;
- **RQ$_1$**: **Alternative hypothesis, H$_1$**: The F3 approach is more efficient than the adhoc one in terms of time spent in framework development. It can be formalized as: **RQ$_1$H$_1$**: $\tau_{F3} < \tau_{adhoc}$.

The second question the experiment had to answer was: **RQ$_2$**: **"Which approach facilitates framework development reducing the number of problems in the frameworks during their development?"**. In order to answer this question, we analyzed the reports of the subjects, in which they documented the problems found (ρ) in their frameworks during development, as well as the source-code of the outcome frameworks. By problems we mean defects and bad smells in the source-code of the frameworks. According to this, the following hypotheses were elaborated:

- **RQ$_2$**, **Null hypothesis, H$_0$**: The F3 approach does not facilitate framework development, as the number of problems during framework development is not reduced. It can be formalized as: **RQ$_2$H$_0$** : $\rho_{F3} \geq \rho_{adhoc}$;
- **RQ$_2$**, **Alternative hypothesis, H$_1$**: The F3 approach facilitates framework development, reducing the number of problems in the frameworks during their development. It can be formalized as: **RQ$_2$H$_1$**: $\rho_{F3} < \rho_{adhoc}$.

3. Variables Selection

The dependent variables of this experiment were "**time spent to develop a framework**" and "**number of problems found in the frameworks**". The independent variables were:

- **Application**: Each subject had to develop two frameworks: Fw1, for the domain of trade and rental transactions; and Fw2 for the domain of automated vehicles. Both Fw1 and Fw2 were composed of 10 features.
- **Development Environment**: Eclipse 4.2.1, Astah Community 6.4.
- **Technologies**: Java version 6.

4. Selection of Subjects

Subjects were selected according to convenience sampling [17]. In this non-probabilistic technique, the selected participants were the closest and most convenient to conduct the experiment. Altogether, 26 Msc and PhD students voluntarily participated in the experiment.

5. Experiment Design

The experiment followed the design of grouping the subjects in homogeneous blocks [17], avoiding that their experience level could directly impact in the results. We used a Participant Characterization Form to determine the experience level of each subject. In this form the subjects had to answer multiple-choice questions about their knowledge regarding Java programming, design patterns and frameworks.

The design type of the experiment was **one factor with two treatments paired** [17]. The **factor** is the approach used to develop a framework and the **treatments** are the adhoc and the F3 approaches. Each subject had to develop two frameworks, one applying the adhoc approach and the other applying the F3 approach. The order in which the subjects applied the treatments had no effect in the result. Therefore, the subjects were divided into two blocks of 13 participants with two tasks, as follows:

- **Block 1**: Task 1, development of Fw1 applying the adhoc approach; and Task 2, development of Fw2 applying the F3 approach;
- **Block 2**: Task 1, development of Fw2 applying the adhoc approach; and Task 2, development of Fw1 applying the F3 approach;

6. Instrumentation

The subjects received all necessary materials to assist them during the execution of the experiment. These documents consist of: textual description and models of the framework domains; manual for creating F3 models; documentation of the F3 patterns; Data Collection Form, in which the subjects had to report the time spent to develop the frameworks and the problems found during their development; one Test Application for each framework, which should be used by the subjects to verify the correctness and the completeness of the outcome frameworks; and Feedback Form, in which the subjects should describe their difficulties and write their opinion after the experiment.

5.2 Operation

After defining and planning the experiment, its operation was carried out in two steps: (1) Preparation and (2) Execution.

1. Preparation

At first, the subjects signed a Consent Form, stating the objectives and confidentiality of the experiment, and filled the Participant Characterization Form in, reporting their experience in the concepts and technologies utilized in the experiment. After this, the subjects had a training in: adhoc framework development, in which they learned design patterns and code structures commonly used in frameworks to identify application-specific elements; and the F3 approach. After training, the subjects were able to carry out the experiment tasks.

2. Execution

Before starting the execution of the experiment, the subjects were positioned in the blocks and received the materials referent to their respective Task 1.

When all subjects were commanded to execute Task 1 (applying the adhoc approach), they started to measure the time. They used the Astah Community to create a class model of the framework and then use the Eclipse IDE to implement its source-code. When they finished framework implementation, they executed its

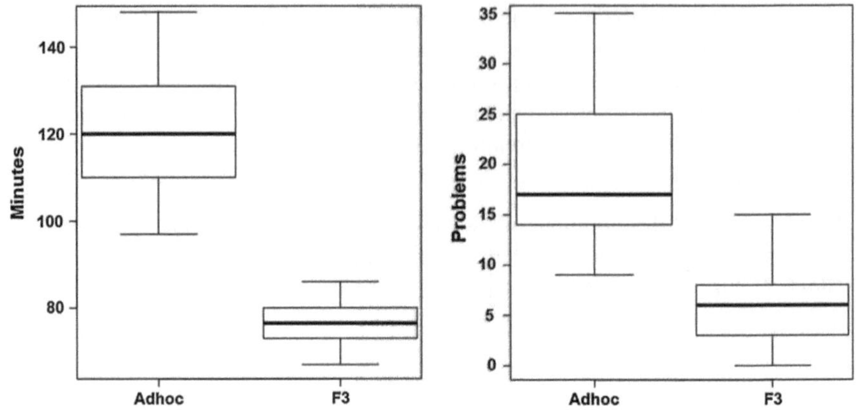

Fig. 8 Dispersion of the total time and number of problems

Test Application to verify whether or not it was developed as expected. If the Test Application showed a message of a problem, the subjects had to report it in the Data Collection Form and fix the problem(s) found. Only when the Test Application returned a successful message, the subject could stop measuring the time. Task 2 (applying the F3 approach) was performed in a similar way to Task 1. In the end, the subjects received the Feedback Form to comment the difficulties and advantages in applying each approach.

5.3 Analysis of Data

The experiment data is presented in Table 2. The groups developed the tasks satisfactorily and the collected data was within the expected limits. It means that the treatments were executed correctly and in accordance with the planning. The analysis of data is presented in the following subsections.

1. Descriptive Statistics

It can be seen in Table 2 that the F3 approach spent less time in framework development than the adhoc approach. Approximately 38.7 % against 61.3 %. The feedback provided by the subjects showed that, although the subjects have spent part of the time trying to identify the F3 patterns that should be used, they saved some time because these patterns assisted them indicating the classes, attributes and operations that should be created. On the other side, when the subjects were developing the frameworks applying the adhoc approach, they spent part of the time trying to find out the code units they should implement. Moreover, most of the subjects reported that they spent a long time maintaining their frameworks, because the Test application returned lots of problem messages. The dispersion of time spent by the subjects are also represented graphically in a boxplot on left side of Fig. 8.

Table 2 Results of the frameworks developed by the subjects

Subject	Time spent (minutes)		Number of problems									
			Incoherence		Structure		Bad smells		Interface		Total	
	Adhoc	F3	Adhoc	F3	Adhoc	F3	Adhoc	F3	Adhoc	F3	Adhoc	F3
S1	108	72	5	1	2	0	2	2	7	0	16	3
S2	113	71	7	1	4	1	2	0	4	1	17	3
S3	139	83	9	3	11	1	3	1	12	2	35	7
S4	124	78	7	1	5	2	2	1	7	2	21	6
S5	101	67	4	0	3	0	1	0	3	0	11	0
S6	133	81	8	4	7	3	3	3	9	3	27	13
S7	131	79	5	3	3	1	2	1	6	2	16	7
S8	116	73	6	1	5	0	3	0	5	1	19	2
S9	109	79	7	1	4	2	2	1	7	2	20	6
S10	106	69	4	2	3	0	1	0	3	1	11	3
S11	119	71	4	1	4	1	2	0	7	0	17	2
S12	148	83	8	3	6	1	3	3	11	4	28	11
S13	110	74	4	1	2	1	3	1	5	0	14	3
S14	107	72	2	1	3	0	3	0	6	1	14	2
S15	117	76	5	3	5	2	2	1	4	2	16	8
S16	97	68	3	1	1	0	2	0	3	0	9	1
S17	137	80	8	5	9	4	3	3	10	3	30	15

continued

Table 2 continued

Subject	Time spent (minutes)		Number of problems									
			Incoherence		Structure		Bad smells		Interface		Total	
	Adhoc	F3	Adhoc	F3	Adhoc	F3	Adhoc	F3	Adhoc	F3	Adhoc	F3
S18	121	75	4	1	6	2	2	2	2	2	14	7
S19	115	73	3	0	4	1	2	0	4	0	13	1
S20	134	81	7	2	6	3	3	1	9	3	25	9
S21	144	86	9	3	7	3	3	3	12	6	31	15
S22	111	76	3	2	4	1	1	1	5	1	13	5
S23	129	83	7	4	8	3	3	2	11	3	29	12
S24	123	79	5	2	5	1	3	1	7	2	20	6
S25	127	77	5	3	3	1	1	0	4	1	13	5
S26	131	78	6	2	4	1	2	2	6	2	18	7
AVG	121.154	76.4231	5.57692	1.961538	4.76923	1.346154	2.26923	1.115385	6.5	1.692308		
%	61.3198	38.6802	73.9796	26.02041	77.9874	22.01258	66.0455	32.95455	79.3427	20.65728		

In Table 2 it is also presented the four types of problems that we analyzed in the outcome frameworks: incoherence, structure, bad smells and interface.

The problem of incoherence indicates that the subjects did not develop the frameworks with the correct features and constraints (mandatory, optional and alternative features) of the domain. In other words, they did not designed and implemented the classes, attributes and operations that could make the framework to behave as expected by its domain. In Table 2 it can be seen that the F3 approach helped the subjects to develop frameworks with less incoherence problems, approximately, 26 % in opposition to 74 % for the adhoc approach.

The problem of structure indicates that the subjects did not implement the frameworks properly, for example, implementing classes with no constructor, non-abstract when they should be or incorrect relationships. In Table 2 it can be seen that the F3 approach helped the subjects to develop frameworks with less structure problems, i.e., 22 % in opposition to 78 %.

The problem of bad smells indicates design weaknesses that do not affect functionality, but make the frameworks harder to maintain. This problem is not a defect, so the Test Applications could not detect it and the subjects did not fixed it. We identified it by analyzing the source-code of the frameworks. In Table 2 we can remark that the use of the F3 approach resulted in a design with higher quality than the use of the adhoc approach, respectively, 33 % against 67 %.

The problem of interface indicates absence of getter/setter operations and the lack of operations that allows the applications to reuse the framework and so on. Usually, this kind of problem is a consequence of problems of structure, hence the number of problems of these two types are quite similar. As it can be observed in Table 2 that the F3 approach helped the subjects to design a better framework interface than when they developed the framework through the adhoc approach, respectively, 21 % against 79 %.

In the last two columns of Table 2 it can be seen that the F3T reduced the total number of problems found in the frameworks developed by the subjects. It is also graphically represented in the boxplot on right side of Fig. 8.

2. Hypotheses Testing

The objective of this section is to verify with any degree of significance, whether it is possible to reject the nulls hypotheses (see Sect. 5.1) in favor of the alternative hypothesis based on the data set obtained. This section is divided into: (1) Hypotheses Testing—Time and (2) Hypotheses Testing—Problems.

1. **Hypotheses Testing—Time**: Since some statistical tests are applicable only if the population follows a normal distribution, we applied the Shapiro-Wilk test and created a Q-Q chart to verify whether or not the experiment data departs from linearity before choosing a proper statistical test. As it can be seen in the upper Q-Q charts in Fig. 9, the experiment data related to the time spent in framework development is normally distributed. Thus, we decided to apply the Paired T-Test to the experiment data. According to *StatSoft*,[3] we carried out this test by

[3] http://www.statsoft.com/textbook/distribution-tables/#t

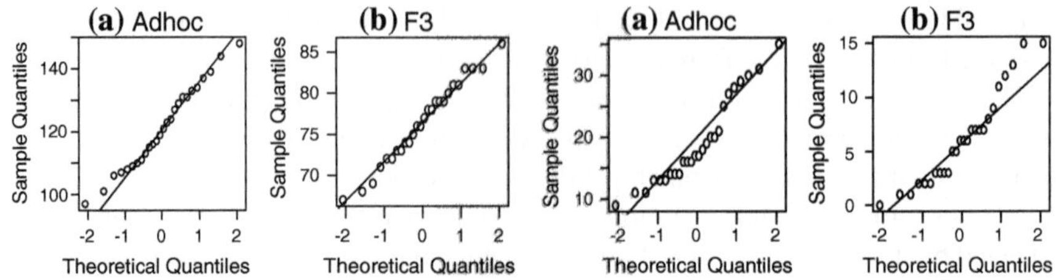

Fig. 9 Q-Q charts of time (*upper*) and problems found (*lower*)

calculating: the difference of time between both approaches, $d = \{36, 39, 56, 46, 34, 52, 52, 43, 30, 37, 48, 65, 36, 35, 41, 29, 57, 46, 42, 53, 58, 35, 46, 44, 50, 53\}$; the standard deviation of this difference, $S_d = 9.357597$; the number of degrees of freedom, $F = N - 1 = 26 - 1 = 25$, where N is the number of subjects; $t_0 = 24.3741$; and $t_{0.05,25} = 1.708141$. Since $t_0 > t_{0.05,25}$ it is possible to reject the null hypothesis with a two sided test at 0.05 level. Therefore, statistically, we can assume that the F3 approach reduces the time spent in framework development when compared with the adhoc approach.

2. **Hypotheses Testing—Problems**: We also used the Shapiro-Wilk test on the data shown in the last two columns of Table 2, which represent the total number of problems found in the outcome frameworks by using the F3 approach and the adhoc one, respectively. As it can be seen in the lower Q-Q charts in Fig. 9, the collected data indicates a normal distribution. Thus, we also used a Paired T-Test by calculating: the difference of number of problems between both approaches, $d = \{ 13, 14, 28, 15, 11, 14, 9, 17, 14, 8, 15, 17, 11, 12, 8, 8, 15, 7, 12, 16, 16, 8, 17, 14, 8, 11 \}$; the standard deviation of this difference, $S_d = 4.463183$; the number of degrees of freedom, $F = N - 1 = 26 - 1 = 25$, where N is the number of subjects; $t_0 = 4.463183$; and $t_{0.05,25} = 1.708141$. Since $t_0 > t_{0.05,25}$, it is possible to reject the null hypothesis with a two sided test at 0.05 level. Therefore, statistically, we can assume that the F3 approach reduces the number of problems found in the outcome frameworks when compared with the adhoc approach.

5.4 Threats to Validity

Internal Validity:

- Experience level of participants: different levels of knowledge of the subjects could affect the collected data. To mitigate this threat, we divided the subjects into two balanced blocks based on their answers in the Participant Characterization Form. All subjects had prior experience in application development reusing frameworks and they were trained in the F3 approach.

- Productivity under evaluation: it might influence the experiment results because subjects often tend to think they are being evaluated. To mitigate this, we explained to the subjects that no one was being evaluated and their participation was considered anonymous.
- Facilities used during the study: different computers and installations could affect the recorded timings. However, the subjects used the same hardware configuration and operating system.

Validity by Construction:

- Hypothesis expectations: the subjects knew the researchers and knew that the F3 approach was supposed to ease framework development before the experiment. These issues could affect the collected data and cause the experiment to be less impartial. In order to keep impartiality, we enforced that the subjects had to keep a steady pace during the whole study.

External Validity:

- Interaction between configuration and treatment: it is possible that the exercises performed in the experiment are not accurate for every framework development in real world applications. Only two frameworks were developed and both had similar complexity. To mitigate this threat, the tasks were designed considering framework domains based on the real world.

Conclusion Validity:

- Measure reliability: it refers to metrics used for measuring development effort. To mitigate this threat we have used only the time spent which was captured in forms fulfilled by the subjects;
- Low statistic power: the ability of a statistic test in reveal reliable data. To mitigate we applied two tests: T-Tests to statistically analyze the time spent to develop the frameworks and Wilcoxon signed-rank test to statistically analyze the number of problems found in the outcome frameworks.

6 Related Works

Xu and Butler [18] proposed an cascaded refactoring method to develop frameworks. In this method, a framework is specified by different models, sorted by abstraction level (from feature model to source-code). Refactorings are performed on these models following their sequence until the framework is completely developed. In the F3 approach the domain is also defined in feature models and framework design and implementation are assisted by patterns, which provide more information to help developers than refactorings.

Zhang et al. [19] proposed the Feature-Oriented Framework Model Language (FOFML) to develop frameworks following the MDA approach. In this language, domain features are defined in a graphical metamodel and domain constraints are specified in a textual role model. Then, framework classes are implemented according to the features and roles defined in these models. In comparison, our approach define domain features and constraints in a single model.

Amatriain and Arumi [20] also proposed a method to develop frameworks through iterative and incremental activities. In their method, the domain of the framework could be defined from existing applications and the framework could be implemented through a series of refactorings over these applications. The advantage of this method is a small initial investment and the reuse of the applications. Although it is not mandatory, the F3 approach can also be applied in iterative and incremental activities, starting from a small domain and then adding features. Applications can also be used to facilitate the identification of the features of the domain. However, the advantage of the F3 approach is the fact that the design and the implementation of the frameworks are performed with the support of patterns specific for framework development.

Oliveira et al. [21] presented the ReuseTool, which assists framework reuse by manipulating UML diagrams. This tool is based in the Reuse Description Language (RDL), which has been created to facilitate the description of framework instantiation processes. The RDL is used to make a description of the framework hot spots. in order to instantiate the framework, application models are created based on this description and their source-code is generated from these models. Thus, the RDL works as a meta language that registers framework hot spots and the ReuseTool provides a more friendly interface for developers to develop applications reusing the frameworks. In comparison, the F3T was designed to support framework development through domain modeling and application development through framework DSML.

Common Variability Language (CVL) is an Object Manager Group (OMG) standardization used for specifying and resolving domain variability [22]. Such as F3 models, CVL is an extended feature model. However, CVL uses a mechanism similar to OCL to implement domain constraints. In comparison, F3 models define domain constrains through relationships and properties. Moreover, since F3 models focus on framework development, the features in this kind of model can contain attributes and operations.

Pure::variants [23] is a tool that supports the development of applications by modeling domain features (Feature Diagram) and the components that implement these features (Family Diagram). Then the applications are developed by selecting a set of features of the domain. Pure::variants generates only application source-code, maintaining all domain artifacts in model-level. Besides, this tool has private license and its free version (Community) has limitations in its functionality. In comparison, the F3T was designed to be free, use only one type of domain model (F3 model) and generate the source-code frameworks as domain artifacts. Moreover, the frameworks developed with he support of the F3T will be able to be reused on the development of applications with or without the support of the F3T.

7 Concluding Remarks and Future Work

In this chapter we proposed the F3 approach to facilitate the development of frameworks. In this approach the framework domain is defined in F3 models, which include elements from feature models and metamodels. Then, F3 patterns guide developers to construct the framework according to its domain. Therefore, the F3 approach provide two major contributions: F3 models allow developers to define framework functionality in a high level of abstraction; and the F3 patterns facilitate framework design and implementation by indicating the structures that must be created to implement its functionality.

Our approach promotes reuse in different levels. F3 models represent domains that can be reused and improved in different frameworks. The F3 patterns represent reuse of experience on framework development. Moreover, the outcome frameworks can be reused in the development of several applications, acting as a core asset for a software product line.

The experiment presented in this chapter indicated that F3 approach facilitates framework development, because it shows developers how to proceed, making them less prone to insert defects and bad smells in the outcome frameworks. Our approach allowed that even subjects with no experience in framework development could execute this task correctly and spending less time.

The F3T supports framework development and reuse through code generating from models. This tool provides an F3 model editor for developers to define the features of the framework domain. Then, framework source-code and DSML can be generated from the F3 models. Besides the gain of efficiency provided by code generation, this tool aims to reduce even more the complexities of framework development and reuse.

In future works we intend to perform more tests considering different domains in order to identify scenarios that we still have not predicted and then create F3 patterns to them. Moreover, we also intend to carry out more tests and experiments to evaluate the F3 approach and its tool.

References

1. Pressman, R.S.: Software Engineering: A Practitioner's Approach, 7th edn. McGraw-Hill Science, New York (2009)
2. Shiva, S.G., Shala, L.A.: Software reuse: research and practice. In: Fourth International Conference on Information Technology, April 2007. pp 603–609
3. Johnson, R.E.: Frameworks = (Components + Patterns). Commun. ACM **40**(10), 39–42 (1997)
4. Srinivasan, S.: Design patterns in object-oriented frameworks. Computer **32**(2), 24–32 (1999)
5. Abi-Antoun, M.: Making frameworks work: a project retrospective. In: Companion to the 22nd ACM SIGPLAN Conference on Object-Oriented Programming Systems and Applications, OOPSLA '07, pp. 1004–1018. ACM, New York (2007)
6. Kirk, D., Roper, M., Wood, M.: Identifying and addressing problems in object-oriented framework reuse. Empirical Softw. Eng. **12**(3), 243–274 (2007)

7. Stanojevic, V., Vlajic, S., Milic, M., Ognjanovic, M.: Guidelines for framework development process. In: Software Engineering Conference in Russia (CEE-SECR), 7th Central and Eastern European, Nov 2011, pp. 1–9
8. Viana, M., Penteado, R., do Prado, A.: Generating applications: framework reuse supported by domain-specific modeling languages. In: 14th International Conference on Enterprise Information Systems (ICEIS'14), June 2012
9. Fowler, M.: Patterns. IEEE Softw. **20**(2), 56–57 (2003)
10. Lee, K., Kang, K.C., Lee, J.: Concepts and guidelines of feature modeling for product line software engineering. In: 7th International Conference on Software Reuse: Methods, Techniques and Tools, pp. 62–77. Springer, London (2002)
11. Kang, K.C., Cohen, S.G., Hess, J.A., Novak, W.E., Peterson, A.S.: Feature-Oriented Domain Analysis (FODA): Feasibility Study. Carnegie-Mellon University Software Engineering Institute,Technical report (Nov 1990)
12. Jezequel, J.M.: Model-driven engineering for software product lines. ISRNSoftw. Eng. **2012**, (2012)
13. Gomaa, H.: Designing Software Product Lines with UML: From Use Cases to Pattern-Based Software Architectures. Addison-Wesley, Upper Saddle River (2004)
14. Bayer, J., Flege, O., Knauber, P., Laqua, R., Muthig, D., Schmid, K., Widen, T., DeBaud, J.: PuLSE: a methodology to develop software product lines. In: Proceedings of the Symposium on Software Reusability, pp. 122–131. ACM (1999)
15. Gronback, R.C.: Eclipse Modeling Project: A Domain-Specific Language (DSL) Toolkit. Addison-Wesley, Upper Saddle River (2009)
16. The Eclipse Foundation. Eclipse Modeling Project. http://www.eclipse.org/modeling Jan (2013)
17. Wohlin, C., Runeson, P., Höst, M., Ohlsson, M.C., Regnell, B., Wesslén, A.: Experimentation in Software Engineering: An Introduction. Kluwer Academic Publishers, Norwell (2000)
18. Xu, L., Butler, G.: Cascaded refactoring for framework development and evolution. In: Software Engineering Conference. pp. 319–330. Australian (2006)
19. Zhang, T., Xiao, X., Wang, H., Qian, L.: A feature-oriented framework model for object-oriented framework: an MDA approach. In: 9th IEEE International Conference on Computer and Information Technology, vol. 2, pp. 199–204 (2009)
20. Amatriain, X., Arumi, P.: Frameworks generate domain-specific languages: a case study in the multimedia domain. IEEE Trans. Softw. Eng. **37**(4), 544–558 (2011)
21. Oliveira, T.C., Alencar, P., Cowan, D.: Design patterns in object-oriented frameworks. Reuse-tool: an extensible tool support for object-oriented framework reuse. J. Syst. Softw. **84**(12), 2234–2252 (2011)
22. Rouille, E., Combemale, B., Barais, O., Touzet, D., Jezequel, J-M.: Leveraging CVL to manage variability in software process lines. In: 19th Asia-Pacific Software Engineering Conference (APSEC). vol. 1, pp. 148–157. (2012)
23. Pure Systems. Pure::Variants. http://www.pure-systems.com/pure_variants. 49.0.html, Feb 2013

About Handling Non-conflicting Additional Information

Éric Grégoire

Abstract The focus in this chapter is on logic-based Artificial Intelligence (A.I.) systems that must accommodate some incoming symbolic knowledge that is not inconsistent with the initial beliefs but that however requires a form of belief change. First, we investigate situations where the incoming knowledge is both more informative and deductively follows from the preexisting beliefs: the system must get rid of the existing logically subsuming information. Likewise, we consider situations where the new knowledge must replace or amend some previous beliefs. When the A.I. system is equipped with standard-logic inference capabilities, merely adding this incoming knowledge into the system is not appropriate. In the chapter, this issue is addressed within a Boolean standard-logic representation of knowledge and reasoning. Especially, we show that a prime implicates representation of beliefs is an appealing specific setting in this respect.

Keywords Artificial intelligence · Knowledge representation and reasoning · Logic · Belief change

1 Introduction

Intelligent agents must often handle a flux of incoming information.[1] In this context, a very fertile domain of research in knowledge representation and reasoning concerns the ways according to which symbolic knowledge and beliefs must change in light of

[1] Since no distinction between belief, knowledge and information is actually needed for the purpose of this chapter, we use all these words interchangeably.

This chapter extends a preliminary version published in IEEE-IRI'2013.

É. Grégoire (✉)
CRIL Université d'Artois, CNRS UMR 8188 rue Jean Souvraz SP18, F-62307 Lens Cedex, France
e-mail: gregoire@cril.univ-artois.fr

T. Bouabana-Tebibel and S. H. Rubin (eds.), *Integration of Reusable Systems*, 285
Advances in Intelligent Systems and Computing 263, DOI: 10.1007/978-3-319-04717-1_13,

new information, taking logic as a representation and reasoning setting. Especially, much attention has been devoted to this issue in the so-called belief revision, belief update [1–5] and knowledge fusion [6, 7] approaches [1, 8–12]. A shared focus by most approaches developed in these domains concerns situations where an incoming piece of information is logically contradicting the pre-existing knowledge. These approaches also recommend to record the incoming piece of information together with the pre-existing knowledge when this does not lead to inconsistency.

In this chapter, we address some frequent situations about non-logical conflicts where the latter recommendation should not apply. In particular, we consider cases where the incoming information brings in more precision or details, or, more generally, is more informative than the preexisting knowledge, without leading to any logical inconsistency. From a logical point of view, this kind of incoming information often takes the form of some logical deductive consequences of the preexisting knowledge. In this respect, adding the incoming information does not drive to logical contradiction. However, inserting it is not sufficient: a form of preemption must also take place so that the incoming information replaces the concerned pre-existing subsuming knowledge. In this chapter, this issue is investigated when the representation setting is standard Boolean logic. Likewise, we consider situations where the incoming knowledge must replace or amend some previous beliefs. Especially, we show that a prime implicates representation of beliefs is an appealing specific setting in this respect.

The chapter is organized as follows. In the next section, some motivating examples are provided. Then, preliminaries about Boolean logic are recalled. Section 4 focuses on prime implicates and implicants, as they play a keystone role in our approach, whereas Sect. 5 defines a notion of a clause being about a concept. In Sect. 6, we show how the various belief change operators can be addressed within a prime implicates representation framework. Section 7 presents a computational method for making sure that clauses are not preempted; some experimental results are provided in Sect. 8. The conclusion motivates some promising paths for further research.

2 Motivating Examples

Let us give some motivating examples, where the incoming symbolic knowledge does not logically contradict the previous premisses, although it requires some belief change.

2.1 Subsumed Knowledge that Must Prevail

Assume that a robot reasons and takes decisions according to standard-logic. Currently, it believes that *the target is located either in room A or in room B*. Assume that a new piece of information arises from its analysis of captors and radar data in such a way that it can no longer exclude the possibility that there might be no target at all. It thus concludes that *the target is either in room A, or in room B or there is no actual target.* Although the robot must give more credit to this conclusion than to its initial knowledge, this new piece of information is not actually new for the robot. Since the robot is logical, the new conclusion was already a consequence of its former knowledge. Indeed, whenever α can be deduced from a set of premisses, any conclusion α *or* β can also be deduced, for any possible β (as well as α *or not* β). In other words, if α is true, so is any subsumed information, like α *or* β. When the robot believed that it was true that *the target is located either in room A or in room B*, it was necessarily also believing that *the target is located either in room A, or in room B or there is no actual target* was also true. Note that both pieces of information do not contradict from a logical point of view using this basic Boolean representation mode.

Thus, if we merely insert the "new" piece of information inside the set of beliefs of the robot without removing the relevant former ones, the robot will still be able to infer that *the target is located either in room A or in room B*. What we need here is the ability to replace the existing subsuming beliefs by the new information in such a way that this latter one prevails.

Unfortunately, this issue is complex both from both conceptual and computational points of view: contrary to this example, deductive links between the initial beliefs and a new piece of information are not as apparent as in this example. They can be hidden through complex reasoning trees. In the worst case, merely proving that some premisses allow a Boolean formula to be derived and thus that this formula is subsumed by those premisses is co-NP-Complete.

It could be argued that the absence of logical contradiction between the old and new information in this example is due to a poor formalization of the available knowledge. Especially, forms of closed world assumption could here exclude any other possible location than *A* or *B* unless logical inconsistency occurs. In the same vein, moving to predicate logic would allow to express in an explicit way that there exists a target. However, the robot should be provided with general capacities to deduce conclusions that are not explicitly present within its premisses. Actually, when the robot is to be provided with minimal genuine deductive power, its ability to infer α *or* β is true from α is true cannot always be blocked by forms of closed-world assumptions. The need for a capacity to make some subsumed knowledge prevail cannot be totally circumvented. Moreover, moving to a logic of a higher degree or to a non-monotonic logic does not solve this issue in the general case either.

2.2 Rules that Must be Weakened

Here is another example, involving more general versus more specific rules, that also illustrates a need to make some deductive consequences prevail over the premisses. Assume that the robot believes that *if the target is in room A then it has to drive directly inside room A*. Assume that it learns the new rule *if the target is in a room A and if the energy level is good then it has to drive directly inside room A*. The second rule is a deductive consequence of the former one and, clearly, it must replace the initial knowledge. Assume that the robot is given now a third rule asserting that *if the target is in room A and if the road is not blocked then it has to drive directly inside room A*. Again, the two last rules are not logically conflicting. Obviously, the new rule should not be inserted as such as a new belief or replace the first rule. What we need here is to fuse the two rules into a new one combining necessary conditions for entering the room: namely, *if the target is in room A and if the road is not blocked and if the energy level is good then it has to drive directly inside room A*. Again, the rule formed in that way needs to be added to the robot's knowledge while making sure that it is not subsumed by shorter rules that are not requiring all those conditions for entering the room.

It is straightforward that not all rules should be merged in that way. Some rules simply assert sufficient conditions for a conclusion to hold whereas in this example rules express a kind of compulsory condition for the conclusion to hold. We thus need to be able to distinguish between these two kinds of rules, merge them when appropriate and make sure that the new rule is not subsumed.

2.3 Replacement and Compulsory Knowledge

To generalize the previous example, assume that our logical robot has some knowledge about a given concept, e.g., conditions that must be true to classify a detected situation as a dangerous one. The robot has thus some information *about* recognizing dangerous situations. Now, some new incoming knowledge about urgency situations that does not logically contradict the robot's initial knowledge might arise and needs to prevail. Especially, this incoming knowledge can be intended to replace the information about what a dangerous situation is. It can also represent some additional compulsory conditions for a situation to be dangerous. In both cases, the new knowledge might not contradict the robot's knowledge from a logical point of view. Again, contrary to belief revision theories, the new information cannot be merely inserted: some belief change in the robot's knowledge must occur so that the epistemological role of the additional information is met.

The study in this chapter investigates these issues and builds on some previous works. First, a complex tentative formal approach to characterize and solve the subsumption issue has been introduced in [13]. Candidate rationality postulates for belief change operators that allow beliefs to be preempted by subsumed ones have been presented in [14]. The formal characterization of this specific handling of subsumption has been extended to a general non-monotonic setting in [15] and applied to the legal domain in [16]. Starting from this, we present in this chapter a new formal solution to handling all the aforementioned kinds of new beliefs that is founded on a prime implicates and prime implicants setting.

3 Logic-Based Representation Setting

The logical setting in this chapter is standard (clausal) Boolean logic. On the one hand, it is the simplest possible framework for presenting and addressing the above subsumption-related issues. On the other hand, recent dramatic progress in Boolean search and reasoning has now revived Boolean logic as a realistic and attractive framework for representing large knowledge bases and solving numerous complex reasoning tasks in artificial intelligence [17].

Let $\mathcal{L}$ be a language of formulas over a finite alphabet $\mathcal{P}$ of Boolean variables, also called *atoms*. Atoms are noted $a, b, c, \ldots$ The $\wedge$, $\vee$, $\neg$, $\Rightarrow$ and $\Leftrightarrow$ symbols represent the standard conjunctive, disjunctive, negation, material implication and equivalence connectives, respectively. A *literal* is an atom or a negated atom. Formulas are built in the usual way from atoms, connectives and parentheses; they are noted α, β, γ, etc. A formula is in conjunctive normal form (CNF) when expressed as a conjunction of *clauses*, where a clause is a disjunction of literals. For convenience, clauses can be represented by their set of involved literals. The empty clause represents *false*. Also for convenience, the set of involved literals of a clause can be enriched by the value *false*, while still representing the clause. Also for convenience, the disjunction forming a clause can be safely enriched by a disjunct representing *false*.

Interpretations are functions assigning either *true* or *false* to every atom. A *model* of a set of formulas Δ is an interpretation that satisfies every formula of Δ. Δ is *consistent* (also said *satisfiable*) when its set of models is not empty. $\Delta \vdash \alpha$ expresses that the formula α can be deduced from Δ, i.e., that it is *true* in all models of Δ.

A *knowledge base* Δ is a consistent finite set of (non-tautological) clauses and the incoming belief γ is a consistent non-tautological clause. We distinguish between Δ, which represents the *explicit* clauses of the base, from the set of all the deductive conclusions of Δ, noted $Th(\Delta)$: $\Delta \vdash \alpha$ iff $\alpha \in Th(\Delta)$.

A word of caution for readers who are familiar with rule-based systems but not with logic. We exploit the sound and complete deductive capabilities of Boolean logic. Especially, we do not only simply allow for mere forward and backward chaining on $\Rightarrow$ as in traditional rule-based systems. For example, from the rule $a \Rightarrow b$ and $\neg b$, we derive $\neg a$ using contraposition. Also, keep in mind that a rule of the form $(a \wedge b \wedge \neg c) \Rightarrow (d \vee e)$ is logically equivalent to $\neg a \vee \neg b \vee c \vee d \vee e$ (which is also represented by $\{\neg a, \neg b, c, d, e\}$) and will be treated as such.

In the following, unless explicitly indicated, when a clause is referred to, it is not tautological. Also, without loss of generality, we assume that Δ is a consistent finite set of non-tautological clauses that represents the current knowledge.

Assume that an incoming non-contradictory and non-tautological clause γ is to be accommodated.

4 Clause About a Concept

A clause $\gamma = a_1 \vee \cdots \vee a_n$ expresses some information about any of its literals, say e.g. a_n. Accordingly, we say that the clause γ is *about* a_n. More generally, we say that γ is about any of its strict sub-clauses. More precisely:

Definition 1. *When γ can be rewritten as $\gamma = \alpha \vee \beta$ where α and β are two clauses with an empty intersection and where α is not the empty clause, γ is said to be about β. α is called the anti-condition for β in γ.*

Note that the condition requiring α not to be the empty clause is a syntactical requirement: α might be some non-empty clause that turns out to be actually *false*, taking other available information into account. When the anti-condition α for β in γ is interpreted to *false*, γ entails β: hence the "anti" prefix. However, by convenience, we will write "condition" instead of "anti-conclusion". Please, note that the definition entails that γ is not about itself: γ is about *false* and about any of its strict subsets. In our framework, the incoming clause γ can be asserted together with the additional information that γ is *specifically intended* to be about one specific given β, or without any information about such a possible intent. In the first case, this does not prevent the above definition from concluding that γ is also about other β when γ is not just a literal.

In the motivating example, assume that the rule *if the target is in room A and if the road is not blocked and if the energy level is ok then it has to drive directly inside room A* is encoded as *target-room-A* $\wedge \neg$ *blocked-road* $\wedge$ *energy-level-ok* $\rightarrow$ *move-to-roomA*, which is expressed as the clause $\neg$ *target-room-A* $\vee$ *blocked-road* $\vee \neg$ *energy-level-ok* $\vee$ *move-to-roomA*. This clause is for example about *move-to-roomA*: in which case, its anti-condition is $\neg$ *target-room-A* $\vee$ *blocked-road* $\vee \neg$ *energy-level-ok*.

5 Prime Implicates Representation of Beliefs

Our developments are deep-rooted in *subsumption*, *strict implicants*, *prime implicants* and *prime implicates*.

Definition 2. *α is a* strict implicant of *β iff $\alpha \vdash \beta$ but $\beta \nvdash \alpha$.*

Definition 3. *Δ* strictly subsumes *a clause β iff $\Delta \vdash \alpha$ for some clausal strict implicant α of β.*

By abuse of words, we will use "subsume" in place of "strictly subsume".

Definition 4. *α is a* prime implicant *of β iff α is a strict implicant of β and there does not exist any strict implicant δ of β such that α is a strict implicant of δ.*

Interestingly, when two non-tautological clauses α and β are under their set-theoretical representation, α is an (a strict) implicant of β iff α is a (strict) subset of β. Moreover, when a consistent non-tautological clause β is made of n literals, the prime implicants of β are the consistent sub-clauses of β made of $n-1$ literals and β is not subsumed by a consistent Δ iff none of the prime implicants of β can be deduced from the same Δ.

The dual concept of *prime implicates* also plays a central role in this chapter.

Definition 5. *A prime implicate of a finite set Δ of formulas is any clause δ that satisfies both conditions below*

(1) $\Delta \vdash \delta$
(2) $\delta' \equiv \delta$ *for every clause* δ' *s.t.* $\Delta \vdash \delta'$ *and* $\delta' \vdash \delta$

Δ_{PI} denotes the set of all prime implicates in Δ.

From now on, we assume that Δ is a consistent non-tautological belief base. Accordingly, δ is a prime implicate of Δ iff δ is a minimal (w.r.t. $\subseteq$) non-tautological clause amongst the set formed of the clauses β such that $\Delta \vdash \beta$.

In our first motivating example, under the set-theoretical representation of clauses, $\alpha = \{RoomA, RoomB\}$ is a prime implicant of $\beta = \{RoomA, RoomB, No\text{-}Target\}$. Thus, $\Delta = Cn(\{\alpha\})$ subsumes β. To make β prevail, we need to deliver a belief base Δ' such that Δ' entails β but does not subsume it. Especially, Δ' cannot contain α and β is a prime implicate of Δ'.

Intuitively, the other motivating examples deal with an additional incoming rule $\delta = \alpha \Rightarrow \beta$ intended to influence existing rules about β in Δ, either by weakening or by replacing their body by means of α. This is done by other forms of preemption that we study in this chapter.

Interestingly, a central concept will be prime implicates, again. Prime implicates have already been investigated in belief revision mainly because they provide a compact and syntax-independent yet complete representation of a belief base (see e.g. [18] and [19]) and because interesting computational tasks (like satisfiability checking and entailment) are tractable in this framework [20]. In the worst case, computing the set of prime implicates of Δ containing a clause β (a task that we will often refer to) is however not in polynomial total time unless P = NP (it is in polynomial total time when for example the clause is positive and Δ is Horn) [21]. Although the compactness and some of the computational features of a prime implicates representation happen to be welcome properties, the motivation for focusing on prime implicates is here different and stems from their intrinsic epistemological nature.

In our setting, we need to make a difference between the basic beliefs of an agent and the beliefs that it is able to deduce from them. Adopting a prime implicates representation allows us to do that. The prime implicates represent these basic beliefs. Amending the beliefs of an agent must address these basic beliefs in the first place, while safely keeping all the deductive capabilities of the agent leading to derived beliefs, as we will illustrate this.

6 Handling the Various Situations

6.1 Preempting Subsuming Knowledge

Let us go back to our first motivating example, where a new piece of knowledge, *the target is located either in room A, or in room B or there is no actual target*, is intended to prevail about the initial information *the target is located either in room A, or in room B*.

An operator that does this must introduce the new knowledge *TargetIsInRoomA* $\vee$ *TargetIsInRoomB* $\vee$ *NoTarget* and remove any of its prime implicants, e.g. *TargetIs-InRoomA* $\vee$ *TargetIsInRoomB*. Interestingly, as formulas of Δ are under CNF format, it is sufficient to remove all the prime implicants of the clause that needs to prevail.

Actually, the problem of making a formula prevail over all its strict implicants in Δ must sometimes be adapted by replacing Δ by one of its subsets, say Δ', in the Definition of strict implicants. Typically, Δ' is selected as containing the permanent information involving generic rules or other stable knowledge, whereas the rest of Δ contains facts that are temporary or related to a specific case or result from the instantiation of the generic rules to a specific situation. When using such Δ' only, we reason about generic rules independently of specific facts and the process of transforming and removing formulas considers the generic rules, only. Note that facts subsume any rule that contains them as part of its conclusion.

6.2 Handling Clauses About a Concept

When a new information $\gamma = \alpha \vee \beta$ about β comes in, following the belief change AGM theory tradition [1], we assume that, by default, γ is merely to be inserted in Δ, unless this leads to a logical contradiction. As emphasized in the introduction, other situations can occur leading to other forms of belief change within Δ. In this case, we require γ to be provided with an indication of its epistemological role.

Mainly, we envision the following main possibilities.

1. Case 1. The new knowledge must be inserted in Δ. This is the by default case.
2. Case 2. The new piece of information must be inserted in Δ and cannot be subsumed.
3. Case 3. It expresses some condition α about a concept β and α must not be subsumed.
4. Case 4. It expresses some conditions α about a concept β that must replace all current conditions that are required to hold in order for β to hold.
5. Case 5. It expresses conditions α that are compulsory for a concept β in the sense that they must occur as one member of the set of compulsory prerequisites for β to hold.

Implementing the above cases must be done in a way that accommodates the agent's ability to reason deductively. This is easily handled from a conceptual point of view when the representation is a prime implicates one.

To understand this, assume that a naive implementation of Case 5 requires α to belong to any clause about β. Clearly, this would be wrong since this would block some legitimate deductions of the agent. For example, the agent must be able to deduce the tautology $\beta \vee \neg\beta$, which does not mention α. Or, if the agent believes γ, it must remain entitled to deduce $\gamma \vee \beta$ (and $\gamma \vee \neg\beta$, as well). Actually, what is needed is that α must belong to any prime implicate that contains β in the resulting Δ. Hence, we assume that Δ is actually recorded under the form of its prime implicates base, namely Δ_{IP}.

Case 2.
This case has been described in the previous paragraph. In the next section, we will describe a computational approach to handle it.

Case 3.
This case is similar to the previous one but restricts the set of prime implicants that must be removed from Δ. Accordingly, we define *restrictive* clauses to encompass the notion of a clause that expresses conditions that must not be subsumed in order for a concept to hold.

Definition 6. *When $\gamma = \alpha \vee \beta$ is about β, γ is restrictive about β in Δ iff*

1. *$\gamma \in \Delta$, and*
2. *$\beta \notin Th(\Delta)$ and, when $\beta \neq false$, $\neg\beta \notin Th(\Delta)$, and*
3. *no strict implicant of γ containing β belongs to $Th(\Delta)$.*

Clearly, when a clause γ about β must be restrictive in Δ, it is sufficient to insert γ within Δ_{IP} and adapt this latter set so that it remains a prime implicates base, by possibly removing other formulas from it.

Case 4.
Clearly, γ must be inserted within Δ_{PI} and all other clauses that are strict supersets of β must be removed from Δ_{PI}.

Case 5.
This concerns situation where γ must be compulsory about β in (the resulting) Δ according to the following sense.

Definition 7. *When $\gamma = \alpha \vee \beta$ is about β. γ is compulsory about β in Δ iff*

1. *$\gamma \in \Delta$, and*
2. *$\beta \notin Th(\Delta)$ and, when $\beta \neq false$, $\neg\beta \notin Th(\Delta)$, and*
3. *when $\beta = false$, no strict implicant of γ belongs to $Th(\Delta)$, and*
4. *$\forall\,(\alpha' \vee \beta) \in Th(\Delta)$, $\alpha \subseteq \alpha'$ unless $\alpha' \in Th(\Delta)$ or $\beta = false$.*

Properties of compulsory clauses are explored in [22]. When dealing with a prime implicates representation, implementing this case amounts to make sure that there is at least one clause in Δ_{PI} that contains γ and that all clauses in Δ_{PI} that contain β also contain α.

7 MUS-Finding Algorithms as Basic Tools

A central computational issue is checking whether a clause γ is subsumed or not in $Th(\Delta)$ and, in the positive case, in determining which clauses should be expelled from Δ in order to break the subsumption links, as in Case 2. A case of subsumption that is easy to detect occurs when one strict implicant of γ belongs to Δ: a first preprocessing step can check whether any of the n strict longest sub-clauses of γ is *explicit* in Δ in $O(nm)$, where m is the total number of clauses in Δ. Obviously, this preprocessing cannot cover deductive paths that also make use of formulas from $Th(\Delta) \setminus \Delta$ to prove subsumption. From a computational point of view, checking whether a formula subsumes another one is actually *coNP*-complete, and thus intractable in the worst case.

However, recent dramatic progress in Boolean and search makes it often possible to get answers within seconds, especially thanks to powerful SAT solvers [17], which check whether a set of Boolean clauses is consistent or not. In this respect, we have experimented an original method to deliver clauses that must be expelled to get rid of subsumption links. It is based on SAT-solvers and on methods [17] for delivering Minimal Unsatisfiable Subsets of clauses (MUSes): a MUS Γ in $Th(\Delta)$ is an inconsistent set Γ of clauses in $Th(\Delta)$ such that each proper subset of Γ is consistent. More precisely: assume it is to be checked whether $Th(\Delta)$ (that contains γ) subsumes γ through a strict implicant γ' of γ. The solver considers $\Delta \cup \{\neg\gamma'\}$, which can only be inconsistent in case of subsumption. Then, it extracts the MUSes, namely the minimal (w.r.t. inclusion) sets of clauses that are inconsistent. Making sure that at least one clause in each of the MUSes is dropped ensures that the subsumption link disappears. Expelling such clauses can be automatic or the knowledge engineer can be asked whether she (he) really wants to drop them, or even be given the choice of selecting the clauses to be dropped in the MUSes. When he (she) prefers keeping these MUSes intact, she (he) is then conducted to revise his (her) former requirement about the subsumption-freeness status of γ. This solver provides efficient results even for large Δ, provided that the number of MUSes and the size of the MUSes remain small [23] (the number of MUSes is exponential in the worst case).

As an alternative to computing all MUSes, the solver also allows a *cover* of MUSes to be computed. A cover of MUSes of Δ is made of several MUSes of Δ such that if the clauses in the cover were dropped from $Th(\Delta)$ then Δ would become consistent. Interestingly, computing a cover of MUSes of Δ can often be done without computing all MUSes of Δ.

8 Experimental Results

All experimentations have been conducted on a PC (IntelCore 2 Quad 2.66GHz-4Gb Ram) under Linux Ubuntu 11.10 (3.0.0-16-generic). The solver is freely available from http://www.cril.univ-artois.fr/~ramon/preempte. In Table 1, a sample of typical experimental results is provided for the central and computationally-heaviest part of the algorithm, namely the MUS detection step. It presents the actual performance

Table 1 Some experimental results on SAT benchmarks (MUS detecting step)

				All MUSes		1 cover of MUSes	
Instances	#c	#var	#MUSes	#sec	#c in MUSes	#sec	#c in cover
Battleship-5-8-unsat	105	40	1	0.23	105	0.18	105
Battleship-6-9-unsat	171	54	1	1.88	171	0.48	171
Battleship-10-10-unsat	550	100	–	*Timeout*		19.58	417
Battleship-11-11-unsat	726	121	–	*Timeout*		82.21	561
Battleship-12-12-unsat	936	144	–	*Timeout*		172.93	711
5cnf_3500_3900_30f1.shuf.	420	30	–	*Timeout*		17.01	194
5cnf_3900_3900_060.shuf.	936	60	–	*Timeout*		33.99	777
Marg3x3add8ch.shuf.-as.sat03-1448	272	41	1	177.15	272	7.10	272
Marg3x3add8.shuf.-as.sat03-1449	224	41	1	8.79	224	3.48	224
Php_010_008.shuf.-as.sat05-1171	370	80	–	*Timeout*		1.14	370
Rand_net60-30-1.shuf.	10,681	3,600	–	*Timeout*		233.46	10,681
C208_FA_UT_3254	6,153	1,876	17,408	2.94	98	17.49	40
C208_FA_UT_3255	6,156	1,876	52,736	6.67	102	18.97	40

of this step on challenging benchmarks from SAT competitions. The columns indicate for each benchmark instance, its name, its number of clauses (#c), of different variables (#v) and the total number of MUSes (#$MUSes$) in the instance (unless the algorithm was not able to extract all MUSes; in that case this is denoted by –). Then, the time spent for finding and extracting all MUSes and one cover of MUSes is given in seconds, together with the numbers of different clauses in these MUSes and in this cover of MUSes, respectively. Timeout was set to 250 s. Not surprisingly, computing a cover of MUSes proved more often time-feasible than computing all MUSes successively. Actually, it was often possible to extract a cover whereas computing all MUSes was not feasible in the preset computing time. Let us stress that the benchmarks involve large MUSes, as this is often a factor of difficulty for SAT solvers. It is well-known that, heuristically, the time spent by MUSes finding algorithms also often increases with the size of MUSes. It is also often acknowledged that in most real-life belief bases, MUSes are of smaller sizes, based on the finding that a new incoming belief generally interacts with a small number of our pre-existing beliefs to form an inconsistent chain of reasoning that is of minimal length (i.e., a MUS). Accordingly, under this latter assumption, we can expect a time-efficiency that should often be better than the one exhibited in the described experimentations.

9 Conclusion and Perspectives

So far, the A.I. research community has mainly concentrated on situations where an incoming piece of information is logically inconsistent with the current beliefs. In other cases, the incoming belief is just expected to increase the set of current ones. Actually, we claim that the belief change domain should also concern other reasoning paradigms involving an additional piece of information to be handled. Especially, there are many situations where an additional piece of symbolic knowledge requires the former beliefs to change, even when no logical inconsistency arises. In this chapter, we have illustrated some of them. A second claim in this chapter is about the usefulness of a prime implicates representation of beliefs. By delimiting a frontier between basic beliefs from what a deductive agent can derive from them, it helps in modeling belief change operators. Accordingly, the various belief change paradigms introduced in this chapter are easily handled using a prime implicates representation. In the future, we plan to investigate detailed properties of these belief change operators and to continue exploring the extent to which they could be implemented and achieved efficiently. We also plan to extend this study to other logical settings, especially non-monotonic ones involving forms of negation as failure or answer-set programming. In this last respect, note that when Δ does not entail any literal -which might often correspond to the targeted framework where we are only interested in generic rules and their interactions- negation as failure amounts to mere logical deduction and Δ_{PI} does not contain any unary clauses. In such a case, this study directly applies, too.

References

1. Fermé, E., Hansson, S.: AGM 25 years. Twenty-five years of research in belief change. J. Philos. Logic. **40**, 295–331 (2011)
2. Alchourrón, C., Gärdenfors, P., Makinson, D.: On the logic of theory change: partial meet contraction and revision functions. J. Symb. Logic. **50**(2), 510–530 (1985)
3. Katsuno, H., Mendelzon, A.: On the difference between updating a knowledge base and revising it. In: Proceedings of KR'91, pp. 387–394 (1991)
4. Gärdenfors, P.: Knowledge in Flux: Modeling the Dynamics of Epistemic States. MIT Press, Cambridge (1988)
5. Hansson, S.O.: A Textbook of Belief Dynamics: Theory Change and Database Updating. Kluwer Academic Publishers, Dordrecht (1999)
6. Konieczny, S., Pino Pérez, R.: On the logic of merging. In: Proceedings of KR'98, pp. 488–498 (1998)
7. Konieczny, S., Grégoire, É.: Logic-based information fusion in artificial intelligence. Inf. Fusion. **7**(1), 4–18 (2006)
8. Doyle, J.: A truth maintenance system. Artif. Intell. **12**, 231–272 (1979)
9. Dalal, M.: Investigations into a theory of knowledge base revision (preliminary report). In: Proceedings of AAAI'88, vol. 2, pp. 475–479 (1988)
10. Revesz, P.Z.: On the semantics of theory change: arbitration between old and new information. In: Proceedings of PODS'93, pp. 71–82 (1993)
11. Subrahmanian, V.S.: Amalgamating knowledge bases. ACM Trans. Database Syst. **19**, 291–331 (1994)
12. Fagin, R., Ullman, J.D, Vardi, M.Y.: On the semantics of updates in databases. In: Proceedings of PODS'83, pp. 352–365 (1983)
13. Besnard, Ph., Grégoire, É., Ramon, S.: Enforcing logically weaker knowledge in classical logic. In: 5th International Conference on Knowledge Science Engineering and Management (KSEM'11), pp. 44–55. LNAI 7091, Springer (2011)

14. Besnard, Ph., Grégoire, É., Ramon, S.: Preemption operators. In: Proceedings of ECAI 2012, pp. 893–894 (2012)
15. Besnard, Ph., Grégoire, É., Ramon, S.: Overriding subsuming rules. In: Proceedings of ECSQARU'11, pp. 532–544. LNAI 6717, Springer (2011)
16. Besnard, Ph., Grégoire, É., Ramon, S.: Logic-based fusion of legal knowledge. In: Proceeings of Fusion 2012, pp. 587–592. IEEE Press, Singapore (2012)
17. http://www.satlive.org satlive is a main website of the research community on SAT
18. Zhuang, Z.Q., Pagnucco, M., Meyer, T.: Implementing iterated belief change via prime implicates. In: Orgun, M.A., Thornton, J. (eds) Australian Conference on Artificial Intelligence, volume 4830 of Lecture Notes in Computer Science, pp. 507–518. Springer (2007)
19. Bienvenu, M., Herzig, A., Qi, G.: Prime implicate-based belief revision operators. In: 20th European Conference on Artificial Intelligence (ECAI 2012), pp. 741–742 (2008)
20. Darwiche, A., Marquis, P.: A knowledge compilation map. J. Artif. Intell. Res. (JAIR) **17**, 229–264 (2002)
21. Eiter, T., Makino, K.: Generating all abductive explanations for queries on propositional horn theories. In: Computer Science Logic, 17th International Workshop, CSL 2003, 12th Annual Conference of the EACSL, and 8th Kurt Gödel Colloquium, KGC 2003, Vienna, Austria, August 25–30, pp. 197–211 (2003)
22. Besnard, Ph., Grégoire, É.: Handling incoming beliefs. In: 6th International Conference on Knowledge Science Engineering and Management (KSEM'13), LNAI, Springer (2013)
23. Grégoire, É., Mazure, B., Piette, C.: Using local search to find MSSes and MUSes. Eur. J. Oper. Res. **199**(3), 640–646 (2009)

A Multi-Layer Moving Target Defense Approach for Protecting Resource-Constrained Distributed Devices

Valentina Casola, Alessandra De Benedictis and Massimiliano Albanese

Abstract Techniques aimed at continuously changing a system's attack surface, usually referred to as Moving Target Defense (MTD), are emerging as powerful tools for thwarting cyber attacks. Such mechanisms increase the uncertainty, complexity, and cost for attackers, limit the exposure of vulnerabilities, and ultimately increase overall resiliency. In this chapter, we propose an MTD approach for protecting resource-constrained distributed devices through fine-grained *reconfiguration* at different architectural layers. We introduce a coverage-based security metric to quantify the level of security provided by each system configuration: such metric, along with other performance metrics, can be adopted to identify the configuration that best meets the current requirements. In order to show the feasibility of our approach in real-world scenarios, we study its application to Wireless Sensor Networks (WSNs), introducing two different reconfiguration mechanisms. Finally, we show how the proposed mechanisms are effective in reducing the probability of successful attacks.

Keywords Moving target defense · Reconfiguration · Proactive security

---

The work presented in this chapter is supported in part by the Army Research Office under award number W911NF-12-1-0448 and MURI award number W911NF-13-1-0421.

V. Casola · A. De Benedictis (✉)
Department of Electrical Engineering and Information Technology,
University of Naples Federico II, Naples, Italy
e-mail: alessandra.debenedictis@unina.it

V. Casola
e-mail: casolav@unina.it

M. Albanese (✉)
Center for Secure Information Systems, George Mason University, Fairfax, VA, USA
e-mail: malbanes@gmu.edu

T. Bouabana-Tebibel and S. H. Rubin (eds.), *Integration of Reusable Systems*,
Advances in Intelligent Systems and Computing 263, DOI: 10.1007/978-3-319-04717-1_14,

1 Introduction

In recent years, we have witnessed a growing interest in techniques aimed at continuously changing a system's attack surface in order to prevent or thwart attacks. This approach to cyber defense is generally referred to as Moving Target Defense (MTD), and it is currently considered one of the *game-changing* themes in cyber security by the Executive Office of the President, National Science and Technology Council [1–3]. As stated in [1], Moving Target Defense *enables us to create, analyze, evaluate, and deploy mechanisms and strategies that are diverse and that continually shift and change over time to increase complexity and cost for attackers, limit the exposure of vulnerabilities and opportunities for attack, and increase system resiliency*.

The MTD paradigm can be successfully adopted to enforce security requirements in networks composed of distributed and mobile devices, that are typically characterized by limited hardware and software resources. Achieving high levels of security in such constrained environments is not a straightforward task, and innovative approaches must be devised. In this chapter we propose an MTD strategy based on fine-grained *reconfiguration* to protect resource-constrained distributed devices, which are characterized by limited processing and storage capabilities, limited battery life, mobility, highly dynamic topology, and frequent failures. Our reconfiguration approach applies to different architectural layers and takes into account not only the hardware and software features of the nodes but also specific security and performance requirements depending on the deployment scenario.

Although changing configuration or system parameters to augment security is a very intuitive principle[1], there is still a lack of metrics to evaluate the security provided by a system and, consequently, quantify the benefits of reconfiguration. To this aim, we introduce a coverage-based security metric to quantify the level of security provided by a given system configuration. Such metric, along with commonly adopted performance and cost metrics, is used to identify the configuration that best meets the current requirements.

In order to show the feasibility of our approach in real applications, we consider Wireless Sensor Networks (WSNs) as a case study. Different mechanisms have been proposed to secure WSNs, but of most such efforts have primarily been aimed at limiting power consumption by reducing the computational and storage requirements. Because of these constraints, the level of security provided by such mechanisms is quite limited, and more complex solutions are not feasible in practice. In this scenario, an MTD approach would make it possible to achieve better security, without requiring computation-intensive solutions, by periodically switching among multiple lightweight cryptosystems. Several reconfiguration mechanisms have been proposed for WSNs [4], mainly based on network reprogramming. They operate at different architectural levels but present similar limitations, as they are battery consuming, introduce a significant overhead, and are potentially not secure.

---

[1] Consider, for instance, the trade-off between the key length in a cryptographic session and the duration of the session itself.

In order to address these limitations, we introduce two novel mechanisms for reconfiguring sensors that provide better performance from several points of view. We carried out a number of experiments by simulating attack scenarios where an attacker is able to gather partial information on the adopted cryptosystem and attempts a brute force attack. We evaluate the effectiveness of the proposed MTD approach by measuring the probability of successfully completing an attack and show how reconfiguration dramatically decreases such probability.

The paper, which extends the work presented in [5], is organized as follows. Section 2 discusses some of the MTD approaches that have been proposed in the literature, whereas Sect. 3 introduces our approach and presents the reconfigurable architectural layers we take into account. Section 3.1 presents a coverage metric to evaluate the level of security provided by a configuration and in Sect. 3.2 the dependency of security on time is discussed. Section 4 illustrates two innovative reconfiguration mechanisms for WSNs, whereas Sect. 5 reports experimental results. Finally, some concluding remarks are given in Sect. 6.

2 Related Work

The idea behind the Moving Target Defense (MTD) is to change one or more properties of a system in order to present attackers with a varying *attack surface*, so that, by the time the attacker gains enough information about the system for planning an attack, the system's attack surface will be different enough to disrupt it [2, 3]. According to the definition in [6], a system's attack surface is *the subset of the system's resources (methods, channels, and data) that can be potentially used by an attacker to launch an attack*. It depends on the system's hardware and software features, and can be changed by dynamically reconfiguring such features at different levels of granularity.

A common MTD practice consists in updating the cryptographic keys used for encryption of communication channels: this introduces some uncertainty for attackers but presents the problem of key distribution, that is a critical phase particularly subject to attacks. More in general, MTD approaches (also referred to as *diversity techniques*) may be applied both at the application level and at a lower level (e.g., code location in memory), as suggested in [7]. Several low-level MTD techniques have been proposed in the literature, based on the idea of automatically generating diverse variants of a program to disrupt vulnerability exploits. A widely deployed example is Address Space Randomization, that was introduced in 2000 by the PAX Team for Linux,[2] and has been implemented in most modern operating systems. The basic idea is to randomize the locations of objects in memory so that an attack depending on the knowledge about the address of these objects will fail.

Instruction Set Randomization [8] is another technique for obfuscating the language understood by a system to protect against code-injection attacks: by

[2] http://pax.grsecurity.net/

randomizing the underlying systems instructions, *foreign* code introduced by an attack would fail to execute correctly, regardless of the injection approach.

Another type of low-level diversification is altering how data is stored in memory: in [9] authors present Data Randomization, a technique that provides probabilistic protection against attacks that exploits memory errors by XOR-ing data with random masks. Data randomization uses static analysis to partition instruction operands into equivalence classes: it places two operands in the same class if they may refer to the same object in an execution that does not violate memory safety. Then it assigns a random mask to each class and it generates code instrumented to XOR data read from or written to memory with the mask of the memory operand's class. Therefore, attacks that violate the results of the static analysis have unpredictable results.

Jackson et al. [10] present a diversity technique based on the generation, during the compilation phase, of multiple functionally equivalent machine codes for the same high-level source: with massive-scale software diversity, every user could get its own diversified program version, so that it is impossible for attackers to run a successful attack.

The advantage of low-level diversity is that it does not require an understanding of the application's behavior and can be done automatically. However, it is only capable of thwarting specific classes of attacks, such as code injection and memory corruption attacks. Looking at higher-level MTD techniques, several approaches have been proposed, aimed at thwarting the attacker's reconnaissance effort: reconnaissance enables adversaries to gather information about the target system including network topology, configurations, network dynamics. This information can be used to identify system vulnerabilities, and to design and execute specific exploits.

In this regard, several approaches for dynamically changing nodes IP addresses for proactive security have been proposed in the literature, [11–13]. In 2001, Kewley et al. [13] presented a technique called DYNAT (Dynamic Network Address Translation), aimed at confusing any adversary sniffing the network by obfuscating host identity information in TCP/IP packet headers when packets enter public parts of the network. Whenever a client host wants to communicate with a protected server host, the addressing information contained in the header of its request packets is translated (encrypted) by a DYNAT shim before routing the packet to the server. A server gateway receives the packets, reverses the translation in the header fields (decryption) and obtains the true host identity information, used to pass the packets to the target server.

Another work funded by DARPA is presented in [12] by Atighetchi et al., that give an overview of current set of network-level defenses in the DARPA APOD (Application That Participate in Their Own Defense) project. Among the proposed network-centric defense mechanisms, the APOD toolkit also provides a *port and address hopping mechanism*, based on constantly changing a service's TCP identity to both hide the service's real identity and confuse the attacker during reconnaissance. Packets intercepted by attackers will reveal random addresses, which are valid only for a small period of time, e.g., 1 min. For a port attack to be successful, the attacker must discover the current ports and execute the attack all within one refresh cycle.

Antonatos et al. [14] introduce a proactive defense mechanism called Network Address Space Randomization (NASR) whose objective is to harden networks against worms that use precomputed hitlists of vulnerable targets, by forcing nodes to frequently change their IP addresses. In order to achieve this goal, the authors implemented an advanced NASR-enabled DHCP server to expire DHCP leases at intervals suitable for effective randomization. As the addresses are actually changed at the end-points of a communication, active connections are disrupted during the update; moreover, NASR is limited in the address space as it uses LAN addresses, and requires changes to the end-host operating system, thus making the deployment costly.

In [15] the authors introduce an MTD technique called OpenFlow Random Host Mutation (OF-RHM): each host is assigned an address range, selected from the entire unused address space in the network, and at each mutation interval, a virtual IP is chosen from this range and associated with the host. A Software-Defined Networking (SDN) approach is adopted for range allocation and mutation coordination: a centralized controller (NOX) properly installs flows in OpenFlow switches to forward requests and perform the address translation actions.

Finally, the MTD defense mechanism proposed in [16] is designed to protect the identity of nodes in Mobile Ad Hoc Networks by turning the classical Sybil attack mechanism into an effective defense mechanism. Legitimate nodes use virtual identities to communicate and periodically change their virtual identity to increase the uncertainty for attackers observing the network. To preserve communication among legitimate nodes, the network layer is modified by introducing a mechanism for mapping virtual identities to real identities, and a protocol for propagating updates of a node's virtual identity to all legitimate nodes.

3 Improving Node Security

In this chapter, we propose an MTD framework for reconfiguring resource-constrained devices at different architectural levels, with the reconfiguration granularity chosen at runtime based on current requirements. By reconfiguring a system, it is possible to increase the overall *security level* it provides. Reconfiguration can be either reactive—i.e., the system is reconfigured in response to a detected or perceived threat or new security requirements—or proactive—the system is periodically reconfigured to limit the amount of time each configuration is exposed to malicious observers. Additionally, reconfiguration should be performed in a way to minimize its impact on the system in terms of resource consumption and performance.

Reconfiguration consists in changing one or more of the system's parameters. In our case study focused on WSNs, we identified two main reconfigurable architectural layers:

- *Security layer*. Security in an embedded network can be achieved by implementing a proper cryptosystem. Security layer reconfiguration can be performed by

switching among different cryptosystems that satisfy specific security requirements while meeting certain performance and energy consumption constraints.
- *Physical layer*. In embedded systems, the software is part of the node's firmware, that is typically preloaded on internal read-only memory chips (ROM), in contrast to a general-purpose computer that loads its programs into random access memory (RAM) at run-time. Firmware provides the control program of the device and represents the skeleton where different libraries for the implementation of the available cryptosystems and APIs can be plugged and activated via proper software switches. Nodes can be equipped with several versions of the firmware in order to perform physical reconfiguration when needed.

Clearly, further parameters could be considered for reconfiguration, such as the application interface, the hardware configuration or the topology, as long as their reconfiguration is feasible from a technical and energy consumption point of view. In order to perform complex tasks, embedded nodes communicate with one another according to specific application interfaces (APIs), defining the format of the exchanged messages and the communication protocols. Reconfiguration could be applied at this level by providing different APIs for the same application. API reconfiguration could be useful to confuse an attacker that is observing the communication protocol in order to find an exploit to interfere with or control the communication.

As for hardware reconfiguration, it is expensive and not feasible on most of the available devices but needed in case of damage. Network topology could be reconfigured in terms of the view offered to external observers. This could be achieved by implementing a mechanism that, for instance, presents virtual identities or introduces additional fake nodes into the network. Such a mechanism would need additional protocols and algorithms that are often too expensive for the considered nodes.

The choice of the reconfiguration level impacts both the system's performance and the provided level of security. From the performance point of view, changing the firmware of all the nodes in the network or a subset of them is much more expensive—in terms of latency and power consumption—than changing the cryptosystem, whose reconfiguration could be handled in software. On the other side, by changing the entire application running on a node, it becomes harder for an attacker to exploit software vulnerabilities and gain complete control of the node.

At the security layer, the cryptosystem itself is designed to cope with a specific set of attacks and provides an intrinsic level of security, depending on the cryptographic scheme, the algorithm, the length of the keys, etc. Reconfiguration of the cryptosystem can increase the level of security in two ways, that is by switching to a cryptosystem that covers a larger set of attacks (e.g., to cope with some detected or perceived threats), or by selecting an equivalent cryptosystem that uses different parameters. Given a certain fixed configuration, the more an attacker is able to observe, the more he will be able to infer information about the system. By continuously changing the system's configuration, the attacker will be presented with different views of the system over time, and will have to restart the reconnaissance effort multiple times in order to identify a viable exploit.

Once the admissible configurations have been identified, the selection of the new configuration is performed by a security-driven scheduler. The scheduler can be either a centralized entity making decisions on the global network configuration, or a decentralized component, independently deployed on each network node, making local reconfiguration decisions. In a *centralized* approach, a central entity triggers a configuration update based on some events (e.g., timer expiration, detected security threat) and transmits its decision to all the nodes that are involved. In a *decentralized* approach, each node is able to schedule, independently from other nodes, when to update its own configuration. Communication among legitimate nodes is preserved adopting additional mechanisms, described in details in the following section.

In the remainder of this chapter, we will discuss the methodology adopted to evaluate the level of security provided by a system configuration, and how to increase it using a reconfiguration approach.

3.1 Security Level Evaluation

The security of complex systems depends on many technical and organizational issues that must be properly addressed. The need for a clear definition and selection of security rules has led system administrators to set up security policies trying to adopt formal approaches to describe system security configurations. In spite of the ambiguity of such policies, a common approach to evaluate a system's security is through evaluation of its security policy. At present, such an evaluation is performed *by hand* whenever enterprises endeavor to extend their trusted domain and cooperate [17]. This approach also includes well known standards as Common Criteria and TCSEC [18, 19], that are very suitable to assess and audit the security level provided by a company, by a specific procedure or, in general, by a system.

The Common Criteria (CC) for Information Technology Security Evaluation (Common Criteria or CC) [18] are an internationally approved set of standard for computer security certification. They are used by Government customers in the USA and the NATO community along with other organizations, particularly in the public sector, to determine the level of security and assurance of various technology products. However, the assurance levels provided by CC (from EAL1 to EAL7) do not measure the security of the system itself, but simply state at what level the system was tested, and do not find a direct application in our approach.

Defining a quantitative measure of the level of security provided by a system is a complex task. Several security metrics have been proposed in the literature, mostly based on the analysis of attack graphs or on risk quantification [20–23]. Other approaches, such as the one adopted by the Common Vulnerability Scoring System (CVSS) [24], try to rate the severity of security vulnerabilities and assign a score to a system based on the vulnerabilities it is subject to.

Other metrics are linked to the adopted configurations [17, 25], they are centered on the mechanisms that are available to enforce a subset of security requirements. We started from these considerations to introduce a metric based on the *coverage* of

Table 1 An example of Attacks Coverage Table

Conf	Attack A	Attack B	Attack C	SL
c_1	×			L_1
c_2	×		×	L_2
c_3	×	×	×	L_3
c_4	×		×	L_2

a configuration respect of a set of known attacks. An attack could have several objectives, such as physically taking possession of a node, interfering with communication at the physical level, exploiting software vulnerabilities to take control of a node, disturb network operation at routing/application level or intercept sensitive data. In this discussion we are interested in attacks aimed at interfering, steering or eavesdropping communications at the application layer among nodes, and at exploiting vulnerabilities of the firmware installed on nodes.

Let *Threats* define the set of threats of interest, belonging to the above discussed set of attacks. A configuration c is said to *cover* a threat $t \in Threats$, if either the cryptosystem implemented at the security layer or the specific firmware version running on the node include mechanisms to protect the node from such threat.

Once the admissible configurations and the attacks of interest have been identified, it is possible to build an *Attack Coverage Table*, that helps define the levels of security provided by each configuration [26]. Table 1 shows an example of attack coverage table relative to configurations $\{c_1, c_2, c_3, c_4\}$, under the hypothesis that three attacks of interest have been identified, namely *Attack A*, *Attack B* and *Attack C*. An increasing level of security (from L_1 to L_4 in the example) can be assigned to configurations, based on the risk associated with the attacks and their coverage properties.

Attack coverage can be defined either as an ON/OFF property (that is an attack is covered or uncovered), or in terms of the degree of satisfaction of specific requirements (e.g. authentication, integrity, confidentiality, key distribution...), using a scoring system (similar to CVSS for vulnerabilities). Coverage with respect to a specific attack could even be defined in terms of the effort an attacker needs to make the attack succeed.

The level of security associated with a configuration could simply depend on the number of covered threats, or it could be set depending on the risk associated with each threat, either in a static way (the risk associated with a threat is set at deployment and remains unchanged for the entire operation of the network) or dynamically (the risk associated with a threat changes dynamically during network operation depending on current conditions and possible detection events).

3.2 Modeling the Security Level

As previously said, each configuration provides a certain level of security, which depends on the implemented cryptosystem (cryptographic scheme, algorithms, and keys) and is characterized by an intrinsic value. Indeed, the longer a system

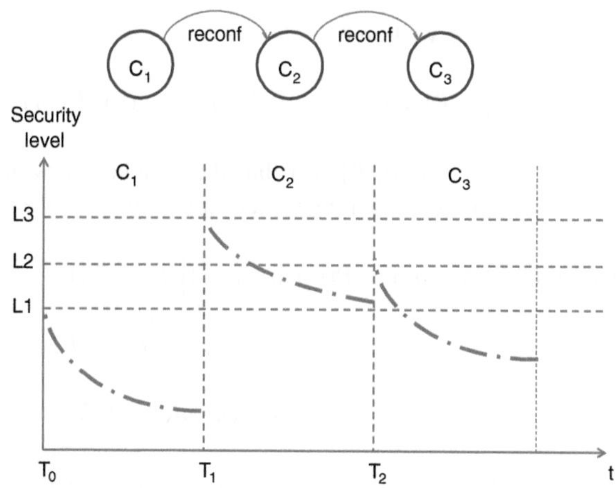

Fig. 1 Reconfigurations and security level

configuration is exposed to malicious observers, the more the actual level of security decreases. For this reason, the security level is a monotonically decreasing function, with its maximum corresponding to the intrinsic security level associated with the specific implemented cryptosystem.

As illustrated in Fig. 1, using reconfiguration, we can prevent the security level from falling below a certain threshold, and periodically reset it to the intrinsic value associated with a new configuration. Dually speaking, we avoid that the probability of successfully completing an attack increases beyond a certain threshold. In fact, such probability depends on the considered type of attack and is usually represented by a monotonically increasing function: the longer an attacker can try to exploit a system, the higher the success probability is. It is easy to demonstrate that, by introducing reconfiguration, we can *break* the monotonicity, such that the probability of successfully completing an attack actually decreases every time the system is reconfigured.

Theorem 1. *Let* $[0, T]$ *be an observation interval, and let* $n \in \mathbb{N}$ *be an integer greater than or equal to 2, representing the number of reconfigurations in* $[0, T]$. *Then the following inequality holds.*

$$Pr(success\ ([0, T], n)) \leq Pr(success\ ([0, T], 0)) \quad (1)$$

where $Pr(success\ (I, x))$ *denotes the probability that the attacker is successful within the temporal interval* I *if* x *reconfigurations are performed during the same interval.* ■

In order to prove Theorem 1, we need to consider that the probability that the attacker will successfully break the cryptosystem between 0 and T when the interval $[0, T]$ is broken down into n validity intervals—and a different cryptosystem is used in each such intervals—can be written as

$$\Pr(success([0,T],n)) = 1 - \Pr(\neg success([0,T],n)) \tag{2}$$

The probability $\Pr(\neg success([0,T],n))$ that the attacker does not succeed by time T is the probability that he does not succeed in any of the n validity intervals.

$$\begin{aligned}\Pr(\neg success([0,T],n)) = \Pr(\neg success & \left(\left[0, \tfrac{1}{n}\cdot T\right]\right) \\ & \wedge \neg success \left(\left[\tfrac{1}{n}\cdot T, \tfrac{2}{n}\cdot T\right]\right) \\ & \wedge \ldots \wedge \neg success \left(\left[\tfrac{n-1}{n}\cdot T, T\right]\right))\end{aligned} \tag{3}$$

The events $\neg success\left(\left[0, \frac{1}{n}\cdot T\right]\right), \ldots, \neg success\left(\left[\frac{n-1}{n}\cdot T, T\right]\right)$ are clearly independent, thus $\Pr(\neg success([0,T],n))$ can be computed as follows.

$$\Pr(\neg success([0,T],n)) = \prod_{i=0}^{n-1}\left(1 - \Pr\left(success\left(\left[\frac{i}{n}\cdot T, \frac{i+1}{n}\cdot T\right]\right)\right)\right) \tag{4}$$

As the probability that the attacker can break the system in a given interval is directly proportional to the length of the interval itself, we can conclude that, for all $i \in [0, n-1]$, $\Pr\left(success\left(\left[\frac{i}{n}\cdot T, \frac{i+1}{n}\cdot T\right]\right)\right) = \frac{\Pr(success([0,T]))}{n}$. This conclusion relies on the simplifying assumption that the different cryptosystems used in our framework are equivalent in terms of attack time. Generalizing this result to the case of heterogeneous cryptosystems is straightforward, but it is omitted for reasons of space. Additionally, the above conclusion assumes that the interval $[0,T]$ is larger than the time needed to complete a full brute force attack.[3] Then, Eq. 4 can be rewritten as follows.

$$\begin{aligned}\Pr(\neg success([0,T],n)) &= \textstyle\prod_{i=0}^{n-1}\left(1 - \frac{\Pr(success([0,T]))}{n}\right) \\ &= \left(1 - \frac{\Pr(success([0,T]))}{n}\right)^n\end{aligned} \tag{5}$$

In order to complete the proof, we need the results of another theorem:

Theorem 2. *Let $x \in [0,1]$ be a real number and let $n \in \mathbb{N}$ be an integer number. The following inequality holds.*

$$\left(1 - \frac{x}{n}\right)^n \geq 1 - x \tag{6}$$

■

[3] If $\Pr(\neg success([0,T])) = 1$, then there may exist a sub-interval $[t_i, t_j]$ of $[0,T]$ such that $\Pr(\neg success([t_i, t_j])) = 1$.

Using the binomial theorem, the expression $\left(1-\frac{x}{n}\right)^n$ can be expanded as follows.

$$\left(1-\frac{x}{n}\right)^n=\sum_{k=0}^{n}\binom{n}{k}\left(-\frac{x}{n}\right)^k=1-x+\sum_{k=2}^{n}\binom{n}{k}\left(-\frac{x}{n}\right)^k \tag{7}$$

To complete the proof, we need to show that the alternating series $\sum_{k=2}^{n}\binom{n}{k}\left(-\frac{x}{n}\right)^k$ is greater than or equal 0. As the first term in the series is positive, we only need to show that all the terms have decreasing absolute values. In order to do so, we now show that the ratio between two consecutive terms is greater than 1.

$$\left|\frac{\binom{n}{k}\left(-\frac{x}{n}\right)^k}{\binom{n}{k+1}\left(-\frac{x}{n}\right)^{k+1}}\right|=\frac{\frac{n!}{k!\cdot(n-k)!}}{\frac{n!}{(k+1)!\cdot(n-k-1)!}\cdot\frac{x}{n}}=\frac{n\cdot(k+1)}{(n-k)\cdot x} \tag{8}$$

It is clear that the quantity at the right end side of Eq. 8 is greater than 1, as $n\cdot(k+1)\geq n$ and $(n-k)\cdot x\leq n$. Using Theorem 2, we can conclude that

$$\left(1-\frac{\Pr\left(success\left([0,T]\right)\right)}{n}\right)^n\geq 1-\Pr\left(success\left([0,T]\right)\right) \tag{9}$$

Combining Eqs. 2, 5, and 9, we can write

$$1-\Pr\left(success\left([0,T],n\right)\right)\geq 1-\Pr\left(success\left([0,T]\right)\right) \tag{10}$$

Equation 1 follows directly from Eq. 10.

In conclusion, Theorem 1 shows that, in theory, the proposed mechanism is effective in reducing the probability that the attacker will successfully discover currently used cryptographic keys in a given amount of time. In other words, it will take more time for the attacker to break the system. Experiments reported in the next section confirm this result.

In the following, we will refer to the *level of security* as a security metric to express how secure is a configuration with respect to the considered attacks. A security value can be assigned, based on the attacks coverage table, both to a single node and to a link, defined as a connection between communicating nodes. Node security is related primarily to the physical layer (e.g., tamper resistant HW, protected external ROM), while subnet security depends on the security layer (e.g., cryptographic algorithm, key length, key agreement mechanisms); as previously discussed, both also depend on the reconfiguration mechanism itself, that is on time.

Assume that the set SEC of available cryptosystems is a totally ordered set: given $s_1, s_2 \in SEC$, there is an ordering relation between them, and $s_1 \leq s_2$ means that the cryptosystem s_1 is not more secure than the cryptosystem s_2. It is possible to have elements in SEC that are equivalent from the security point of view, adopting for instance the same algorithm but using different parameters (e.g. different keys).

Assume the sequence of the M configurations adopted by a node n is given by $\langle C_1(n), \ldots, C_M(n)\rangle$, and the sequence of time instants in which such configuration were activated is $\langle T_1(n), \ldots, T_M(n)\rangle$.

Let the configurations $C_i(p) = (im_i(p), api_i(p), s_i(p))$ and $C_i(q) = (im_i(q), api_i(q), s_i(q))$ be the i-th active configurations respectively on node p and q. Note that in order for the nodes to be able to communicate, they should either share the same security and API configurations, or they should be provided with a mechanism to always know what is the configuration currently used by other legitimate nodes. Let $T_i(p, q)$ identify the initial time instant when the status of p and q is such that they are able to communicate. In the following, we will refer to a $link$ as a directed edge (p, q) connecting two nodes involved in a communication, with packets traveling from p to q. A link configuration is defined as $C_i(p, q) = (C_i(p), C_i(q))$.

Let us refer to $SL_{(p,q)}(t)$ as the level of security, at time t, of a link (p, q). It is the level of security associated with the cryptosystem used to secure data flow from p to q, denoted with $s_i(p, q)$. With $SL_p(t)$ we identify the level of security of node p, depending on its physical configuration $im_i(p)$ and on time.

Definition 1 (Level of security of a link). *The level of security $SL_{(p,q)}(t)$ of a link (p, q), provided by $C_i(p, q)$ at time $t \in [T_i(p, q), T_{i+1}(p, q)]$, can be expressed as a function of the specific cryptosystem adopted $s_i(p, q)$ and the time elapsed since the current configuration was activated.*

$$SL_{(p,q)}(t) = f(s_i(p, q), t - T_i(p, q)) \tag{11}$$

Definition 2 (Level of security of a node). *The level of security $SL_p(t)$ of a node p can be expressed as a function of the specific firmware adopted $im_i(p)$ and the time elapsed since the current physical configuration was activated.*

$$SL_p(t) = f(im_i(p), t - T_i(p)) \tag{12}$$

Finally, we can define the Security Level associated to a configuration as:

Definition 3 (Level of security of the network). *Assuming that the network is partitioned in different subnets, each composed of nodes communicating with one another with a certain interface (security and application layer), the overall level of security of the network depends both on the security of nodes composing the network, and of the different subnets, other than on time.*

$$SL_{net}(t) = A \cdot \sum_{i=0}^{N-1} \sum_{j=0, j \neq i}^{N-1} \alpha_{ij} \cdot SL_{(i,j)}(t) \cdot x_{ij} + B \cdot \sum_{i=0}^{N-1} \beta_i \cdot SL_i(t) \tag{13}$$

where

- *A and B represent the relative importance of the set of links and the set of network nodes respectively, and satisfy the following constraint: $A + B = 1$.*

- *the* α_{ij} *constants represent link weights, and the* β_i *constants represent node weights and are useful to give more importance to critical nodes or portions of the network. They are subject to the following constraints:*

$$\begin{aligned} &\sum_{i=0}^{N-1} \sum_{j=0, j \neq i}^{N-1} \alpha_{ij} = 1 \\ &\sum_{i=0}^{N-1} \beta_i = 1 \end{aligned} \tag{14}$$

- *the* x_{ij} *variables represent the existence of links and are defined as follows:*

$$x_{ij} = \begin{cases} 1 \; if \; i \neq j \; and \; \exists \; a \; link \; between \; node \; i \; and \; j \\ 0 \; if \; i \neq j \; and \; \nexists \; a \; link \; between \; node \; i \; and \; j \end{cases} \tag{15}$$

4 WSN Reconfiguration: A Case Study

A WSN is an embedded network composed of a base station—able to perform multi-node data fusion and complex application logic, and often provided with a consistent source of energy—and several motes, which merely perform local processing on sensed data. Nodes communicate by exchanging messages over a radio channel: the base station sends queries to motes in order to sample physical variables (e.g., humidity), whereas motes simply reply to these queries by sending unicast messages to the base station.

Security is a fundamental concern in WSNs, as they are widely adopted in several critical application domains. Nevertheless, because of their peculiar features—constrained processing and storage capabilities, limited battery life, highly dynamic topology and mobility, frequent failures—providing security is not a straightforward task. The introduction of security mechanisms has a strong impact on performance and resource consumption, that often represent a limiting factor. For this reason, although the adoption of a complex cryptosystem (e.g., based on public key primitives) for all network activities could be desirable from a security point of view, it is not feasible in practice. The proposed reconfiguration approach is able to overcome these concerns, as it allows to maintain an acceptable level of security in the network by leveraging not only the intrinsic features of the adopted cryptosystems, but also other features, such as the physical configuration and the application interfaces, other than the reconfiguration mechanism itself. This way, the use of simpler cryptosystems for short periods of time can be preferable to the adoption of a single strong but computation-intensive cryptosystem.

In this discussion, we refer to TinyOS, the most commonly adopted operating system for WSNs. TinyOS applications and the OS itself are built by connecting components that represent functional building blocks, such as communication protocols, device drivers, or data analysis modules. During the default compilation process of TinyOS, these building blocks are converted into a monolithic, static binary, to enable code optimization and ensure a small memory footprint. This means that the OS and its applications' executables lack modularity, and it is not possible to dynamically replace a single component at runtime. Security mechanisms could be implemented either as independent TinyOS components or as different static libraries wired in the same component, whose functions are invoked by applications to ensure security requirements. Reconfiguration of both the security layer and the application interfaces could be easily achieved by including the implementation of all the available solutions into the firmware installed on the device, and activating the desired configuration through software switches and ad hoc protocols. Firmware reconfiguration can be performed by adopting node reprogramming techniques, that will be illustrated in details later. Two innovative approaches to reconfiguration are presented in the following subsections, along with some implementation details.

4.1 Security Layer Reconfiguration

Assume that, in order to enforce security, queries are signed by the base station for authentication purposes, and reply messages are encrypted for ensuring confidentiality and integrity. The security layer performing these operations can be designed to implement different cryptographic protocols, depending on the required security level and available resources. The basic idea of the proposed approach is to dynamically change the security layer, by switching between two or more different implementations. We assume that each node is provided with a pool of different cryptosystem implementations, which are identified by a unique ID.

To give a concrete example, we refer to the two cryptosystems presented in [27], based respectively on Elliptic Curve Cryptography (WM-ECC libraries) and Identity-based cryptographic techniques (TinyPairing libraries). WM-ECC [28] provides key agreement algorithms and digital signature that can be used to authenticate packets in the sensor network. It provides support for all the ECC operations and we used it to implement a hybrid cryptosystem [27] based on a public key function for ensuring authentication of the base station, and on a key agreement protocol for establishing a symmetric key, to be used for encryption/decryption of data packets sent by the motes. TinyPairing [29] is an open-source pairing-based cryptographic library for wireless sensors, providing an interesting solution to the key management problem, that still represents an open issue in WSN security research.

From a security point of view, the cryptosystem based on WM-ECC does not authenticate public keys, thus allowing man-in-the-middle attacks in the key exchange phase. Moreover, sensitive data is encrypted with a symmetric cipher, and this increases overall vulnerability of the network. Instead, TinyPairing adopts an asymmetric scheme and is much more secure in the initialization phase as it does not use a key exchange mechanism. As stated in Sect. 3.1, the intrinsic level of security of the two configurations can be represented by an attack coverage table, identifying, for each configuration, what attacks it is able to thwart or, dually, what requirements it is able to satisfy. Table 2 shows an example of attack coverage table for the two

Table 2 Attack Coverage Table for the considered cryptosystems

Configuration	Man-in-the-middle	Eavesdropping	Brute force	Replay attack
WM-ECC	Non-auth key	Yes	Weak symm	No
TinyPairing	Auth key	Yes	Asymm	No

cryptosystems (configurations) considered here. In this case, coverage is not defined as a binary property, but through a qualitative score capturing the level of protection provided by the configuration with respect to each considered attack.

In the simplest reconfiguration scenario, each node can decide independently when to update, and an identifier of the cryptosystem used to encrypt a message is encoded in the message itself, so that each receiving node, sharing the same reconfiguration strategy, is able to properly handle it.

Figure 2 illustrates a typical scenario for security protocol reconfiguration. In the INIT phase, the base station and the motes agree on the parameters of N different cryptosystems. The details of the initialization phase depend on the specific cryptosystems (the parameters can be public points for an ECC based cryptosystem, or system parameters for an identity based [27]). Initialization should be performed in a secure environment, either in the pre-deployment phase or later. After initializing the N available cryptosystems, each node can independently choose the valid cryptosystem to adopt for performing cryptographic operations in the current validity interval. In particular, the base station choses the cryptosystem it will use to digitally sign the outgoing queries and ensure authentication (CRYPTO(i) in figure). Any mote receiving the query message will use the cryptosystem whose ID is included in the message itself to verify the signature. Similarly, after verifying the signature, any mote encrypts data using the locally selected cryptosystem (CRYPTO(j) and CRYPTO(w) in figure), and the base station will use the ID included in the reply messages to decrypt them.

As illustrated in Fig. 2, the parameters of the N cryptosystems (i.e., the cryptographic keys) could be either preloaded on all network nodes in the INIT phase, or dynamically determined in each reconfiguration phase according to available key agreement mechanisms. All the cryptographic keys can be stored in each node for the entire lifetime of the network, and they can be used as master keys for generating new keys.

As discussed in Sect. 3, the introduction of reconfiguration mechanisms impacts a system's performance by introducing some overhead depending on the mechanism itself and the particular architectural level it is applied to. Each reconfiguration solution that encompasses switching among different implementations of the same functionality is characterized by an unavoidable increase in memory storage, at least to include the different available versions and the additional mechanisms to implement the reconfiguration itself. Referring to the application security reconfiguration mechanism proposed in this section, each node will be loaded with an application image including the binaries of all the available security libraries and their initialization

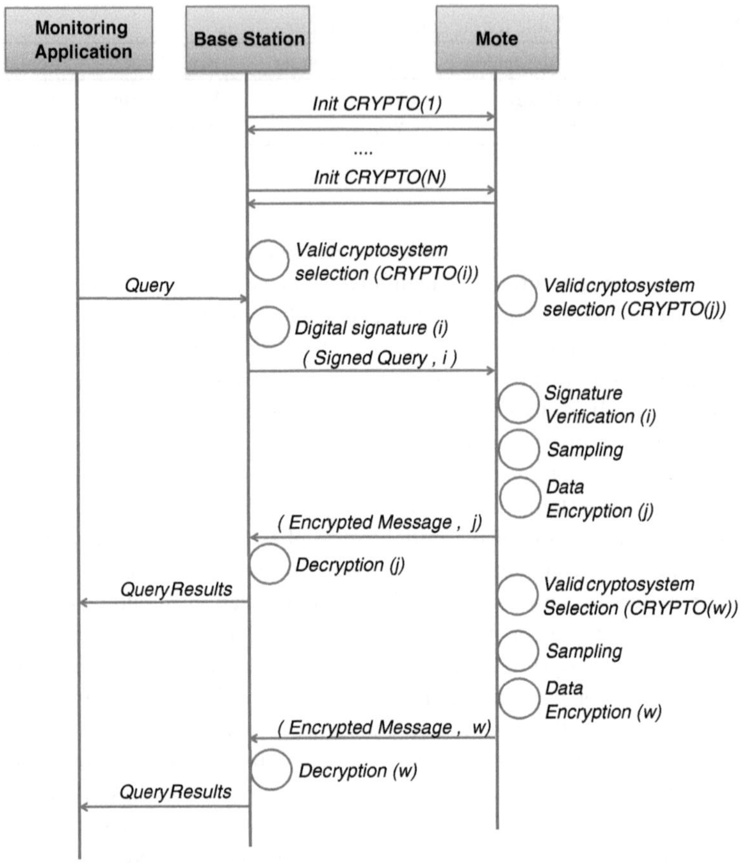

Fig. 2 Security protocol reconfiguration

parameters, along with the software switches needed to dynamically select different security primitives. Clearly, this could represent a problem for constrained devices such as sensor nodes, that are typically equipped with a small flash memory and RAM, therefore a more prudent design of security libraries should be devised in order to save as much memory as possible. As for energy consumption and required CPU computational effort, the proposed solution does not affect them, since cryptographic operations belonging to different libraries are activated by simple software switches. Moreover, there is no latency to swap from a cryptosystem to another and we do not need to stop the monitoring application during the reconfiguration.

Let us now take a look at possible security weaknesses of this strategy: an attacker who is aware of the message format may try to manipulate some fields of query and data packets, such as those coding the cryptosystem ID and its parameters, so that nodes are no longer able to communicate. As for query messages, their payload is signed with the base station's private key, so that, if any field is altered during transmission, the signature verification at mote's side will not succeed, and the message will be discarded. This aspect of the protocol could be exploited to execute a denial of service attack, with motes not able to verify the authenticity of queries and thus refusing to provide the required data. To detect this type of attack, a timeout is set by the base station every time a query is sent. If no reply is received before the timer expires, the query is sent again to cope with possible message losses. If no reply is received after a few attempts, an alert is raised. The cryptosystem ID is also encoded in each response message, as it is necessary to decrypt the message. An attacker could alter it as messages are not authenticated, but then the base station will not be able to decrypt them, and will discard them. This situation may cause the loss of some response messages. However, as a typical sensor network is composed of many redundant motes, this situation is not critical.

4.2 Physical Layer Reconfiguration

Several existing approaches for sensor network reprogramming perform a *full-image replacement*, consisting in completely replacing the image of the application running on a node. Deluge [30] is a reliable data dissemination protocol for propagating large data objects (larger than a node's memory) from one or more source nodes to many other nodes over a multi-hop network. As Dutta et al. pointed out in [31], this approach is unsafe and too battery-consuming. We implemented a different approach to remotely reconfigure each node in the network. We decoupled the reconfiguration mechanisms from the components to enforce the new configuration according to a scheduling policy.

To this aim, we designed a reconfiguration application by augmenting several components of the Deluge framework. In particular, we implemented new reconfiguration functionalities to enable a single node to swap to a new image that was previously preloaded on its storage. The reconfiguration application is defined by wiring new components specifically designed to manage external reconfiguration commands, and components designed to manage the images loaded on the node storage. The proposed reconfiguration application consists of three main components, namely (i) a bootloader component, (ii) a reprogramming component, and (iii) a management component (Fig. 3).

The *bootloader component* is a persistent layer in the architecture, which can enforce the chosen reconfiguration mechanisms. This component is intended for TinyOS and provides needed functionalities to program the node with an already stored program image. The parameters passed to this component are specified in the external command and indicate the location of the binary in the external flash memory to program the node's microcontroller. When reprogramming is requested, the bootloader will erase the program flash and write the new binary to it. On completion, it jumps to the first instruction of the new application.

The *reprogramming component* is the core of the *reconfiguration application*. In our implementation, it accepts commands from the base station, but can be extended

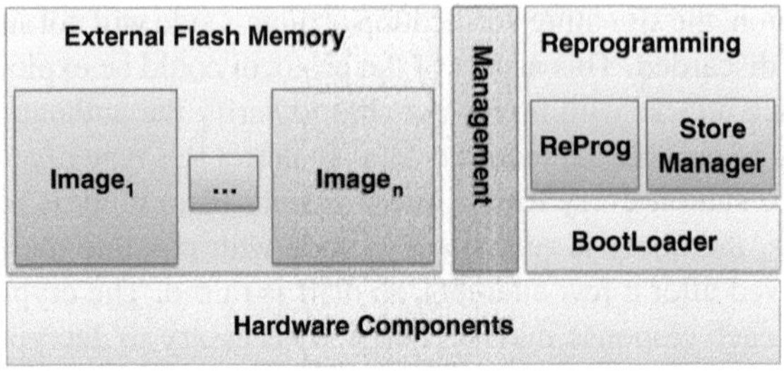

Fig. 3 Reconfiguration Application components

to implement a decentralized reconfiguration approach. This component is built by connecting two primary subcomponents: the *ReProg* and the *StorageManager*. The ReProg component is an extension of the NetProg component of Deluge T2. It handles a reprogramming request from the network by providing a dedicated API to initialize a reconfiguration process. When a node wants to perform a reconfiguration, it only has to invoke this API by specifying the name of the new binary in the flash memory to load. Subsequently, the ReProg sets the environment variables needed by the bootloader component and reboots the node. The StorageManager component deals with image name resolution, mapping names of program images to their respective physical addresses in the external flash memory.

The *management component* has a master (base station) and a mote side. It is used to initialize the mote and deploy different images. Usually this operation is done in a secure environment and it is accessible only during the initialization. The master-side *management component* has been derived from the *tos-deluge* application of the Deluge T2 Framework, and it is called mote-manager. This component allows to inject one or more images into the mote by writing directly into nodes' external flash memory volumes. It is also possible to erase a volume and ping the status of a mote to get information about already injected images.

Finally, the reconfiguration application runs on a workstation connected to the base station, which implements the reprogramming scheduler. As discussed in the next section, the reprogramming frequency and the new configuration to load can be chosen to balance overhead and attack probability.

Clearly, physical layer reconfiguration introduces a greater overhead than security layer reconfiguration, as it is based on full image replacement. As previously discussed, due to resources limitations, a node's memory cannot be preloaded with many different application images. Consider that the monitoring application we implemented occupies about 16 KB of ROM and 700 Bytes of RAM on a telosB platform, equipped with a 48 KB ROM memory and a 10 KB RAM.

The proposed solution introduces considerable advantages in terms of overall performance with respect to WSN reprogramming approaches based on code dissemination. In fact, the reconfiguration time is now not dependent on the image size and the network topology as the images are not sent over the network but preloaded via a serial interface. This approach avoids any security risk in the dissemination phase, and reduces the battery consumption as it only needs the introduction of a simple command message to swap from an image to another one. As the available application images are all loaded on the node's external memory, the swapping latency is considerably reduced with respect to the code dissemination case. We experimented a reduction of one order of magnitude with respect to the original Deluge approach: from 50 s to send a 40 Kb image implementing a monitoring application secured with WM-ECC, to about 6 s to perform the swap.

One known drawback of this approach is the need to stop the monitoring application and any ongoing query in order to swap to another image. Nevertheless, as previously discussed, any available approach that is based on full image replacement presents even worst issues. A possible solution to mitigate this problem could be that of delaying the reconfiguration operation if the node is involved in some communication, or pausing an ongoing communication until reconfiguration is completed. It is up to network administrators to decide frequencies and strategies for node reconfiguration. As for possible security weaknesses of this solution, some attacks can be considered undermining this reconfiguration mechanism. First of all, preloading nodes with all images exposes them to physical compromise. We can assume that, in presence of strict security requirements, nodes are equipped with tamper-resistant packages so that they cannot be compromised. Moreover, an attacker may try to replay control packets sent by the base station and containing a reconfiguration command, in order to control communication or simply perform a denial of service attack by forcing motes to continuously swap images. This risk can be prevented by introducing a sequence number for reconfiguration commands and proper channel encryption.

5 MTD Evaluation

In order to evaluate the effectiveness of security-driven reconfiguration—even under adverse conditions—we assume that an attacker is able to understand when the adopted cryptosystem changes and what type of cryptosystem is used at each time (e.g., by observing control messages sent over the network by the base station in the node reconfiguration strategy, or control flags found in data packets in the protocol reconfiguration strategy).

Many types of cryptographic attacks can be considered. In our case, an attacker can only observe encrypted packets traveling on the network and containing information about sensed data, and can perform a brute force attack on captured packets by systematically testing every possible key for the current (known) cryptosystem—assuming he is able to determine when the attack is successful. In the *worst case* (for the defender), the attacker knows the encryption algorithm and the key length associated with the algorithm, therefore he can systematically try all possible keys of that length. In the *intermediate case*, the attacker knows the encryption algorithm but does not know the key length associated with it, thus he systematically tries all possible keys for a given set of key lengths. In the *best case*, the attacker does not

Table 3 Characteristics of cryptosystems

Cryptosystem	Key lenght (bits)	Time (ms)	Max attack time (ms)
WM_ECC_sk	80	0.001251	1.5123E +21
WM_ECC_rc5	160	0.001221	1.7845E +45
TinyPairing	208	13.019531	5.3560E +63

know anything about the adopted cryptosystem, thus he tries all possible keys for a given set of key lengths and a given set of cryptosystems.

We evaluated our approach with respect to the cryptosystems described in [27], whose characteristics are summarized in Table 3.

The WM_ECC_sk and WM_ECC_rc5 cryptosystems are both based on the WM-ECC library, used to execute key exchange and digital signature operations. They both perform symmetric encryption using respectively a Skipjack cipher with a 80 bit key and an RC5 cipher with a key of 160 bits. The TinyPairing cryptosystems is based on TinyPairing and uses a 208 bit key. In Table 3, the time needed to test a single key is reported for each cryptosystem, along with the maximum attack time, that is the time necessary to test all the possible keys. The reported execution times (third column) refer to the execution of the decryption operation on TelosB devices, equipped with a 4.15 MHz MSP430 microcontroller, a CC2420 radio chip, a 10 KB internal RAM, and a 48 KB program flash memory.

The maximum attack times reported in the fourth column of Table 3 have been computed analytically based on the measured time needed to perform a single decryption operation. It is important to point out that these attack times are significantly high due to the nature of the attacks we considered. In practice, attacks may be more sophisticated and efficient than brute force attacks. However, this does not affect the validity of the proposed MTD approach as we are interested in illustrating how the probability of successfully completing an attack decreases, compared to a static configuration scenario.

We carried out our experiments considering both worst and intermediate cases, and analyzed the cumulative distribution function (cdf) of the *attack time*. In both cases, we simulated an attacker sequentially exploring the key space. We considered an observation interval as long as the attack time of the most complex cryptosystem, TinyPairing, and validity intervals of decreasing length. A validity interval is the time interval in which a single system configuration is active. At the end of the i-th validity interval, the new configuration to activate should be chosen based on a specific strategy (e.g., related to security or battery consumption requirements) in order to maximize reconfiguration benefits. However, for the sake of simplicity, the results shown in this section have been obtained by performing random choices among the available cryptosystems at each validity interval. In particular, during an observation interval, we randomly generated 1,000 different sequences of valid cryptosystems and recorded the time of successful attacks to build the cdf. Clearly, the sequence length depends on the chosen validity interval.

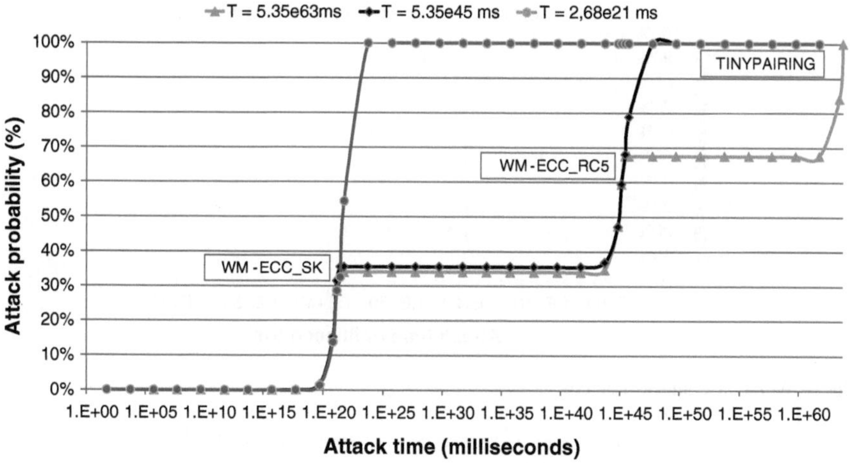

Fig. 4 Worst case attack time cdf for large validity intervals

In the first experiment, we chose three validity interval lengths in such a way to be comparable to the maximum attack times of the three different cryptosystems. The resulting attack time's cdf in the worst case is shown in Fig. 4: the labels in the figure—note that the x-axis is on a logarithmic scale—identify three inflection points in the middle of the maximum attack times of each cryptosystem. These correspond to the maximum values of the attack time probability distribution functions (pdf) for each cryptosystem.

When analyzing the chart, a seemingly counterintuitive behavior can be identified: when considering smaller validity intervals the attacker seems to benefit.

The chart can be explained as follows. The WM_ECC_sk cryptosystem can be certainly broken in 1.5123E+21 ms (worst case for WM_ECC_sk) as shown in Table 3. This means that, if randomly selecting one cryptosystem among the 3 available, and choosing a validity interval greater than this threshold, the cryptosystem will always be broken. As each cryptosystem has a 33 % probability of being selected at next reconfiguration time—based on our assumptions—in 33 % of cases the system will be broken. Similar considerations can be made for the other two cryptosystems, explaining the other inflection points.

As illustrated in Fig. 5, when reducing the length of the validity interval—with validity intervals larger than the maximum attack time of the weakest cryptosystem—the attack time increases, with the percentage of successful attacks reducing dramatically. The same behavior is highlighted in Fig. 6, which shows how the probability of completing a successful attack within a time t varies as the length of the validity interval changes: as soon as the validity interval drops below the maximum attack time of the weakest cryptosystem, the rate at which probability decreases becomes higher.

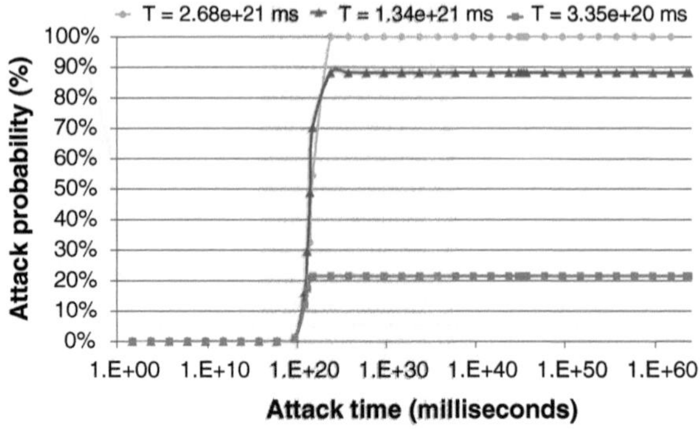

Fig. 5 Worst case attack time cdf

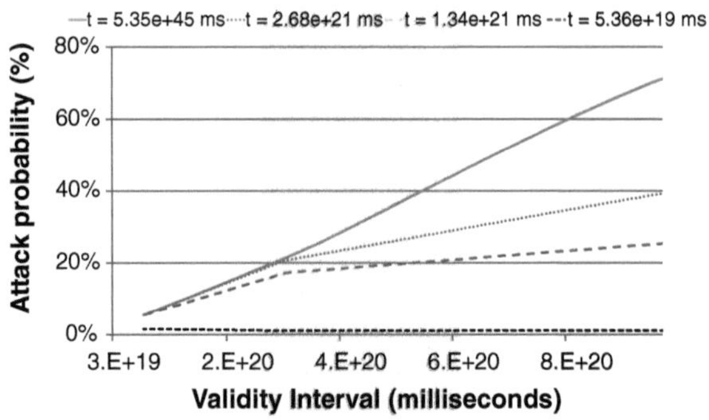

Fig. 6 Probability of successful attack

Similar results can be obtained when reconfiguration is performed by selecting an equivalent cryptosystem that uses different parameters (i.e different keys). Figure 7a shows the attack time's cdf in the worst case when reconfiguration is performed by switching among three cryptosystems that implement the WM-ECC library with the Skipjack cipher, but have different keys. When reducing the validity interval, the probability of successfully completing an attack significantly decrease as the intrinsic security level is restored every time a new key is adopted. For comparison purposes, Fig. 7b shows the attack time's cdf when three different cryptosystems are used. As expected, increased diversity results in a lower probability of attack.

Figure 8 compares the attack time's cdf for the intermediate and the worst cases, under the assumption that the attacker performs a brute force attack using the set of key lengths in Table 4. The validity interval of 5,36E+45 milliseconds is long enough to break both WM_ECC_sk and WM_ECC_rc5. As shown, the attacker's success probability is smaller in the intermediate case. Clearly, when the attacker's uncertainty about the used cryptosystem is higher, more key lengths will be tested, making the proposed approach even more effective.

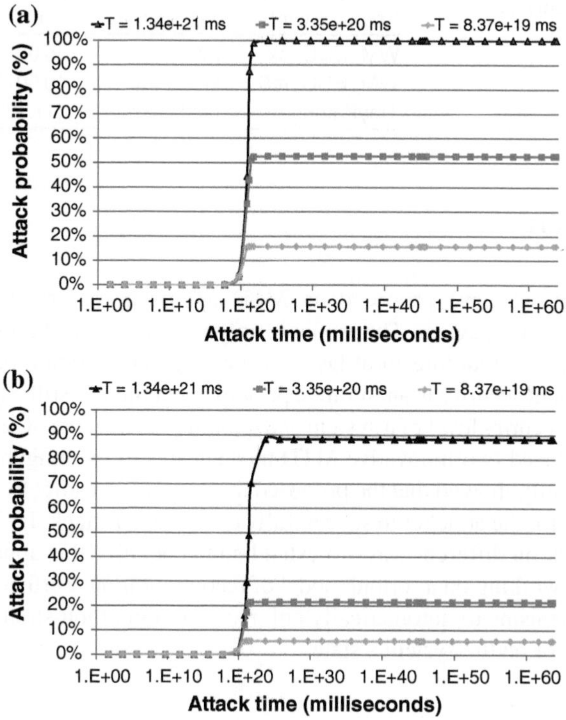

Fig. 7 Worst case attack time cdf. **a** Same cryptosystems with different keys. **b** Three different cryptosystems

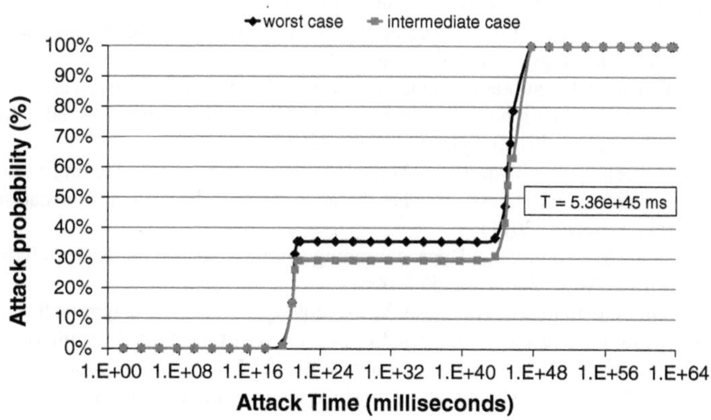

Fig. 8 Worst case vs. intermediate case

Table 4 Key lengths set

Cryptosystem	Key len (bits)	Time (ms)
WM_ECC_sk	[80]	[0.001251]
WM_ECC_rc5	[120,160]	[0.001120,0.001221]
TinyPairing	[180,208]	[11.023211,13.019531]

6 Conclusions

In this chapter, we have proposed an MTD approach for protecting resource-constrained distribute devices. The proposed approach is based on fine-grained *reconfiguration* at different architectural layers. Changing configuration or system parameters to augment security is an intuitive principle, but there is still a lack of metrics to evaluate the security level of a system and quantify the benefits of reconfiguration. We have introduced two innovative MTD mechanisms to reconfigure the network, and experimentally showed that the proposed mechanisms are effective in increasing the complexity for the attacker to successfully complete an attack. In the near future, we plan to work on different ways to extend and generalize this approach. Indeed, we are already working on a formal model of reconfiguration. Furthermore, we plan to define mechanisms to automatically enforce reconfiguration strategies based on external events or on the system's state.

References

1. Executive Office of the President, National Science and Technology Council: Trustworthy cyberspace: Strategic plan for the federal cybersecurity research and development program. http://www.whitehouse.gov/. Accessed Dec 2011
2. Jajodia, S., Ghosh, A.K., Subrahmanian, V.S., Swarup, V., Wang, C., Wang, X.S. (eds.): Moving target defense II: Application of game theory and adversarial modeling. 1st edn. Advances in Information Security, vol. 100, Springer, Berlin (2013)
3. Jajodia, S., Ghosh, A.K., Swarup, V., Wang, C., Wang, X.S. (eds.): Moving target defense: Creating asymmetric uncertainty for cyber threats. 1st edn. Advances in Information Security, vol. 54, Springer, Berlin (2011)
4. Wang, Q., Zhu, Y., Cheng, L.: Reprogramming wireless sensor networks: challenges and approaches. IEEE Netw. **20**(3), 48–55 (2006)
5. Casola, V., De Benedictis, A., Albanese, M.: A moving target defense approach for protecting resource-constrained distributed devices. In: Proceedings of the 14th IEEE International Conference on Information Reuse and Integration (IEEE IRI 2013), San Francisco, CA, Aug 2013
6. Manadhata, P.K., Wing, J.M.: An attack surface metric. IEEE Trans. Software Eng. **37**(3), 371–386 (2011)
7. Evans, D., Nguyen-Tuong, A., Knight, J.C.: Effectiveness of moving target defenses. In: Moving Target Defense: Creating Asymmetric Uncertainty for Cyber Threats, pp. 29–48. Springer, New York (2011)
8. Kc, G.S., Keromytis, A.D., Prevelakis, V.: Countering code-injection attacks with instruction-set randomization. In: Proceedings of the 10th ACM conference on Computer and communications security. CCS '03, ACM, New York. pp. 272–280 (2003)

9. Cadar, C., Akritidis, P., Costa, M., Martin, J.P., Castro, M.: Data randomization. Technical report, Microsoft Research (2008)
10. Jackson, T., Salamat, B., Homescu, A., Manivannan, K., Wagner, G., Gal, A., Brunthaler, S., Wimmer, C., Franz, M.: Compiler-generated software diversity. In: Moving Target Defense: Creating Asymmetric Uncertainty for Cyber Threats, pp. 77–98. Springer, New York (2011)
11. Antonatos, S., Akritidis, P., Markatos, E.P., Anagnostakis, K.G.: Defending against hitlist worms using network address space randomization. Comput. Netw. **51**(12), 3471–3490 (2007)
12. Atighetchi, M., Pal, P., Webber, F., Jones, C.: Adaptive use of network-centric mechanisms in cyber-defense. In: Proceedings of the Sixth IEEE International Symposium on Object-Oriented Real-Time Distributed Computing (ISORC 2003), pp. 183–192 May 2003
13. Kewley, D., Fink, R., Lowry, J., Dean, M.: Dynamic approaches to thwart adversary intelligence gathering. In: Proceedings of the DARPA Information Survivability Conference & Exposition (DISCEX 2011). Vol. 1, pp. 176–185. Anaheim, CA, June 2011
14. Antonatos, S., Akritidis, P., Markatos, E., Anagnostakis, K.: Defending against hitlist worms using network address space randomization. Comput. Netw. **51**(12), 3471–3490 (2007)
15. Jafarian, J.H., Al-Shaer, E., Duan, Q.: Openflow random host mutation: transparent moving target defense using software defined networking. In: Proceedings of the First Workshop on Hot Topics in Software Defined Networks. HotSDN '12, pp. 127–132. ACM, New York (2012)
16. Albanese, M., De Benedictis, A., Jajodia, S., Sun, K.: A moving target defense mechanism for MANETs based on identity virtualization. In: Proceedings of the First IEEE Conference on Communications and Network Security (IEEE CNS 2013), Washington, DC, Oct 2013
17. Casola, V., Mazzeo, A., Mazzocca, N., Vittorini, V.: A policy-based methodology for security evaluation: a security metric for public key infrastructures. J. Comput. Secur. **15**(2), 197–229 (2007)
18. Common criteria project: Common criteria for information technology security evaluation 2.1. Technical report, US NIST (1999)
19. Trusted computer system evaluation criteria. Technical Report DoD 5200.28-STD, US Department Of, Defense (1985)
20. Li, X., Parker, T.P., Xu, S.: A stochastic model for quantitative security analyses of networked systems. IEEE Trans. Dependable Sec. Comput. **8**(1), 28–43 (2011)
21. Barth, A., Rubinstein, B.I.P., Sundararajan, M., Mitchell, J.C., Song, D., Bartlett, P.L.: A learning-based approach to reactive security. IEEE Trans. Dependable Sec. Comput. **9**(4), 482–493 (2012)
22. Ahmed, M.S., Al-Shaer, E., Khan, L.: A novel quantitative approach for measuring network security. In: INFOCOM. pp. 1957–1965 (2008)
23. Pamula, J., Jajodia, S., Ammann, P., Swarup, V.: A weakest-adversary security metric for network configuration security analysis. In: QoP. pp. 31–38 A novel quantitative approach for measuring network security. In: INFOCOM. 1957–1965 (2008)
24. Mell, P., Scarfone, K., Romanosky, S.: NIST IR 7435: The common vulnerability scoring system (CVSS) and its applicability to federal agency systems, Aug (2007)
25. Casola, V., Preziosi, R., Rak, M., Troiano, L.: A reference model for security level evaluation: policy and fuzzy techniques. J. Univers. Comput. Sci. **11**(1), 150–174 (2005)
26. Foley, S.N., Fitzgerald, W., Bistarelli, S., OSullivan, B., Foghl, M.: Principles of secure network configuration: towards a formal basis for self-configuration. Lecture Notes in Computer Science, vol. 4268. Springer, Berlin Heidelberg (2006)
27. Casola, V., De Benedictis, A., Drago, A., Mazzocca, N.: Analysis and comparison of security protocols in wireless sensor networks. In: Proceedings of the 30th IEEE Symposium on Reliable Distributed Systems Workshops (SRDSW 2011), pp. 52–56. Madrid, Spain (Oct 2011)
28. Wang, H., Sheng, B., Tan, C., Li, Q.: WM-ECC: An elliptic curve cryptography suite on sensor motes. Technical Report WMCS-2007-11, College of William and Mary (Oct 2007)
29. Xiong, X., Wong, D.S., Deng, X.: TinyPairing: a fast and lightweight pairing-based cryptographic library for wireless sensor networks. In: Proceedings of the IEEE Wireless Communications and Networking Conference (WCNC 2010), Apr 2010

30. Hui, J.W., Culler, D.: The dynamic behavior of a data dissemination protocol for network programming at scale. In: Proceedings of the 2nd International Conference on Embedded Networked Sensor Systems (SenSys 2004), pp. 81–94. Baltimore, MD (2004)
31. Dutta, P.K., Hui, J.W., Chu, D.C., Culler, D.E.: Securing the deluge network programming system. In: Proceedings of the Fifth International Conference on Information Processing in Sensor Networks (IPSN 2006), pp. 326–333. Apr 2006

Protocol Integration for Trust-Based Communication

Fatma Laidoui and Thouraya Bouabana-Tebibel

Abstract In mobile ad hoc networks (MANETs), nodes have to cooperate in order to accomplish routing tasks. Nevertheless, they have limited resources, and may behave in a selfish way. On the other hand, the networking infrastructure supporting routing is quite weak faced with such misbehaviors. In this chapter, we propose a trust model for reactive routing in MANETs. The proposed solution applies to any source routing protocol. It is based on mechanisms inspired by the CONFIDANT protocol to install and update trust in the network. The model also integrates new protocols to improve trust in the selected routes. It may adapt to topology changes caused by the mobility of nodes and takes into account new routes learned after the route request phase. Finally, it improves the choice of the safest route towards the destination. Fundamental and elaborate tests prove the efficiency of the solution.

Keywords Mobile ad hoc network · DSR · CONFIDANT · Trust · Security · Reactive routing

1 Introduction

Mobile ad hoc networks (MANETs) have known a fulgurating success, more particularly with the appearance of wireless technology. In that environment, mobile hosts must behave like routers to forward, hop by hop, data packets through the network. The existing routing protocols in literature suppose that the environment is ideal to

F. Laidoui
Laboratoire de Communication des Systèmes Informatiques, Ecole Militaire Polytechnique—EMP, Algiers, Algeria
e-mail: fat.laidoui@gmail.com

T. Bouabana-Tebibel (✉)
Laboratoire de Communication des Systèmes Informatiques, École nationale Supérieure d'Informatique—ESI, Algiers, Algeria
e-mail: t_tebibel@esi.dz

T. Bouabana-Tebibel and S. H. Rubin (eds.), *Integration of Reusable Systems*, Advances in Intelligent Systems and Computing 263, DOI: 10.1007/978-3-319-04717-1_15,

information exchange and free from malicious attacks. This induced multiple vulnerabilities on the routing process that threat the reliability of data transmission. Among the most common attacks, we cite: (1) Packets dropping where the malicious node absorbs the exchanged packets; this attack may paralyze the whole network due to the loss of connectivity between nodes, if the number of deleted packets is high. (2) No-cooperation or selfishness where nodes change their behavior from normal to selfish by refusing to participate in the routing process with the aim of saving their own resources. Such a behavior is motivated by the fact that cooperation generates lots of messages between nodes, which induces high costs of energy.

Many solutions are proposed in the literature to protect networks against no cooperation of nodes. Those solutions are essentially based on the overhearing of neighbors' transmissions [1]. They, nevertheless, suffer from wrong accusations caused by collusions and interferences, and need complementary mechanisms such as acknowledgments and reputation mechanisms.

Secure routing schemes based on reputation [2–4] focus on the cooperation between nodes, through packet forwarding, which is considered as the principal function to assure routing in MANETs. In systems based on reputation, the key idea rests on direct neighborhood observation in order to decide whether a node is malicious or not. Accordingly, the reputation of a node increases when it accomplishes correctly its tasks and decreases otherwise. In such systems, if a node declares another node as suspicious because of its bad reputation, this information is propagated through the network and the misbehaving node is quarantined from the rest of the network. However, direct observations of the neighborhood do not allow objective measurements of nodes behavior. It is, often, necessary that each node takes into account the opinion of the other nodes about the trustworthiness of the neighborhood, and this constitutes the major approach of systems based on trust [5–8]. These latter systems include two components: a routing process and a trust model. The principal goal of the trust model is to make routing decisions. A node classifies another node as honest or misbehaving one according to the information gathered about direct observations and that are sent by the others. This is done in order to cope with uncertainty in trustworthiness.

In this work, we are interested in securing source routing protocols using trust mechanisms. We select as source routing protocol the Dynamic Source Routing protocol (DSR) [9, 10]. Protocols based on trust are mainly conceived to secure against attacks for which cryptographic functions reveal to be ineffective, such as no-cooperation of nodes and packets dropping. We focus on no cooperation causing the dropping of control packets RREQ (Route REQuest) and RREP (Route REPly). We assure the propagation of trust in the network by using mechanisms drawn from the CONFIDANT protocol [6]. However, CONFIDANT is based on the classical source routing route discovery phase, and may miss new routes which are learned late in the process of route discovery. These routes may be more secure, and should, consequently, be evaluated. We propose a method that takes into consideration that issue. On the other hand, to make routing decisions about trust in routes, we use the MNRS model [11] which we have improved with new parameters in order to enhance the selection of trustworthy routes.

This chapter is organized as follows. In Sect. 2, we discuss works related to ours. Section 3 describes the background of our solution. We present, in Sect. 4, the proposed secure protocol. In Sect. 5, we discuss the performance of the proposed protocol compared with conventional ones. Finally, we conclude with some observations based on the obtained results and we propose some recommendations for future research directions.

2 Related Work

To combat no cooperation behaviors, a common idea is to motivate the nodes to participate in the routing process [12]. This idea is based on either virtual currency (nuglets) or credit. Its drawback is that it sometimes privileges a rich malicious node chosen in route discovery than poor ones, because the metric of optimal route is the length. Besides, solutions based on credit require a central authority to distribute credits. In [13], the authors propose to send a message via different available routes in order to attenuate dropping of data packets. However, this idea is difficult to implement. In addition, it floods the network with several copies of a message. The study proposed in [14] allows monitoring, detecting, and isolating droppers without using promiscuous listening. It also can distinguish between selfish and malicious nodes. However, the authors did not present results to show the effectiveness of their solution, nor they compare it with existing trust based routing protocols. Authors in [15] proposed a trust evaluation method that offers a security mechanism for data protection and secure routing. This mechanism uses global information to provide trust in the network, which delays the response time. More recently, a simple model is proposed in [16] to secure routing based on reputation. The mechanism reveals to be efficient for isolating the network from selfish nodes. The principal drawback of the solution is the irreversible character of punishment. When a node is classified as selfish, it is punished for not participating in the routing process until the entire network is reformed. This may be unfair as the cause of selfishness may be the result of transmission errors or an inappropriate node location which makes it difficult to participate in packet forwarding.

Other proposed solutions, in the literature, are based on end-to-end feedbacks, where the destination acknowledges packet reception to the source. The drawback of this technique is that it only detects routes containing the misbehaving nodes, but cannot detect the malicious nodes. That's why, many sophisticated solutions have been developed [17] but they considerably overhead the network by acknowledgements. As a compromise, the authors in [18] combined end-to-end feedbacks with the Probing technique to decrease the overhead due to acknowledgements. Later, they claim in [19] that the biggest threats appear to be join-leave attacks, used to isolate honest peers in the system, and against which no provably robust mechanisms are known so far. In their proposition, they showed that, in a high level, a scalable DHT (Distributed Hash Table) can be designed as provably robust against adaptive adversarial join-leave attacks.

In regard to DSR security, authors in [11] proposed an approach to avoid misbehaving nodes using reputation mechanisms. In the traditional DSR protocol, when a node receives a RREQ packet, it drops it if it has been previously processed. A misbehaving node takes advantage of this action and forwards the RREQ fast so that RREQ from other nodes are dropped and the discovered path includes itself. To overcome this issue, a new approach for RREQ packet broadcasting is proposed in [20]. In Trusted Dynamic Source Routing (TDSR) [20], the trust between nodes is represented by a trust score composed of direct and indirect trust. Trust relationships and routing decisions are based on node experienced, observed, or reported routing as well as forwarding behavior of other nodes. However, TDSR might not determine the direct trust of neighboring nodes in presence of collisions.

3 Background

The security extension that we propose for DSR is based on the CONFIDANT protocol to ensure trust. We present in this section, the basic concepts on which both DSR and CONFIDANT are based.

3.1 Dynamic Source Routing Protocol (DSR)

The Dynamic Source Routing protocol (DSR) [9, 10] is a reactive routing protocol which uses the source routing concept, i.e. the whole route to cross is inscribed in the data packet header. DSR includes two main phases: route discovery and route maintenance. When a node S wants to send data packets to another node D and does not know any route to that destination, it launches a route discovery process, by broadcasting a RREQ packet to its neighbors. When a neighbor node receives RREQ, it checks if it knows a route towards D. If no route is known, it inscribes itself in the RREQ packet and broadcasts the packet to its neighbors. When node D or another node knows a route towards D, it unicasts a RREP packet back to S, based on the nodes which have been crossed during the route request phase and inscribed in the RREQ. If a node detects a broken link, it sends a RERR (Route ERRor) packet towards S with a notification about the node responsible for the failure. When S receives RERR, it deletes from its cache all routes that include the broken link. If the source has another alternative route towards the same destination, it uses it and continues packets forwarding. Otherwise, it launches a new request to find another route towards D.

3.2 CONFIDANT Protocol

CONFIDANT is composed of four modules [6]: the monitor, reputation manager, trust manager and path manager. The function of the monitor is to collect information about direct neighbors. It can detect a bad behavior of a node and distinguish it from the good one. It does so using a watchdog mechanism. The reputation manager module manages a reputation table saved at each node. This table associates every neighbor node identity with a reputation value. The reputation value reflects direct observations noticed by a node about another node. It does not enable a node to decide about the trust of another node. Indeed, a node may have a bad reputation due to link errors caused by nodes mobility and not because of its selfishness. That's why CONFIDANT needs a trust manager. The latter module estimates the trust value of a node and saves it in a trust table. The trust value reflects indirect observations about a node. It is estimated by accumulating observations sent by the neighbors. As for the path manager module, it cleans paths in use and nodes cache from any selfish node. In regard to the update of reputation and trust values, the authors of CONFIDANT use a Bayesian model with the Beta distribution Beta(α, β). This model is very used in the literature for representing reputation and trust values [3, 7, 21, 22]. The advantage of using the Beta function is that it only needs two parameters which are continuously updated. These two parameters reflect the current belief. In our work, we use Beta distribution in which α and β are initialized to one ($\alpha = \beta = 1$), which means that initially, we have no information about the nodes behavior [15].

4 The RTDSR Protocol

Our proposed protocol, named RTDSR (Reinforcement of Trust in DSR), is a security extension for DSR, that integrates CONFIDANT [6] tasks, Fading [7] and Second Chance [23] mechanisms, as well as the MNRS model [11].

CONFIDANT provides trust between nodes by analyzing packets forwarding and neighborhood discovery. The Fading mechanism gives less weight to observations received in the past. The Second Chance mechanism rehabilitates nodes previously considered as misbehaving ones. As for the MNRS model, it allows making decision about the most trustworthy path.

Our first contribution is to integrate, in a complementary and consistency way, these four techniques to secure DSR. CONFIDANT modules are integrated into the route request and route reply steps of the DSR discovery phase in order to ensure trust. The CONFIDANT reputation manager checks if the new calculated reputation value respects the tolerate reputation threshold, every time it evaluates it. If the checking fails, we use the Second Chance mechanism to check whether the node has consumed all its chances. If so, it is blacklisted. Otherwise, the CONFIDANT trust manager is invoked to estimate the trust in the node and make a decision about its honesty. The node is blacklisted if its trust value is less than a trust threshold. For each decision made on either reputation or trust in a given node, the Fading principle is applied to favor the last evaluation by providing it with a higher weight.

In addition to these components integration, we inject two further modifications on the composite model. The first one consists in improving trust in the route by taking into account new paths that are not included in the *route request phase*. CONFIDANT protocol aims to prevent malicious nodes from participating in the routing process.

We reinforce this approach by tracking down the misbehaving nodes applying trust and reputation concepts on the whole process of route discovery which includes both of route request and route reply phases. This reinforcement constitutes our second contribution.

We notice that in CONFIDANT mechanism, routes learned after the *route request phase* are ignored because of the unicast forward of RREPs, which is used to answer the source request. However, these new routes may be more secure than the current one. We suggest to take them into consideration. Thus, at the *route reply phase,* when an intermediate node receives RREP, it first checks if it has learned a more secure route towards the destination. If this is the case, it substitutes the current sub-route (from this node towards the destination) by the most secure one.

The second modification that we bring to the composite solution enhances trust in a route by considering new parameters. The revision is made in the MNRS model. In MNRS, trust in the route is calculated using the average value of trust in all nodes of the route in both directions (source-destination, destination-source). We propose to add to this estimation the lowest trust value of the route nodes. We believe that this new parameter allows to better evaluate the trust in a route. Indeed, a node with a very little trust value is probably a misbehaving node. Thus, it should disqualify the route even if the latter average trust value is higher than the other routes. This revision constitutes our third contribution.

Thus, RTDSR intervenes, on one hand, in the route request and route reply phases of the DSR protocol using CONFIDANT mechanisms which it improves. It resorts, on the other hand, to the MNRS model in order to select routes using a new phase that we call a route selection phase. These three phases are presented below.

4.1 Route Request Phase

When a source wants to send a data packet to a destination and does not know any route to that destination, it launches a route request phase. We propose to include in the broadcasted RREQ packet two new fields, namely p_{trust} and min_{trust}. p_{trust} is the accumulated trust value of nodes from the source to the current node. It is initialized to zero by the source. This metric is inspired by the MNRS model. As for min_{trust}, it is the lowest trust value of the nodes of the path and is initialized to 0.5 at the source [7]. It means that we have no idea about the path trust.

When an intermediate node *B* receives RREQ from the node *A*, its CONFIDANT *path manager* module checks the trust in the forwarder. If the latter is a dishonest node, the packet is ignored. Otherwise, if the receiver does not know any route towards the destination, it inscribes its identifier in the RREQ and updates the p_{trust} and min_{trust} values as follows:

$$\mathrm{p_{trust}(new)} = \mathrm{p_{trust}} + \mathrm{T_{BA}} \tag{1}$$

$$\mathrm{min_{trust}(new)} = \mathrm{Min(min_{trust}, T_{BA})} \tag{2}$$

T_{BA} is assigned by node B to node A and means how can B trusts A. It is provided by the CONFIDANT *trust manager.* RREQ is, afterwards, broadcasted to all one hop neighbors.

When the destination or any intermediate node that knows a route towards the desired destination receives RREQ, it checks the trust in the RREQ forwarder, and answers the source by launching the *route reply phase*. Thus, it adds p_{trust}, n_{trust} and min_{trust} to RREP. The values of p_{trust} and min_{trust} are those inscribed in RREQ. n_{trust} is a value that has the same role as p_{trust}. It reflects the trust degree of the route from the destination to the current node. RREP is then sent to the source in unicast using the reverse route inscribed in RREQ. We note that this route remains the same and does not change during all the *route reply phase* of DSR when associated with CONFIDANT.

4.2 Route Reply Phase

When the destination or any intermediate node that knows a route towards the destination receives RREQ, it answers the source with RREP. We distinguish between two different cases within this phase.

Case 1: no member of the reverse route knows a new route from itself towards the destination.

In this case, RREP packet is forwarded in unicast until it reaches the source. Each node A, member of the crossed route, checks the trust in the previous node B. If it finds that the node is not honest, the packet is ignored. The honesty of the node is evaluated using the CONFIDANT protocol, namely the monitor, reputation and trust modules. Otherwise, it updates n_{trust} and min_{trust} as follows:

$$\mathrm{n_{trust}(new)} = \mathrm{n_{trust}} + \mathrm{T_{AB}} \tag{3}$$

$$\mathrm{min_{trust}(new)} = \mathrm{Min(min_{trust}, T_{AB})} \tag{4}$$

Case 2: a node member of the reverse route learns a new route, which is safer, from itself towards the destination.

The main idea behind this case is to provide more chances to select the most trustworthy route if possible new routes towards the destination, which are safer, are learned after the route request phase has been executed. Such a case may occur when multiple sources within a network launch, at the same time, a route request phase towards the same destination. In that situation, a given node may learn several routes towards the destination, but these routes are not all sent to the different sources. Every source only receives replies to its broadcasted RREQ packets. Figure 1 illustrates this case. It shows two sources, namely source1 and *source2*, discovering a route towards a destination node *Dest*. The route reply phase of source1 provides the sub-route (*Dest*, *B*, *A*), whereas that of *source2* provides (*Dest*, *F*, *A*). This way, node A which

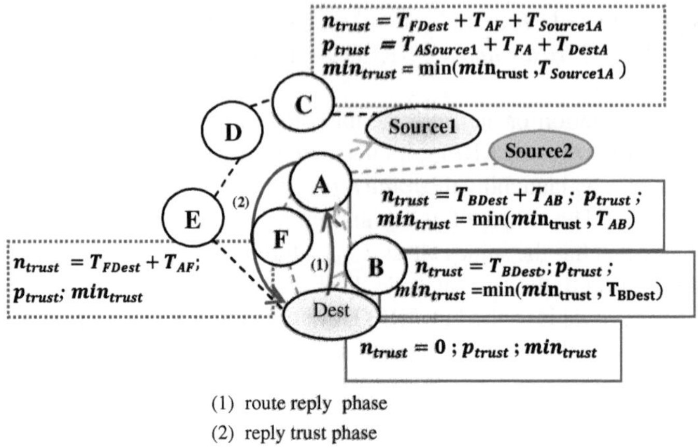

Fig. 1 New learned routes

receives *source2* RREP before *source1* RREP learns a new route before this latter receipt. To check which of the two routes is safer, it performs a new phase called *the route trust phase*.

4.3 Route Trust Phase

When an intermediate node receives a RREP packet, it checks if it has learned new sub-routes towards the destination, and if so, it controls whether they are more secure. The checking is done only if the packet has not crossed yet more than the half of the route. To check whether the learned route is safer, the receiver compares the current n_{trust} with that corresponding to the learned sub-route. If n_{trust} of the learned sub-route is lower, then it treats RREP packet as described in case 1; else it substitutes the current sub-route with the new one and recalculates some trust parameters.

The condition according to which new learned routes are taken into account only if the packet hasn't crossed more than the half route, is done in the aim not to force nodes to recalculate the current route when the packet is near to its source node. Such revaluation may delay unnecessarily the packets delivery process.

On the other hand, since the route towards the destination changes, it is necessary that the p_{trust} value be updated. For this purpose, the current node unicasts a special request *REQtrust* (REQuest for p_{trust}) to the destination. The REQtrust packet contains the field p_{trust} which is initialized to 0 ($p_{trust} = 0$). When each node member of the learned sub-route receives this request, it updates p_{trust} by adding its trust value according to (1), updates min_{trust} according to (2) and forwards the request to the next hop. When REQtrust reaches the destination, the latter inscribes p_{trust} and min_{trust} in a *REPtrust* (*Reply for REQtrust*) packet and unicasts that packet to the source of REQtrust. When the source of REQtrust receives REPtrust, it selects the safest sub-route towards the destination, updates p_{trust}, min_{trust} and continues the forwarding of RREP to the next hop. That selection is assured by the *route selection phase.* Figure 1 illustrates this selection.

4.4 Route Selection Phase

If several routes are discovered by either a source node or an intermediate node which learns new sub-routes, these nodes must decide about the safest route. In the same manner as the MNRS model, we use p_{trust}, n_{trust} and the route node number to make a decision. However, to enhance the security of the selected route, we consider another parameter which is the lowest trust value of the route nodes. This constraint is justified by the fact that the presence of a node with a very low trust value in a route increases the probability of insecurity of this route. This constraint is expressed by considering min_{trust} value in the decision process that will be unfolded into two steps.

First, we estimate the global trust value for each discovered route as follow:

$$\mathrm{path\_truste_i} = \mathrm{c1} * \left(\frac{\mathrm{p_{trust}} + \mathrm{n_{trust}}}{2}\right) * \mathrm{w_i} + \mathrm{c2} * \mathrm{min_{trust}} \tag{5}$$

$$w_i = \frac{\frac{1}{n_i}}{\sum \frac{1}{n_{i(i=1\,\grave{a}\,n)}}} \tag{6}$$

$path\_truste_i$ is the global trust value of the ith path, n_i is the nodes number of the i'th path, n is the total number of discovered routes from the source towards the destination, w_i is a weight assigned to the ith path, $c1$ is a coefficient that reflects how important is the selection of the safest route based on the MNRS model, $c2$ is a coefficient that reflects how significant is the new added constraint min_{trust} to select the safest route.

In our present work, we give a higher weight to the new added constraint and perform evaluations with $c1 = 0.25$ and $c2 = 0.75$.

Finally, the second step consists in selecting the safest route. This one corresponds to the highest value pah_trust. The value of pah_trust can be estimated by the following formula:

$$\mathrm{path\_trust} = \max(\mathrm{path\_trust_i});\ \mathrm{i} = 1, \mathrm{n} \tag{7}$$

5 Simulation Results

Simulations were performed using the NS2 [24] network simulator. We chose NS2 because of its popularity among academic researchers. In addition, it already supports a verified version of DSR. Sixty nodes were randomly placed within a

Table 1 Simulation parameters

Parameter	Value
Topology area	1000×1000 m^2
Nodes transmission range	250 m
Total simulation time	400 s
Transmission rate	2 packets/s
Pause time	100 s
Speed	10 m/s
Application traffic	CBR
Movement model	Random waypoint
Types of attack	Control packet drop

$1000 \times 1000\,\text{m}^2$ area. The simulation time was 400 s for each simulation. The other simulation parameters are defined in Table 1.

In order to evaluate RTDSR efficiency, we compared it to DSR, CONFIDANT and MNRS protocols. We first study in Sect. 5.1 the impact of mobility on RTDSR's performance. We test, afterwards, RTDSR's resilience against data packets dropping. In Sect. 5.2, we examine the effect of the number of sources in the network on the performance of RTDSR.

We evaluate, for this purpose, the following metrics:

- **Packet Delivery Fraction (PDF)** is the ratio of packets successfully received.
- **Average End to End Delay (AEED)** is the average time required to transmit a data packet from a source to a destination node.
- **Throughput** is the ratio of the total data that flow through the network.

5.1 Fundamental Tests

For each evaluated metric, namely the PDF, AEED and throughput, we compare the performance of the protocols DSR, CONFIDANT, MNRS and RTDSR relative to two different variables. We first consider the impact of mobility in presence of 50 % of misbehaving nodes. We afterwards set the nodes speed to 10 m/s and test the resilience of each protocol against packets dropping, with 10, 20, 30, 40 and 50 % of misbehaving nodes. The misbehaving nodes were selected randomly. In all cases, the number of sources is equal to 10.

Packet Delivery Fraction

Figure 2 represents PDF measures for DSR, CONFIDANT, MNRS and RTDSR. Through the curves, we see that RTDSR packet delivery fraction is more important compared with that of DSR, CONFIDANT and MNRS for the two measured variables. This is due to the mechanisms supported by RTDSR to improve the selection of secure routes. Fewer packets pass through misbehaving nodes, which increases the number of packets correctly received by their destination.

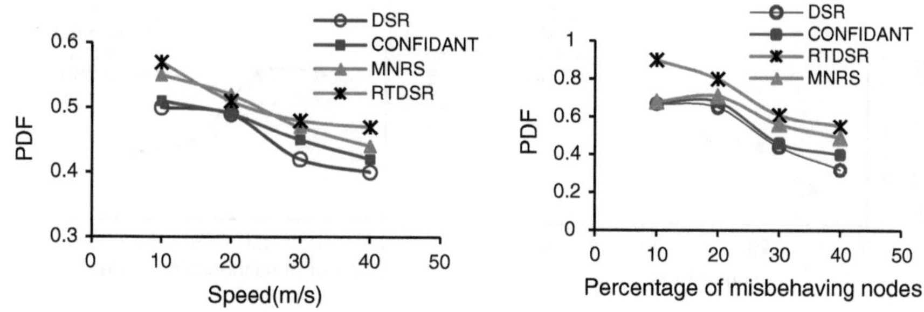

Fig. 2 PDF measures for DSR, CONFIDANT, MNRS and RTDSR

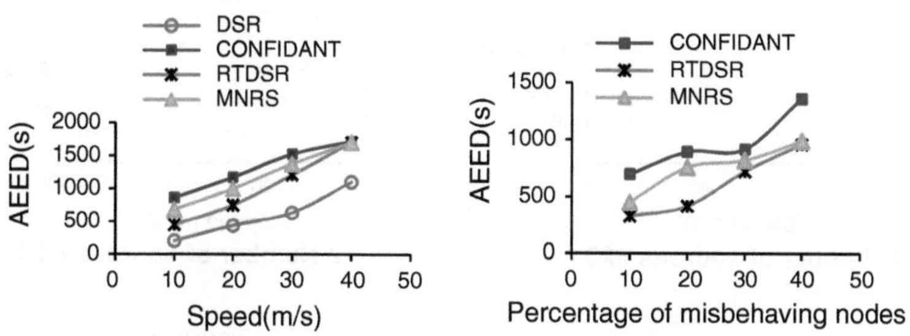

Fig. 3 AEED measures for DSR, CONFIDANT, MNRS and RTDSR

Average End to End Delay

Figure 3 depicts the variation of AEED for the four protocols. The results show that RTDSR has the lowest value of AEED compared to CONFIDANT and MNRS protocols. We can explain this arguing that RTDSR assures the selection of more secure routes against packet dropping. This implies less retransmissions, and consequently less time to deliver packets. DSR protocol has been evaluated in absence of misbehaving nodes. It has the lowest AEED value, because it does not use any secure mechanism to select routes, and hence spends no time for evaluation.

Throughput

In Fig. 4, we analyze the throughput for the four protocols. The simulation results show that RTDSR has the highest value in all conditions. This means that RTDSR is the most efficient protocol for establishing routes exempt of misbehaving nodes.

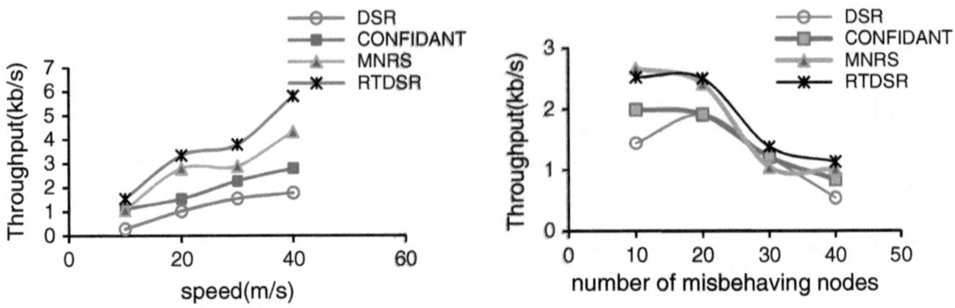

Fig. 4 Throughput measures for DSR, CONFIDANT, MNRS and RTDSR

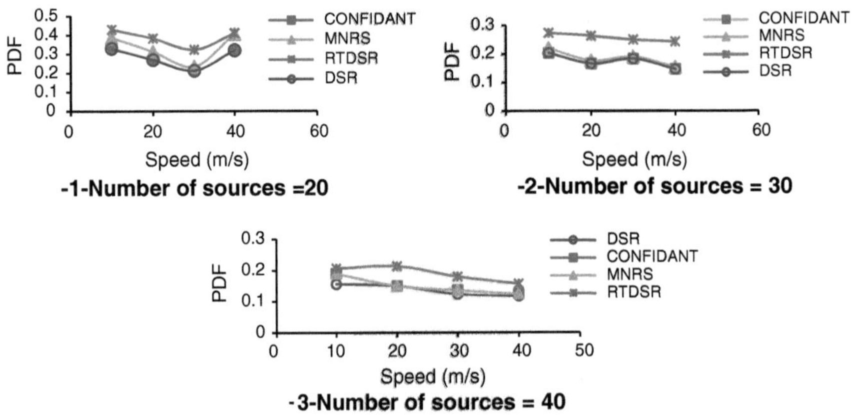

Fig. 5 PDF obtained by varying the number of sources in the network

5.2 Elaborate Tests

In this section, we study the impact of sources number on PDF, Throughput and AEED metrics for DSR, CONFIDANT, MNRS and RTDSR protocols. For this purpose, we conducted our experiment with 20, 30 and 40 sources of 60 nodes. The analysis is firstly performed by varying nodes mobility. Afterward, we set the nodes speed to 10 m/s and changed the percentage of misbehaving nodes using 10, 20, 30, 40 and 50 %.

(a) Mobility

Packet delivery fraction
In Fig. 5, the experiments are conducted with 20, 30 and 40 sources of 60 nodes present in the network, by varying the number of sources. The curves demonstrate that RTDSR remains the most efficient protocol in spite of the network congestion and its mobility. Besides, the results show that PDF decreases when the number of sources increases, which is due to the network congestion.

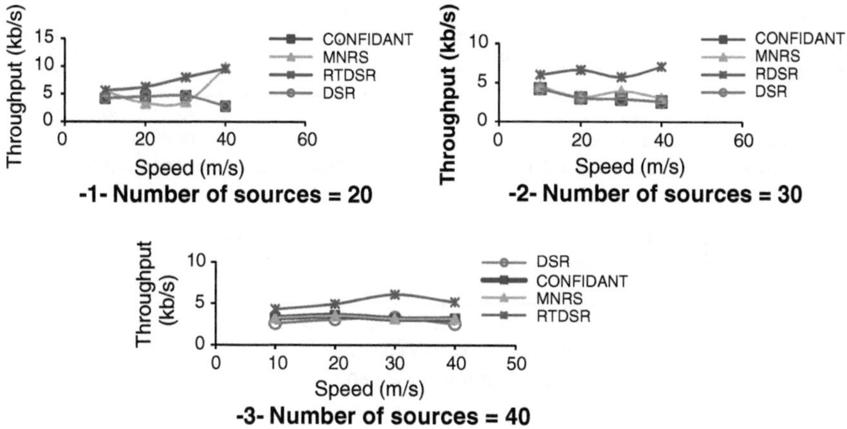

Fig. 6 Throughput obtained by varying the number of sources in the network

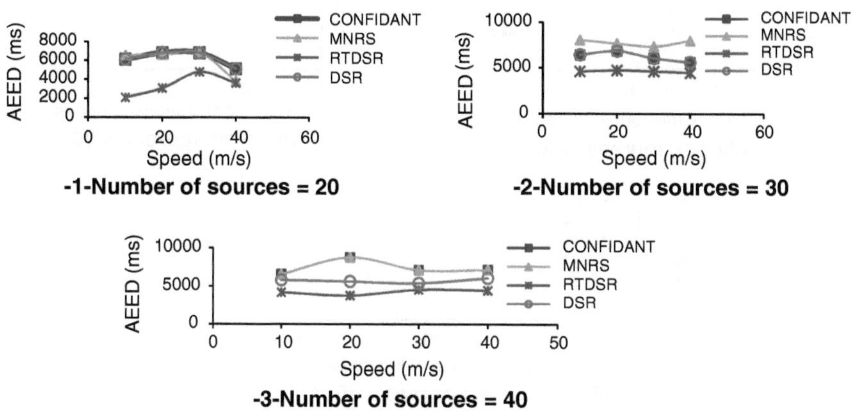

Fig. 7 AEED obtained by varying the number of sources in the network

Throughput
When we compare the performance of RTDSR in term of throughput with DSR, CONFIDANT and MNRS, we notice in Fig. 6 that it has the highest value, which means that it is the most resilient against dropping attacks.

Average End to End Delay
Figure 7 shows that RTDSR has the lowest value in comparison with DSR, CONFIDANT and MNRS, which proves that it is the most rigid against misbehaving nodes. This is due to its capacity to find routes more quickly than the other protocols.

(b) Percentage of misbehaving nodes

Packet Delivery Fraction
Curves in Fig. 8 trace the variation of PDF for the four protocols relative to the percentage of misbehaving nodes. This variation is done in a network containing

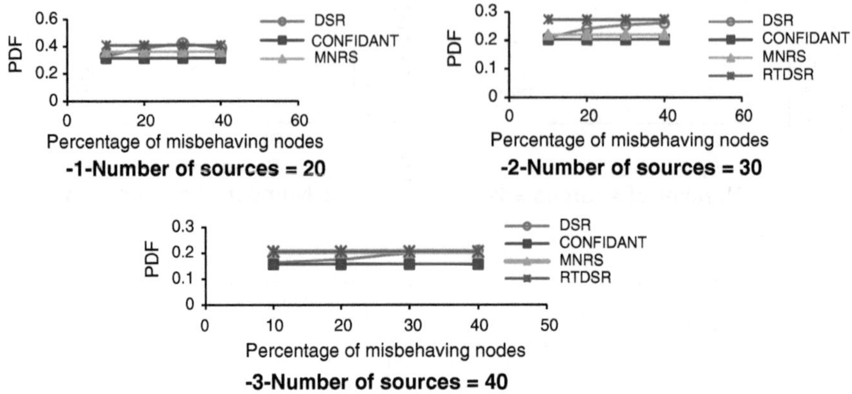

Fig. 8 PDF obtained by varying the number of sources in the network

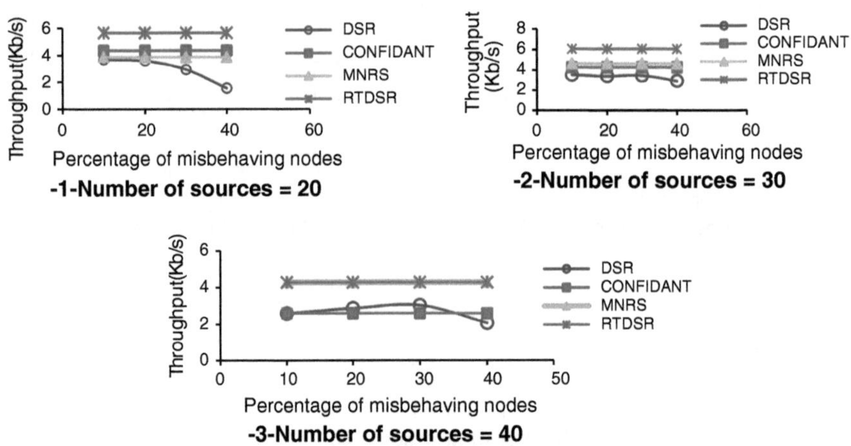

Fig. 9 Throughput obtained by varying the percentage of misbehaving nodes

respectively 20, 30 and 40 sources. In all schemas, RTDSR has the highest value, which proves its robustness against misbehaving nodes.

Throughput
Figure 9 shows the impact of the number of sources present in the network on the throughput for the four protocols by varying the percentage of misbehaving nodes. In the whole of the experiments, RTDSR is still the most efficient protocol.

6 Conclusion

In this chapter, we proposed a model based on trust concepts to secure source routing protocols. The proposed solution improves trust provided by the CONFIDANT protocol using a new mechanism that takes into consideration new routes learned by nodes after the *route request phase*. To evaluate trust in routes, we used the MNRS model which we rewrote using a new constraint on nodes. This constraint contributes to depreciate routes composed of nodes with the lowest value of trust, which increases the probability of selecting routes free of misbehaving nodes. Simulation results based on fundamental and elaborate tests show that RTDSR is resilient to packets dropping attacks and maintains good performances. We propose for future works to formally validate RTDSR using the AVISPA tool.

References

1. Kargl, F., Klenk, A., Schlott, S., Weber, M.: Advanced detection of selfish or malicious nodes in ad hoc networks. In: Proceedings of 1st European Workshop on Security in Ad-hoc and Sensor Networks, ESAS 2004, Heidelberg, Germany, pp. 152–165 (2004)
2. Bhalaji, N., Shanmugam, A.: A trust based technique to isolate non-forwarding nodes in DSR based mobile ad hoc networks. In: The third International Conference on Network Security and Applications, Chennai, India (2010)
3. Buchegger, S., Mundinger, J., Le Boudec, J.Y.: Reputation systems for self-organized networks. IEEE Technol. Soc. Mag. **27**(1), 41–47 (2008)
4. Michiardi, P., Molva, R.: Core: a collaborative reputation mechanism to enforce node cooperation in mobile ad hoc networks. In: Proceedings of IFIP Conference on Communications and Multi-media Security, Kluwer, B.V. Deventer, The Netherlands vol. 228, pp. 107–121 (2002)
5. Arya, K.V., Vashistha, P., Gupta, V.: Three phase technique for intrusion detection in mobile ad hoc network. In: DICTAP 2011, vol. 166, pp. 675–684 (2011)
6. Buchegger, S., Le Boudec, J.Y.: Performance analysis of the confidant protocol. In: ACM 3rd International Symposium on Mobile Ad hoc Networking and Computing-MobiHoc'02, pp. 226–236 (2002)
7. Buchegger, S., Le Boudec, J.Y.: A robust reputation system for P2P and mobile ad hoc networks. In: Proceedings of P2PEcon 2004 (2004)
8. Yang, X., Gao, Y.: A routing protocol based on trust for MANETs. In: Zhuge H., Fox G.C. (eds.) Proc. 4th International Conference on Grid and Cooperative Computing GCC 2005, vol. 3795, pp. 959–964. LNCS, Beijing, China, Nov 30–Dec 3 (2005)
9. Johnson, D. B., Maltz, D.A., Broch, J.: DSR: the dynamic source routing protocol for multi-hop wireless ad hoc networks. In: Perkins C.E. (ed.) Ad Hoc Networking. Addison-Wesley, Reading (2001)
10. Johnson, D.B., Hu, Y-C., Maltz, D.A.: The dynamic source routing protocol (DSR) for mobile ad hoc networks. IETF RFC4728 (2007)
11. Suresh, A., Duraiswamy, K.: Routing protocol with node reputation scheme. J. Comput. Sci. **7**(2), 242–249 (2011)
12. Balakrishnan, K., Deng, J., Varshney, P.K.: TWOACK: preventing selfishness in mobile ad hoc networks. In: Proceedings of Wireless Communication and Networking Conference, New Orleans, LA, USA (2005)
13. Djenouri, D., Ouali, N., Mahmoudi, A., Badache, N.: Random feedbacks for selfish nodes detection in mobile ad hoc networks. In: The 5th IEEE International Workshop on IP Operations and Management-IPOM 2005, Barcelona, Spain (2005)
14. Abd El-Haleem, A.M., Ali, I.A.: TRIUMF: trust-based routing protocol with controlled degree of selfishness for securing MANET against packet dropping attack. Int. J. Comput. Sci. Issues, **8**(4), 1 (2011)
15. Zhao, L., Delgado-Frias, J.G.: MARS: Misbehavior Detection in Ad Hoc Network. IEEE GLOBECOM, Washington (2007)

16. Sukumran, S., Jaganathan, V., Korath, A.: Reputation based dynamic source routing protocol for MANET. Int. J. Comput. Appl. **47**(4), 42–46 (2012)
17. Choudhury, S., Roy, S.D., Singh, S.A.: Trust management in ad hoc network for secure DSR routing, novel algorithms and techniques. In: Sobh, T., Elleithy, K., Mahmood, A., Karim, M.A. (eds.) Telecommunications, Automation and Industrial Electronics, pp. 496–500. Springer, The Netherlands (2008)
18. Awerbuch, B., Curtmola, R., Holmer, D., Nita-Rotaru, C., Rubens, H.: ODSBR: an on-demand secure byzantine resilient routing protocol for wireless ad hoc networks. ACM Trans. Inf. Syst. Sec. **10**, 3 (2007)
19. Awerbuch, B., Scheideler, C.: Robust random number generation for peer-to-peer systems. Theor. Comput. Sci. Elsevier **410**(6–7), 453–466.28 (2009)
20. Yong, C., Chuanhe, H., Wenming, S.: Trusted dynamic source routing protocol. In: Wireless Communications, Networking and Mobile Computing (WiCom) pp. 1632–1636 (2007)
21. Quercia, D., Hailes, S., Capra, L.: B-trust: Bayesian trust framework for pervasive computing. In: Proceedings of the 4th International Conference on Trust Management (iTrust), Pisa, Italy, pp. 298–312, May (2006)
22. Marias, G.F., Georgiadis, P., Flitzanis, D., Mandalas, K.: Cooperation enforcement schemes for MANETs: a survey. Wirel. Commun. Mob. Comput. Wirel. Netw. Sec. **6**(3), 319–332 (2006)
23. Bansal, S., Baker, M.: Observation-based cooperation enforcement in ad-hoc networks. Stanford University, Technical Report (2003)
24. NS Manual, VINT Project. 2011, www.isi.edu/nsnam/ns/doc/ns_doc.pdf

Author Index

T. Bouabana-Tebibel and S. H. Rubin (eds.), *Integration of Reusable Systems*,
Advances in Intelligent Systems and Computing 263, DOI: 10.1007/978-3-319-04717-1,

MIX
Papier aus verantwortungsvollen Quellen
Paper from responsible sources
FSC® C105338

If you have any concerns about our products,
you can contact us on
ProductSafety@springernature.com

In case Publisher is established outside the EU,
the EU authorized representative is:
Springer Nature Customer Service Center GmbH
Europaplatz 3, 69115 Heidelberg, Germany

Printed by Libri Plureos GmbH
in Hamburg, Germany